Topics in Applied Physics   Volume 66

Topics in Applied Physics   Founded by Helmut K. V. Lotsch

Volumes 1–56 are listed on the back inside cover

Light Scattering in Solids V

Superlattices and Other Microstructures

Edited by M. Cardona and G. Güntherodt

With Contributions by
G. Abstreiter   M. Cardona   M. H. Grimsditch
P. Grünberg   G. Güntherodt   B. Jusserand
R. Merlin   D. L. Mills   A. Pinczuk   J. C. Tsang

With 184 Figures

Springer-Verlag Berlin Heidelberg GmbH

Professor Dr., Dres. h.c. *Manuel Cardona*

Max-Planck-Institut für Festkörperforschung, Heisenbergstrasse 1,
D-7000 Stuttgart 80, Fed. Rep. of Germany

Professor Dr. *Gernot Güntherodt*

2. Physikalisches Institut der Rheinisch-Westfälischen Technischen Hochschule,
Templergraben 55, D-5100 Aachen, Fed. Rep. of Germany

ISBN 978-3-662-31112-7   ISBN 978-3-540-45998-9 (eBook)
DOI 10.1007/978-3-540-45998-9

2154/3150-543210 – Printed on acid-free paper

Preface

This volume is the fifth of a series [Topics in Applied Physics (TAP), Vols. 8, 50, 51, 54, 66] devoted to the scattering of light by solids. The first and second volumes (TAP 8 and 50) emphasize general concepts and theoretical background, the third (TAP 51), investigations of specific materials and also Brillouin scattering, and TAP 54 discusses light scattering by electronic excitations, surface enhanced Raman scattering and effects of pressure on phonon spectra. A detailed list of contents of these volumes can be found in the second edition (paperback) of TAP 8.

The present volume is devoted to light scattering by interfaces and artificial microstructures (such as the so-called superlattices). The growth and characterization of such structures has now reached a high degree of sophistication including developments which culminated in the discovery of the ordinary and fractional quantum Hall effects and related theoretical work. Light scattering spectroscopy is an ideal technique for the investigation and characterization of such structures, since the incident laser light can be focused to dimensions of the order of those of the microstructures. This volume contains theoretical background on the excitations accessible to light scattering (Chap. 2), and a review of scattering by phonons (Chap. 3) and by electronic excitations (Chap. 4). Chapter 5 discusses artificial quasiperiodic (Fibonacci) and aperiodic (Thue-Morse) superlattices. Chapter 6 presents the state of the art of multichannel detectors, which can be used to great advantage for recording very weak scattering signals such as those obtained at a single interface. Chapter 7 treats metallic superlattices and Chap. 8 scattering by magnetic excitations.

The editors would like to thank all authors, especially those who were diligent enough to comply with the original, seldom adhered to, deadline.

This volume is dedicated to the various groups who contributed to the preparation of the samples discussed in its chapters. Without their high degree of expertise, sophistication and dedication, the advances described here would not have been possible. Last but not least thanks are due to S. Birtel and I. Dahl for expert secretarial help and skillful typing.

Stuttgart and Aachen,   *Manuel Cardona*
November 1988   *Gernot Güntherodt*

Contents

Contributors

Abstreiter, Gerhard
Walter Schottky Institut, Technische Universität München
D-8046 Garching, Fed. Rep. of Germany

Cardona, Manuel
Max-Planck-Institut für Festkörperforschung, Heisenbergstrasse 1
D-7000 Stuttgart 80, Fed. Rep. of Germany

Grimsditch, Marcos H.
Materials Science Division, Argonne National Laboratory
Argonne, IL 60439, USA

Grünberg, Peter
Institut für Festkörperforschung, Kernforschungsanlage Jülich
Postfach 1913, D-5170 Jülich, Fed. Rep. of Germany

Güntherodt, Gernot
2. Physikalisches Institut der Rheinisch-Westfälischen Technischen
Hochschule, Templergraben 55, D-5100 Aachen, Fed. Rep. of Germany

Jusserand, Bernard
Centre National d'Etudes des Télecommunications, 196 Rue de Bagneux
F-92220 Bagneux, France

Merlin, Roberto
Department of Physics, The University of Michigan
Ann Arbor, MI 48109-1120, USA

Mills, Douglas L.
Department of Physics, University of California
Irvine, CA 92717, USA

Pinczuk, Aron
AT & T Bell Laboratories, Murray Hill, NJ 07974, USA

Tsang, James C.
IBM T. J. Watson Research Center, P.O. Box 218
Yorktown Heights, NY 10598, USA

Sir C. V. Raman *1888–†1970
(Nobel laureate 1930)

1. Introduction

Manuel Cardona and Gernot Güntherodt

With 2 Figures

> – Tant és més aconsolada la mia ànima – dix Tirant – que jo puga
> donar coses tals que sien de molta estima, car lo donador no deu
> donar coses que sien de poca condició, mas donar coses que
> apareguen a les gents ésser de gran estima o floresquen en honor e
> fama. Jo do açò en nom meu, e faço-ho per fer-ne servir a la
> majestat del senyor Emperador.
>
> *J. Martorell, M. J. de Galba:* Tirant lo Blanch (N. Spindeler,
> València, 1490)

This is the fifth volume of the Series "Light Scattering in Solids" which appears in the collection "Topics in Applied Physics". The first volume of the series (1975) was reissued in 1982 as a second corrected and annotated edition, with cross references to other volumes of the series [1.1]. It includes the list of contents of Volumes I–IV. Volumes II [1.2] and III [1.3] also appeared in 1982 while Volume IV was published in 1984 [1.4]. Volume VI, including work on organic polymers (polyacetylene), semimagnetic semiconductors, silver halides, heavy fermion systems, high-T_c superconductors, and a chapter on the formal theory of light scattering in solids is in the planning stage.

We should mention that the preparation of the present volume coincides with the 100th birthday of Prof. Chandrasekhara Venkata Raman and the 60th anniversary of the discovery in Calcutta of the effect which bears his name and which earned him the Nobel Prize (awarded two years after the discovery, in 1930). We would like to dedicate this volume to his memory.

1.1 Contents of Previous Volumes and Related Recent Developments

Volume I of this series was written at a time when the power of light scattering for studying elementary excitatons in solids had been amply established. This power had been greatly expanded by the capability to perform backscattering experiments in opaque materials (semiconductors, metals) using lasers as sources of incident light. However, a large data base was not yet available.

Chapter 1 of that volume contains a historical introduction by M. Cardona.

Chapter 2, by A. Pinczuk and E. Burstein, discusses general macro- and microscopic aspects of the theory of the scattering efficiency of phonons and electronic excitations. Emphasis is given to so-called quasistatic or adiabatic formulations in which it is assumed that the frequency of the created (or annihilated) excitations is small compared to the relevant electronic frequencies. Such formulations break down for laser (or scattered) frequencies close to sharp electronic interband excitations. In that article, however, prescriptions are given to transform quasistatic expressions (usually given in terms of electronic susceptibilities and their derivatives with respect to the amplitude of the Raman excitations) into generalized expressions in which the adiabatic restriction has been lifted.

Chapter 3, by R. M. Martin and L. M. Falicov, dwells on more detailed aspects of resonant Raman scattering including its time dependence and the phenomenon of double resonance. At that time, the latter had been observed only in second order scattering. Very recently, several doubly resonant spectra, in which not only the incident but also the scattered photon is resonant with interband transitions, have been obtained in superlattices [1.5] and in bulk semiconductors under stress [1.6]. More recently, even triple resonances have been observed for superlattices [1.7] and for bulk GaAs under stress [1.8].

Chapter 4, by M. V. Klein, was devoted to electronic Raman scattering, a subject which up to then had received relatively little experimental (but a lot of theoretical) attention. The author treats the theory of scattering by single-particle excitations, plasmons, plasmon–L0-phonon coupled modes, acoustic plasmons, electron intervalley fluctuations, and excitations involving impurity levels. Subsequent rapid experimental developments in these fields and in related areas involving two-dimensional samples (heterojunctions, multiple quantum wells, superlattices, surfaces) were later collected in Volume IV. Today, many of these spectra are used as standard techniques for the characterization of semiconductors [1.9].

Chapter 5, by M. H. Brodsky, is devoted to vibrational spectra of amorphous semiconductors. It was brought up to date in Chap. 2 of [1.2] and complemented in [1.10, 12]. Raman spectroscopy has, in the meantime, become a standard tool for the scientific and industrial characterization of amorphous semiconductors [1.9–12].

Chapter 6, by A. S. Pine, contains a discussion of Brillouin scattering as applied to semiconductors. It was written prior to the discovery of polariton-mediated resonant Brillouin scattering (Chap. 7 of [1.3]) and to the introduction of multipass and tandem Fabry-Pérot spectrometers by J. R. Sandercock (Chap. 6 of [1.3]), at a time when a rather reduced data base was available (in particular, no data on opaque materials).

Chapter 7, by Y. R. Shen, covers stimulated Raman scattering and some related coherent nonlinear phenomena. Subsequent developments in the non-linear optics field, including the incoherent Hyper-Raman effect, can be found in Chap. 4 of [1.2]. We have tried to keep the format of subsequent volumes similar to that of Volume I.

Volume II [1.2] contains an extensive article by M. Cardona on the theory of Raman scattering, mainly by phonons in semiconductors, and a discussion of extant experimental data (Chap. 2). At the time of its writing, the available data base had grown considerably. It included scattering efficiencies in absolute units and a large number of resonant and nonresonant data for phonons in many semiconductors. As shown in Chap. 2 of [1.2], microscopic theory based on electronic band structures was able to account for the absolute cross sections. Most of the observed resonances did not, at that time, give separate incoming (laser frequency equal to electronic gap) and outgoing (scattering frequency equal to electronic gap) components because of the lack of resolution, related either to instrumentation or to lifetime broadening of electronic states. Hence, the theory in Chap. 2 of [1.2] implied the quasistatic approximation. It can, however, be easily generalized, following the prescriptions given in Chap. 2 of [1.1], to describe separate incoming and outgoing resonances. Such separate resonances have become standard at low temperatures for the lowest absorption edges of bulk semiconductors [1.11] and semiconductor superlattices [1.13] (see also Chap. 3 of this volume). In Chap. 2 of [1.2] the dipole-forbidden, Fröhlich-interaction-induced scattering by L0-phonons in polar materials was discussed to have three contributions: q-induced (quadrupole), E-(electric field)-induced, and ionized-impurity-induced. The existence of a q-induced component has been widely documented in the meantime through coherent interference with allowed components [1.11, 14]. The E-induced components, also observed, are discussed at length in Chap. 2 of [1.4]. Impurity-induced components are also documented in [1.11, 13].

Chapter 3 of [1.2], by R. K. Chang and M. B. Long, is devoted to the experimentally important topic of multichannel detection. While early work (Raman) was performed using "multichannel" photographic plates which enable *simultaneous* detection of the whole scattered spectrum, developments in photoemissive electron multipliers, permitting single photon counting, displaced photographic detection. Single channel photomultipliers, however, imposed the constraint of single frequency channel detection, and rejection, at one given time, of most of the scattered fequencies. Chapter 3 of [1.2] discusses early methods of multichannel detectors (optical multichannel analyzers, OMA). The most popular of such devices at the time consisted of a photocathode and a microchannel plate electron multiplier followed by a Si diode array. Since then, other systems have appeared on the market, including position-dependent detectors (MEPSICRON) and charge coupled devices (CCD). They are discussed in Chap. 6 of the present volume.

Chapter 4 of [1.2], by H. Vogt, is devoted to non-linear Raman and related techniques. Such techniques, especially the *Coherent* anti-Stokes Raman spectroscopy (CARS) and the *incoherent* hyper-Raman spectroscopy (HRS) have mushroomed (for a list of ubiquitous acronyms see Table 4.1 of [1.2]) due, in particular, to the development of high power lasers and multichannel detectors. CARS spectroscopy has also been reviewed in [1.15]. A list of recent references to Hyper-Raman work can be found in [1.16].

Volume III discusses light scattering spectra of typical families of materials and for several classes of phenomena. Chapter 2, by M. S. Dresselhaus and G. Dresselhaus, a husband-and-wife team, discussed graphite intercalation compounds. These quasi-two-dimensional structures have many features in common with the superlattices discussed in the present volume (Chap. 2). In particular, the lowering in the number of translational symmetry operations induces additional Raman lines. Chapter 3, by D.J. Lockwood, discusses electronic and magnetic excitations (magnons) in the large family of transition metal halides which includes perovskite, rutile, and layered $CdCl_2$-type structures. It was complemented by two articles in [1.4]: a theoretical (Chap. 4) and an experimental (Chap. 5) dealing with phonons, electron-phonon interactions, spin-phonon excitations and localized electronic excitations in in the rare earth chalcogenides. Chapter 4, by W. Hayes, treats the problem of light scattering by superionic conductors, a family of materials of great technological interest (solid electrolytes, solid state batteries). They possess partly disordered structures which lead to broad vibrational spectra, somewhat similar to those found in amorphous materials.

Chapter 5 discusses Raman spectra of phonons and related anomalies in metallic transition metal compounds, some of which (V_3Si and NbC) are superconductors. These were probably the metals (and the highest T_c materials) most investigated by Raman spectroscopy until the recent advent of hight-T_c superconductors. The article contains very detailed and rigorous theoretical considerations whose unifying idea is that the same electron-phonon processes which produce Raman scattering and phonon anomalies (e. g., Fano line shapes, see Chap. 2 of [1.4]) may also be responsible for superconductivity and charge density wave (CDW) transitions. The theory in this article should be useful for the interpretation of phonon anomalies recently observed in hight-T_c superconductors [1.17] and also for phonon anomalies encountered in heavily doped semiconductors (Chap. 2 of [1.4]).

Chapter 6, by J. R. Sandercock, describes the high resolution, high contrast Fabry-Pérot spectrometers developed by the author and based on the multiple pass and tandem principle. They are the clue to Brillouin scattering measurements in opaque materials using the backscattering configuration. In this way, metals, opaque semiconductors, and thin films have been investigated. This work includes acoustic surface waves (Rayleigh modes) and surface spin waves (Damon-Eshbach modes).

Chapter 7, by C. Weisbuch and R. G. Ulbrich, is devoted to Brillouin scattering (i.e., scattering by acoustic phonons) in strong resonance with very sharp excitons in semiconductors (discovered first by the authors in 1977 for GaAs). In this "strong coupling" case the photon cannot be treated in perturbation theory: one must first solve exactly the coupled exciton-phonon problem in terms of exact mixed eigenstates called polaritons. The scattered particles (phonons) couple different polariton branches and can have large values for the wavevectors *perpendicular* to the crystal surface. Hence, large frequency shifts result which can be easily observed with conventional Raman spectrometers (a Fabry-Pérot interferometer is not needed). The phenomenon was *predicted* in 1972 by Brenig, Zeyher, and Birman and it took five years to observe it. This represents one of the rare and most beautiful recent examples of predictive solid state theory.

Volume IV has a structure similar to Volume III, in fact it resulted as a spill-over of the articles requested for that volume. Thus, it also contains case studies for families of materials and classes of scattering phenomena.

Chapter 2, by G. Abstreiter, M. Cardona, and A. Pinczuk, is devoted to electronic excitations in bulk semiconductors and in two-dimensional electron gases (at heterojunctions, Schottky barriers, quantum wells, and some super-lattices). At the time of writing, the work on two-dimensional systems was at its beginning. It had been recognized by *Burstein* et al. [1.18] that light scattering is an ideal tool for the investigation of such two-dimensional systems. In the meantime, so much progress has taken place that the writing of a new article in the present volume (Chap. 4), completely devoted to scattering by electronic excitations in superlattices, has become necessary. Much of the required background can still be found in Chap. 2 of [1.4].

Chapter 3, by S. Geschwind and R. Romestain, is devoted to spin-flip resonant Raman scattering, as illustrated by the great wealth of observations performed by the authors for CdS. The method can be applied to study phenomena as varied as the metal-insulator transition in impurity bands, spin relaxation times and terms linear in k in the energy bands.

Chapter 4, by G. Güntherodt and R. Zeyher, is concerned with the effects of magnetic order and also spin disorder on the scattering by phonons in magnetic semiconductors, as evidenced mainly by the rare earth monochalcogenides, cadmium chromium spinels and vanadium dihalides. Both phenomenological and microscopic theories are presented and illustrated with experimental data for the various magnetic phases of those materials.

Chapter 5, by G. Güntherodt and R. Merlin, reviews the very prolific research area of Raman scattering in rare earth chalcogenides, a family consisting of magnetic semiconductors, magnetic or superconducting metals, and mixed valence compounds. In all these materials the electron-lattice (i.e., phonon) coupling can be consistently described by the concept of phonon-induced local charge deformabilities.

Chapter 6 by A. Otto, treats the important topic of surface enhanced Raman scattering (SERS), a subject which was then at the peak of its activity. Basic research in this field has more recently subsided as the technique has achieved the stage of a standard analytical tool. The emphasis in Otto's review is to disentangle experimentally the various mechanisms which contribute to the striking SERS phenomenon (enhancement by a factor of up to 10^6 of the scattering cross section of molecules adsorbed on certain metal surfaces), including electromagnetic resonances produced by "macroscopic roughness" and "chemical effects" of roughness present on a microscopic scale. In many cases, both mechanisms seem to lead to the same enhancement factor, probably on the order of 10^3 each.

Chapter 7, by R. Zeyher and K. Arya, presents a theoretical formulation of the SERS problem based, in part, on classical electromagnetic resonances and on the quantum-mechanical theory of chemisorption.

Finally, Chap. 8 by B. A. Weinstein and R. Zallen discusses the most widely investigated class of so-called morphic effects, namely the effect of *hydrostatic pressure* on Raman spectra of phonons. The field had reached maturity after the development of the ruby manometer for the diamond anvil cell in the early 1970s. While the effect of *uniaxial stress* on phonon frequencies is not explicitly discussed, a comprehensive bibliography covering work in this field done before 1984 completes the chapter. We should close by remarking that the dependence of phonons on strain (or stress), both uniaxial and hydrostatic, has recently regained interest [1.19], with the development of lattice mismatched, strained semiconductor superlattices (see Sect. 3.4.4 of the present volume).

1.2 Contents of This Volume

The original call of J. W. Goethe in his death bed for *more light* (mehr Licht!), repeated on p. 1 of [1.1], has partially lost its anguish. Powerful gas and solid state cw lasers, covering the region from the near IR to the near UV are now commercially available. Tunable lasers of the dye and color center varieties have become standard tools in light scattering. Excellent double and triple monochromators can be purchased at modest prices. Photomultipliers with InGaAsP photocathodes, and also cooled germanium detectors, have extended the range of photon counting to the near infrared (~ 0.7 eV $\cong 1.8$ μm). And finally, multichannel detectors are cutting down measuring times by about two orders of magnitude, with possible improvement in the quantum efficiency/dark signal ratio in the case of CCDs. These advances in instrumentation make light scattering an ideal standard tool to tackle problems of new materials, such as recently demonstrated in the case of the new high-T_c superconductors [1.17]. More than 60 papers (necessarily of variable quality) on Raman spectroscopy of these materials have appeared in the past year, a fact which will be fully documented in the forthcoming book of the series.

The present volume is largely devoted to light scattering by two-dimensional systems (surfaces and adsorbed layers, heterojunctions, multiple and single quantum wells, superlattices, thin films). Some of this work would not have been possible only a few years ago without the developments in instrumentation just mentioned.

Maybe the development most relevant to the present work is that concerning multichannel detectors. Thus, we have devoted a whole chapter (Chap. 6) to a comparative discussion of the commercially available multichannel detection systems (without explicitly naming the manufacturers). The capabilities of these detectors are illustrated by studies of monolayer-like systems deposited on bulk substrates. Last, but not least, we should mention that much of the work reported here would not have been possible without the know-how accumulated during the past 20 years in the growth of superlattices, both semiconducting and metallic. While some of the samples discussed here were grown by metal-organic chemical vapor deposition (MOCVD), the large majority of the semiconductor superlattices were grown by molecular beam epitaxy (MBE). The principles of this technique were incipient in the early work of *Günther* [1.20] but the method was mainly developed by *Arthur* [1.21] and *Cho* [1.22] nearly 20 years ago. The development of the technique closely paralleled advances in ultrahigh-vacuum technology. Commercial MBE equipment, supplied by several manufacturers of UHV components, has been available for the past 10 years. Pioneering work on metallic superlattices by I. K. Schuller and C. M. Falco started out at Argonne National Laboratory in the late 1970s by using sputtering techniques, and will be discussed in Chap. 7. Presently, the topic has been taken up by many other groups in the US, Japan, and Europe, with the emphasis on metal-MBE.

As an example of the high quality of present day superlattices, we show in Fig. 1.1 a transmission electron micrograph of a GaAs-AlAs superlattice and, in Fig. 1.2, a lattice image of the same superlattice which corresponds to a resolution of better than 3 Å.

Chapter 2 by D. L. Mills presents the theoretical underpinning of many of the elementary excitations discussed in this volume. They can be brought under a common denominator on the basis of a macroscopic response function and appropriate boundary conditions. In the case of electrical excitations, the response function is the dielectric function. At frequencies at which the dielectric constant of one of the media forming the interface is negative, excitations localized near the interface, i.e., decaying exponentially away from it, and propagating along it, result. They give rise to interface excitations which propagate along and perpendicular to the interfaces for an infinite superlattice, surface interface excitations for a semi-infinite superlattice, and standing wave excitations for a finite superlattice. The case of magnetic excitations (spin waves) is also treated. It gives rise to similar types of excitations, except for some phenomena typical of the interface between magnetic and nonmagnetic layers, and of the symmetry properties of the magnetic field and magnetization with respect to spatial inversion and time reversal.

Chapter 3, by B. Jusserand and M. Cardona, is entirely devoted to scattering by phonons in semiconductor superlattices and quantum wells. Emphasis is placed on the concepts of zone folding and dispersion relation averaging, applicable to acoustic phonons, and mode confinement, ususally applicable to optic phonons. The formation of a superlattice induces scattering by bulk-like phonons but with wavevectors throughout the whole bulk Brillouin zone (i.e. forbidden in bulk samples). A macroscopic treatment of the electrostatic fields which accompany optical phonons in polar materials yields the interface modes already discussed in Chap. 2. Theoretical dispersion relations so obtained are compared with experimental data and with the results of full microscopic calculations. The various electron-phonon coupling mechanisms responsible for the scattering and related resonance phenomena, are also treated. The effects of

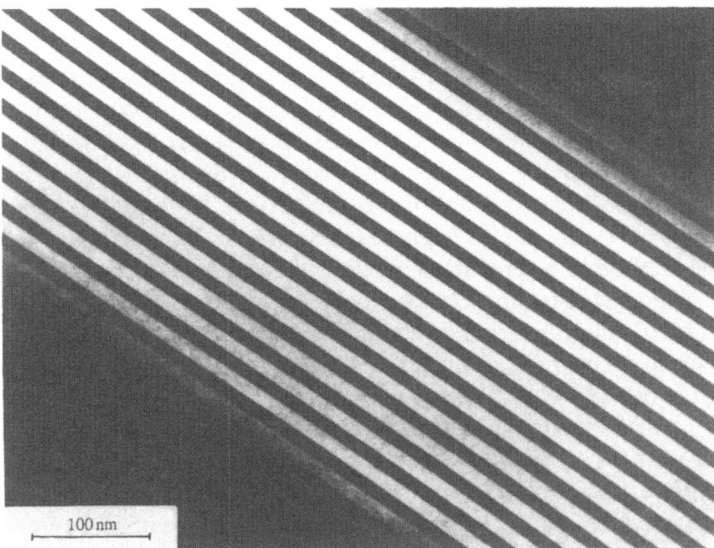

Fig. 1.1. Dark-field transmission electron micrograph of a GaAs/AlAs superlattice taken with [110] beam incidence. Courtesy of H. Oppholzer (Siemens AG)

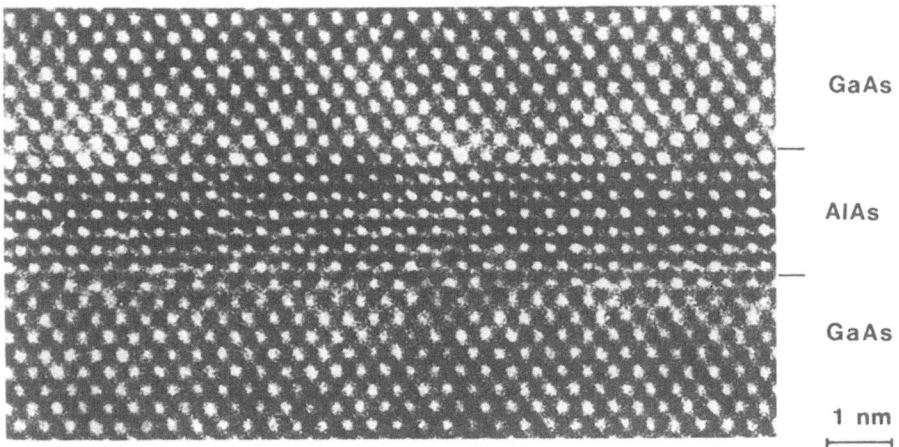

Fig. 1.2. Transmission electron lattice image of GaAs/AlAs superlattice at [110] beam incidence, indicating the perfect atomic arrangement at the interfaces. Courtesy of K. Ploog (Max-Planck Institut, Stuttgart)

built-in strain on the vibrational frequencies of superlattices composed of materials with different lattice parameters are discussed.

Chapter 4, by G. Abstreiter and A. Pinczuk, considers the spectroscopy of excitations of free carriers in semiconductor quantum wells. These electron systems of reduced dimensionality exist in heterojunctions and superlattices and in metal-insulator-semiconductor structures. In 1978, the inelastic light scattering method was proposed for studies of two-dimensional electron systems, and the first observations, by the authors of Chap. 4, were reported in 1979. Since then, inelastic light scattering has become a very versatile experimental probe of the excitations of free carriers in semiconductor systems of reduced dimensionality. Abstreiter, Cardona, and Pinczuk reviewed the early work in [1.4]. Chapter 4 of the present volume considers the recent developments in inter-subband spectroscopy, collective excitations in multilayers, and excitations of the two-dimensional electron gas in high magnetic fields. It highlights the applications of the method to several areas of current research of quantum wells and superlattices. This includes topics like the complex structure of the valence states of semiconductor quantum wells and space-charge layers, the spectroscopy and lifetimes of electron states, and the single particle and collective excitations in high magnetic fields. The collective intra-subband excitations can be treated with a formalism similar to that of Chap. 2: plasmon-like interface modes result. Because of the strong dispersion of these modes for in-plane wavevectors tending to zero, it is possible to observe such dispersion by rotating the sample with respect to the incident and scattered light beams. Copious experimental data are available involving standing waves and surface modes for finite thickness superlattices. Inter-subband excitations can also have single particle and collective character, discrimination being possible by the use of different polarizations. A number of additional phenomena can be observed in taylor-made wells with particular (asymmetric) shapes. The chapter also discusses space-charge-induced potential wells, such as δ-function doped and "nipi" structures.

During the past three years, considerable interest has been devoted to the possible existence of quasiperiodic crystals which require a number of translational basis vectors for their description higher than their dimensionality [1.23]. This interest was triggered by the discovery of materials with quasicrystal X-ray scattering patterns (e.g., Al+14% Mn [1.24]). Computer controlled molecular beam epitaxy can be used to grow one-dimensional quasicrystals, the most interesting of which are based on the principle of the Fibonacci sequence (1, 2, 3, 5, 8, 13...). Such crystals were first grown and investigated by Bhattacharya, Merlin and coworkers. They probably represent the simplest possible type of quasicrystals.

Chapter 5, by R. Merlin, discusses light scattering by phonons, plasmons, and other elementary excitations in Fibonacci superlattices. It also presents theoretical and experimental results for a certain type of deterministic, non-periodic superlattices (Thue-Morse superlattices) and for random superlattices.

The contents of Chap. 6 by J. C. Tsang, multichannel detectors and applications, have been discussed at the beginning of this section.

Chapter 7, by M. H. Grimsditch, is devoted to elastic and magnetic properties of metallic superlattices as observed in Brillouin scattering. The first part of this chapter treats anomalies in the composite elastic constants of metallic superlattices (e.g. alternating layers of metal A and B) which have been observed by means of Brillouin scattering using tandem Fabry-Pérot interferometers. The second part is devoted to the observation of spin wave excitations in metallic superlattices with alternating magnetic and nonmagnetic layers.

Chapter 8, by P. Grünberg, is completely devoted to the theory and experimental observation of spin-wave excitations in single thin films, double layers, and multilayered structures. Emphasis is on the effects of dipolar interlayer coupling, intra- and interlayer exchange coupling and on the determination of the magnetic parameters, such as surface and volume anisotropy constants, exchange stiffness parameters and exchange coupling length between magnetic layers, via a nonmagnetic or antiferromagnetic spacer layer of variable thickness.

References

1.1 M. Cardona (ed.): *Light Scattering in Solids*, 1st ed.; *Light Scattering in Solid I: Introductory Concepts*, Topics Appl. Phys., Vol. 8, 2nd ed. (Springer, Berlin, Heidelberg 1975 and 1982)

1.2 M. Cardona, G. Güntherodt (eds.): *Light Scattering in Solids II: Basis Concepts and Instrumentation*, Topics Appl. Phys., Vol. 50 (Springer, Berlin, Heidelberg 1982)

1.3 M. Cardona, G. Güntherodt (eds.): *Light Scattering in Solids III: Recent Results*, Topics Appl. Phys., Vol. 51 (Springer, Berlin, Heidelberg 1982)

1.4 M. Cardona, G. Güntherodt: *Light Scattering in Solids IV: Electronic Scattering, Spin Effects, SERS, and Morphic Effects*, Topics Appl. Phys., Vol. 54 (Springer, Berlin, Heidelberg 1984)

1.5 R.C. Miller, D.A. Kleinmann, C.W. Tu, S.K. Sputz: Phys. Rev. B **34**, 7444 (1986); D.A. Kleinman, R.C. Miller, A.C. Gossard: Phys. Rev. B **35**, 664 (1987)

1.6 F. Cerdeira, E. Anastassakis, M. Cardona: Phys. Rev. Lett. **57**, 3209 (1986)

1.7 A. Alexandrou, M. Cardona, K. Ploog: Phys. Rev. B **38**, 2196 (1988)

1.8 A. Alexandrou, M. Cardona: Solid State Commun. **64**, 1029 (1987)

1.9 M. Cardona: Proceedings of Intl. Symp. on the Physics of Semiconductors and Applications, 1984 Seoul (Korean Physical Society 1984) p. 43; G. Abstreiter, E. Bauser, A. Fischer, K. Ploog: Appl. Phys. **16**, 345 (1978); P.M. Fauchet: IEEE Circuits and Devices Mag. **2**, 37 (1986); I.P. Herman, F. Magnotta, D.E. Kotecki: J. Vac. Sci. Tech. **4**, 859 (1986); K.K. Tiong, P.M. Amirtharaj, F.H. Pollak, D.E. Aspnes: Appl. Phys. Lett. **44**, 122 (1984); M. Holtz, R. Zallen, A.E. Geissberger, R.A. Sadler: J. Appl. Phys. **59**, 1946 (1986)

1.10 M.H. Brodsky (ed.): *Amorphous Semiconductors*, Topics Appl. Phys., Vol. 36 (Springer, Berlin, Heidelberg 1979)

1.11 W. Kauschke, M. Cardona: Phys. Rev. B **33**, 5473 (1986)

1.12 M. Cardona: In *Phonon Physics* (World, Singapore 1985) p. 2; J. Joannopoulos, G. Lukovsky (eds.): *The Physics of Hydrogenated Amorphous Silicon II*, Topics of Appl. Phys., Vol. 56 (Springer, New York 1984)

1.13 A.K. Sood, J. Menéndez, M. Cardona, K. Ploog: Phys. Rev. Lett. **54**, 2111 (1985)

1.14 J. Menéndez, M. Cardona: Phys. Rev. B **31**, 3696 (1985)

1.15 A.B. Harvey (ed.): *Chemical Applications of Nonlinear Raman Spectroscopy* (Academic, New York 1981); M.D. Levenson: *Introduction to Nonlinear Laser Spectroscopy* (Academic, New York 1982) Y.R. Shen: *The Principles of Nonlinear Optics* (Wiley, New York 1984)

1.16 S. Shin, M. Ishigame: Phys. Rev. B 37, 2718 (1988); "Observation of the wave-vector dependence of the central mode by HRS and the A_u-optical-phonon-central-mode coupling in CsH_2PO_4"
L.D. Ziegler, Y.C. Chung, Y.P. Zhang: J. Chem. Phys. 87, 4498 (1987); "Resonant rotational HRS intensities of symmetric top molecules"
S. Shin, A. Ishida, T. Yamakami, T. Fujimara, M. Ishigame: Phys. Rev. B 35, 4455 (1987); "Central mode in HR spectra of the quasi-one-dimensional hydrogen-bonded ferroelectric CsH_2PO_4"
L.E. Zubkova, A.A. Mokhnatyuk, Yu.N. Polivanou, K.A. Prokhorov, R.Sh. Sayakhov: JETP Lett. 45, 59 (1987); "HRS of light by hot phonon polaritons"
H. Vogt: Phys. Rev B 36, 5001 (1987); "Absolute cross sections for HRS by soft modes in oxygen-octahedra ferroelectrics"
S. Shin, M. Ishigame: Phys. Rev. B 34, 8875 (1986); "Defect-induced HR spectra in cubic zirconia"
V.N. Denisov, B.N. Mavrin, V.B. Podobedov, Kh.E. Sterin: Sov. Phys. JETP 63, 338 (1986); "Two-frequency excitation, anisotropy of the intensity, and effects of asymmetry of the tensor of HR light scattering in crystalline quartz"
H. Vogt, M.D. Fontanta, G.E. Kugel, P. Günter: Phys. Rev. B 34, 410 (1986) "Low-frequency dielectric response in cubic $KNbO_3$ studied by HRS"
1.17 R.M. Macfarlane, H. Rosen, H. Seki: Solid State Commun. 63, 831 (1987);
A. Wittlin, R. Liu, M. Cardona, L. Genzel, W. König, W. Bauhofer, Hj. Mattausch, A. Simon, F. García-Alvarado: Solid State Commun. 64, 477 (1987);
R. Liu, C. Thomsen, W. Kress, M. Cardona, B. Gegenheimer, F.W. de Wette, J. Prade, A.D. Kulkarni, U. Schröder: Phys. Rev. B 37, 7971 (1988)
1.18 E. Burstein, A. Pinczuk, S. Buchner: In Physics of Semiconductors 1958, ed. by B.L.H. Wilson, (Institute of Physics, London 1979) p. 123
1.19 P. Wickboldt, E. Anastassakis, M. Cardona: Phys. Rev. B 35, 1362 (1987)
1.20 K.G. Günther: Z. Naturforsch. A 13, 1081 (1958)
1.21 J.R. Arthur: J. Phys. Chem. Solids 28, 2257 (1968); J. Appl. Phys. 39, 4032 (1968)
1.22 A.Y. Cho: J. Appl. Phys. 41, 2780 (1970); J. Vac. Sci. Technol. 8, 531 (1971)
1.23 D. Levine, P.J. Steinhardt: Phys. Rev. Lett. 53, 2477 (1984)
1.24 D. Schechtman, I. Bleck, D. Gratias, J.W. Cahn: Phys. Rev. Lett. 53, 1951 (1984)

2. Collective Excitations in Superlattice Structures

Douglas L. Mills

With 7 Figures

Superlattice structures support a rich variety of collective excitations whose dispersion relation is influenced strongly by the nature of the constituent films and their thicknesses. Examples are provided by plasmons and spin waves, which exist in the form of collective excitations of the structure as a whole. We review the literature of this field along with the basic concepts, with emphasis on theoretical discussions illustrated with experimental examples. In addition, we summarize recent theoretical studies of gap solitons in superlattices.

2.1 Introductory Remarks

Solid state physics remains a lively and active field from the intellectual point of view, largely because new classes of materials are being continually discovered or subjected to detailed study, and they are understood only after qualitatively new concepts are introduced. During the past decade, a diverse variety of artificially synthesized solid structures have been studied. Typically these are multilayer materials, often superlattices which consist of alternating layers of two materials A and B. The superlattice is thus periodic in the direction normal to the interfaces between the two constituents, with period $d = d_A + d_B$, where d_A and d_B are the thicknesses of the films of material A and material B respectively. Multilayers need not be fabricated in the form of periodic superlattices. An exciting area of current study is quasiperiodic structures, where the layer thicknesses are arranged to form a Fibonacci sequence (see Chap. 5).

Multilayer films and structures have been fabricated and studied for decades, of course. The new feature of the current era is the extraordinary high quality of the interfaces. By growth techniques such as molecular beam epitaxy (MBE), one may create superlattices where each film is a nearly perfect crystal, the atomic planes of material B are in registry with those in material A, and the interface between two films is nearly perfect on the atomic scale. The individual film thicknesses in the new materials may range from a few atomic layers to a few hundred Angstroms; multilayer structures synthesized in earlier eras typically consisted of polycrystalline films with thicknesses in the micron regime, and the interfaces were far from sharp on the atomic scale. At the time of this writing, the art of growing semiconductor superlattices of high quality is well developed, though we continue to see new combinations of materials incorporated in them.

Metallic superlattices, some of which incorporate magnetically ordered films (rare earth ferromagnets and spiral spin materials, for example), can now be prepared with quality comparable to semiconductor structures.

The new superlattices are endowed with properties that are unique, and distinct from those of either constituent. For instance, there is a diverse variety of collective excitations of these materials which involve coherent motions of electrons, electron spin, or nuclei (phonons) throughout the whole structure. It follows that the dispersion relations of these modes depend on parameters such as d_A and d_B, the two film thicknesses. The dispersion relation of these excitations and even the existence of certain classes of modes is then subject to design and manipulation. It is a well-established principle of solid state physics that the linear response of a material to an external probe is controlled by the nature of its spectrum of elementary excitations, so that we have the possibility of designing new materials with desired response characteristics.

This chapter is devoted to a discussion of the nature of collective excitations in superlattice structures. We consider the idealized superlattice of infinite extent, and also the effect of its surfaces, which as we shall see, leads to interesting modes localized to the near vicinity of the surface. We shall concentrate our attention to modes whose very existence depends on the presence of an underlying periodic structure, rather than those whose properties are modified by it. For instance, a perfect crystal admits acoustical waves, and the dispersion relation of such waves will be very similar to those in a crystal, save for modification of their velocity, and the appearance of "mini-gaps" in the dispersion relation for propagation normal to the interface. Since these issues may be addressed by straightforward application of the theory of wave propagation in periodic media, we focus our attention on new classes of modes encountered in superlattices. Our emphasis is primarily on the theoretical description of these modes, since other authors in the present volume explore various experimental studies of them. We will, however, refer to experimental work at various points.

As the reader will appreciate, from the point of view of propagation of collective excitations with wavelength long compared to a lattice constant, the superlattice is a physical realization of the one-dimensional Kronig-Penney model that is discussed in elementary texts on solid state physics [2.1]. In essence, the waves propagate in a structure of alternating one-dimensional square wells with thickness d_A and d_B, respectively. Here, our attention is confined to the ideal superlattice, either infinite or semi-infinite in extent, as remarked earlier. Defects may be deliberately introduced in the structure by varying film thicknesses. In this manner, one may create a random, one dimensional Kronig-Penney structure. Thus, one may explore localization [2.2] phenomena in one dimension, in a structure whose underlying parameters are precisely known. Also, as mentioned earlier, quasi-periodic, one-dimensional structures may be grown in this manner. While we regard these as fascinating topics and exciting areas of current and future research, a discussion of them lies outside the scope of the present chapter. In Chap. 5 of this volume, Merlin presents a discussion of quasi-periodic superlattices.

2.2 Optical Phonons and Plasmons in Superlattices

2.2.1 Introduction

We begin by considering the most straightforward example, the long wavelength collective excitations in superlattices which contain ionic or semiconducting materials that admit long wavelength optical phonons, or doped semiconductors or metals which support plasmons by virtue of the free carriers. An important class of collective excitations in such structures may be discussed within a rather simple and general framework. We begin with general remarks.

A full theory of such modes should include an atomistic description of the atomic vibrations and a fully quantum mechanical description of the electrons. Such an approach also requires a microscopic description of the interface. However, long wavelength optical phonons in ionic crystals and plasmons in conducting media generate macroscopic electric fields that may be described by use of Maxwell's equations, in combination with the appropriate frequency dependent dielectric constant (or tensor) of the medium of interest. Such treatments successfully describe the longitudinal phonons or plasmons, along with the polaritons of bulk media [2.3] and surface polaritons of films and semiinfinite crystals [2.4].

In superlattices, an important class of collective excitations has its origin in modes localized at the various interfaces, which interact with each other by virtue of the finite film thickness, couple together, and thus form a collective excitation of the structure as a whole. If the films have thicknesses in the range of a few tens or few hundreds of Angstroms, the dominant sources of coupling are the macroscopic electric fields set up by the atomic or electronic motions near the interface. These are described nicely by the macroscopic theory mentioned in the previous paragraph. We discuss the macroscopic approach here, and we shall see that through it we also obtain key results which emerge from microscopic theories of collective excitations in superlattices. We refer the reader to a paper by *Camley* and the present author, for a more detailed exposition of the approach sketched below, and additional explicit results [2.5].

A full treatment of the problem requires the use of the complete Maxwell's equations, with the influence of retardation included fully. Here we ignore retardation – an approximation valid for many purposes. Retardation effects are crucial when the ratio c/l, where l is a relavant linear dimension (wavelength, measure of structure size) and c the velocity of light, is comparable to the frequency ω of the mode. When $\omega \ll c/l$, we may ignore retardation. If we choose l to be the thickness of a film in our model superlattice, and $l = 100$ Å, then $c/l = 3 \times 10^{16}$ s^{-1}, which lies in the ultraviolet. Thus, retardation influences the modes we discuss only modestly.

2.2.2 The Isolated Film

We begin by considering an isolated film, described by a frequency dependent dielectric constant $\varepsilon(\omega)$. This discussion will be presented in detail, because much of what follows is obtained through very similar reasoning (see also Sect. 3.2.4). If we have an ionic insulator, cubic in structure, with a single infrared active transverse optical (TO) phonon of frequency ω_{TO}, then

$$\varepsilon(\omega) = \varepsilon_\infty + \frac{\Omega_p^2}{\omega_{TO}^2 - \omega^2} \tag{2.1}$$

with ε_∞ the background dielectric constant with origin in interband electronic transitions, and Ω_p^2 a measure of the oscillator strength of the TO phonon. If we have a doped non-polar semiconductor, or a metal, the free carrier contribution is represented by writing

$$\varepsilon(\omega) = \varepsilon_\infty - \frac{\omega_p^2}{\omega^2} \tag{2.2}$$

with ω_p the free carrier plasma frequency. In a doped ionic semiconductor such as n-type GaAs, the contribution from the TO phonon (2.1) and the free carriers are present simultaneously.

Lattice motions or density fluctuations in the free carriers create an electric field $E(x, t)$ which may be expressed in terms of the scalar potential ϕ in the absence of retardation. For a mode with frequency ω, we have

$$E(x, t) = -\nabla \phi \exp(-i\omega t). \tag{2.3}$$

Within the medium, $\nabla \cdot D = 0$, so we have

$$\varepsilon(\omega) \nabla^2 \phi = 0 \tag{2.4a}$$

while in the vacuum above or below the film,

$$\nabla^2 \phi = 0. \tag{2.4b}$$

In the film, and in the superlattices to be discussed later, we have translational invariance in the two directions parallel to the surface, here taken to be the xy plane. Thus, we seek solutions with

$$\phi(x) = \Phi(z) \exp(i k_\parallel \cdot x_\parallel), \tag{2.5}$$

with $x_\parallel$ the projection of x onto the xy plane. Equations (2.4) are then replaced by

$$\varepsilon(\omega) \left[\frac{d^2 \Phi}{dz^2} - k_\parallel^2 \Phi \right] = 0 \tag{2.6a}$$

as the equation obeyed by the electrostatic potential within the film, and

$$\left[\frac{d^2\Phi}{dz^2} - k_{\parallel}^2\right] = 0 \tag{2.6b}$$

outside.

For the model dielectric constant in (2.1), there is a frequency ω_{L0} for which $\varepsilon(\omega_{L0})$ vanishes. This is in fact the longitudinal optical phonon frequency of the bulk material. One shows easily that

$$\omega_{L0} = (\varepsilon_s/\varepsilon_\infty)^{1/2} \omega_{T0} , \tag{2.7}$$

with ε_s the static dielectric constant, given by setting ω to zero on the right hand side of (2.1). This is the well-known Lyddane-Sachs-Teller relation. The dielectric constant in (2.2) vanishes at the screened plasma frequency $\tilde{\omega}_p = \omega_p/\sqrt{\varepsilon_\infty}$. If free carriers are present in a polar material such as GaAs, there are two frequencies where $\varepsilon(\omega)$ vanishes. These are the frequencies of the L+ and L− modes [2.6], which are coupled LO phonon-plasmon modes.

For such special frequencies, (2.6a) is satisfied for any choice of $\Phi(z)$. Let the film extend from $z=0$ to $z=d$. Outside the film, we must have

$$\Phi(z) = \Phi^> \exp(-k_{\parallel}[z-d]) , \quad z > d \tag{2.8}$$
$$\Phi(z) = \Phi^< \exp(+k_{\parallel}z) , \quad z < 0$$

But normal components of D are conserved at each interface, so we require

$$\varepsilon(\omega)\left.\frac{d\Phi}{dz}\right|_{z=0+} = +k_{\parallel}\Phi^< \quad \text{and} \tag{2.9a}$$

$$\varepsilon(\omega)\left.\frac{d\Phi}{dz}\right|_{z=d-} = -k_{\parallel}\Phi^> . \tag{2.9b}$$

For our special frequencies, $\varepsilon(\omega)=0$ once again, and it follows that $\Phi^< = \Phi^< = 0$. Thus the finite slab supports excitations at frequencies identical to longitudinal modes of the bulk material, and these slab modes have the property that the macroscopic electric field *outside* the slab vanishes identically.

We may have solutions with

$$\Phi(z) = \Phi_s \sin(k_z z) + \Phi_c \cos(k_z z) \tag{2.10}$$

at these frequencies. These are standing wave excitations in the slab. Continuity of tangential E requires continuity of $\Phi(z)$ at $z=0$ and $z=d$. Application of the condition at $z=0$ requires

$$\Phi_c = 0 \tag{2.11a}$$

and its application at $z=d$ gives

$$k_z = n\pi/d , \quad n=1,2,3 \dots . \tag{2.11b}$$

Note that (2.10a,b) may have to be modified as a result of the mechanical boundary condition (Sect. 2.3).

If $\varepsilon(\omega) \neq 0$, then within the slab, we have two independent solutions of (2.6a). We write these in the form

$$\Phi(z) = \Phi_+ \cosh[k_{\parallel}(z-d/2)] \quad \text{and} \tag{2.12a}$$

$$\Phi(z) = \Phi_- \sinh[k_{\parallel}(z-d/2)] , \tag{2.12b}$$

where the subscripts on the right refer to the parity of the potential with respect to reflection through the midplane of the slab. Conservation of tangential E gives

$$\Phi^> = \Phi^< = \Phi_+ \cosh[k_{\parallel}d/2] \quad \text{and} \tag{2.13a}$$

$$\Phi^> = -\Phi^< = \Phi_- \sinh[k_{\parallel}d/2] . \tag{2.13b}$$

Continuity of normal D leads to a second constraint:

$$\Phi^> = \Phi^< = -\varepsilon(\omega)\Phi_+ \sinh(k_{\parallel}d/2) \tag{2.14a}$$

$$\Phi^> = -\Phi^< = -\varepsilon(\omega)\Phi_- \cosh(k_{\parallel}d/2) \tag{2.14b}$$

Equations (2.13) and (2.14) impose a constraint on $\varepsilon(\omega)$ that yields a dispersion relation $\omega_{\pm}(k_{\parallel})$ for the two modes. We have

$$\varepsilon(\omega_+(k_{\parallel})) = -\coth(k_{\parallel}d/2) \quad \text{and} \tag{2.15a}$$

$$\varepsilon(\omega_-(k_{\parallel})) - = -\tanh(k_{\parallel}d/2) . \tag{2.15b}$$

Quite clearly, (2.15) yield solutions only if there are frequency regimes where $\varepsilon(\omega) < 0$. For our two model dielectric constants, $\varepsilon(\omega)$ is negative when $\omega_{TO} < \omega < \omega_{LO}$ (2.1), or $0 < \omega < \omega_p/\sqrt{\varepsilon_\infty}$ (2.2).

In each case, an explicit dispersion relation is obtained. For an ionic slab described by the dielectric response function in (2.1), we have

$$\omega_+^2(k_{\parallel}) = \omega_{TO}^2 + \frac{\omega_{LO}^2 - \omega_{TO}^2}{1 + (1/\varepsilon_\infty)\coth(k_{\parallel}d/2)} \quad \text{and} \tag{2.16a}$$

$$\omega_-^2(k_{\parallel}) = \omega_{TO}^2 + \frac{\omega_{LO}^2 - \omega_{TO}^2}{1 + (1/\varepsilon_\infty)\tanh(k_{\parallel}d/2)} , \tag{2.16b}$$

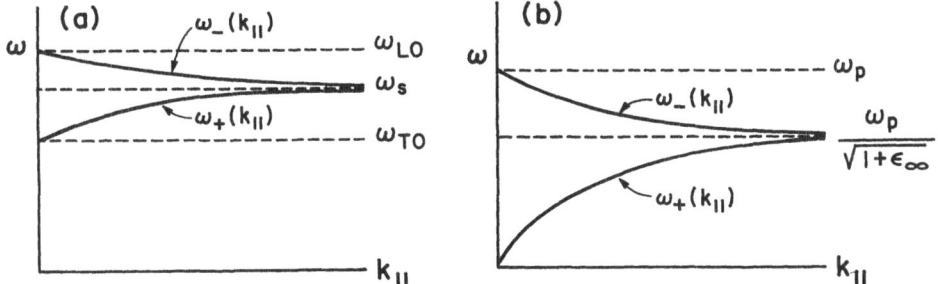

Fig. 2.1a, b. The dispersion relation of the two surface modes of a finite dielectric film. **a** Ionic film described by the dielectric function in (2.1), and **b** conducting film described by the dielectric function in (2.2). See also Fig. 3.17

while for the conducting film described by (2.2),

$$\omega_+^2(k_\parallel) = \frac{\omega_p^2}{\varepsilon_\infty + \coth(k_\parallel d/2)} \quad \text{and} \tag{2.17a}$$

$$\omega_-^2(k_\parallel) = \frac{\omega_p^2}{\varepsilon_\infty + \tanh(k_\parallel d/2)} . \tag{2.17b}$$

A sketch of these dispersion relations is provided in Fig. 2.1. As $k_\parallel d \to \infty$, in both cases $\omega_+(k_\parallel)$ and $\omega_-(k_\parallel)$ approach a common limit, while as $k_\parallel \to 0$, the odd parity mode is driven to the limit ω_{LO} for the ionic slab, and the bulk plasma frequency $\omega_p/\sqrt{\varepsilon_\infty}$ for the conducting slab.

If one traces through the field amplitudes, one sees that the odd parity mode involves, as $k_\parallel \to 0$, a rigid body excitation of polarization normal to the film surfaces, with the generation of a depolarization field which provides a restoring force. As $k_\parallel \to 0$, the even parity mode is driven to ω_{TO} for the ionic slab, from (2.16a). This describes an ionic slab excited with polarization field parallel to its surface; there is no macroscopic field generated in this case, so the frequency drops to ω_{TO}. When $k_\parallel d \ll 1$ for the conducting slab, we have

$$\omega_+(k_\parallel) = \omega_p(2k_\parallel d)^{1/2} , \tag{2.18}$$

which is the dispersion relation of a plasmon of long wavelength in a two dimensional plasma. As $k_\parallel \to 0$, the frequency drops to zero, since no macroscopic field is generated within the plasma in this limit.

The physical interpretation of the modes just discussed becomes evident by examining the limit $k_\parallel d \gg 1$, where $\omega_+(k_\parallel)$ and $\omega_-(k_\parallel)$ become degenerate. The potential $\Phi(z)$ is non-zero only in the vicinity of each interface. We have two surface modes, one localized near each of the two film surfaces. When the film has finite thickness, the two modes interact, and we form even and odd parity combinations of the potential in the manner familiar from the quantum mechanics of weakly interacting bound states. As we pass into the regime

$k_\parallel d \ll 1$, the character of the modes changes, but we may still view the resulting entities as two interacting surface waves, one on each surface.

2.2.3 The Infinite Superlattice

From the concepts just set down, we can now address the nature of the collective excitations in a superlattice. The structure we consider first is the perfect superlattice of infinite extent, which consists of alternating layers of dielectric films, A and B, as illustrated in Fig. 2.2. The films have dielectric constants $\varepsilon_A(\omega)$ and $\varepsilon_B(\omega)$, with thicknesses d_A and d_B, respectively. The length of the unit cell is thus $d = d_A + d_B$.

We have seen that within the isolated film, we have standing wave excitations at those frequencies where the dielectric constant vanishes; the electric fields associated with such modes are entirely confined to the interior of the film, within the framework of dielectric theory. If we consider one film, say film A within the superlattice, we also have standing wave excitations at those frequencies where $\varepsilon_A(\omega) = 0$. These modes are uncoupled to those in the adjacent films, since the electric field generated by the excitations in film A vanishes identically outside it. Similarly, within the B films, we have a set of standing wave modes where $\varepsilon_B(\omega) = 0$. The character of these standing waves is uninfluenced by incorporation of the films within the superlattice, within the framework of our dielectric theory which neglects the effect of dispersion of the optical phonon branches on the mode structure. A simple model which incorporates this dispersion has been presented in [2.16].

We saw in the discussion of the isolated film that in spectral regions where its dielectric constant is negative, each surface supports a surface mode, localized to its vicinity when $k_\parallel d \gg 1$. Each interface in the superlattice will also support a surface mode, if the relationship between $\varepsilon_A(\omega)$ and $\varepsilon_B(\omega)$ is appropriate. Analysis of a single interface between two semi-infinite dielectrics shows that when $\varepsilon_A(\omega)$ and $\varepsilon_B(\omega)$ have opposite signs, the interface supports surface waves.

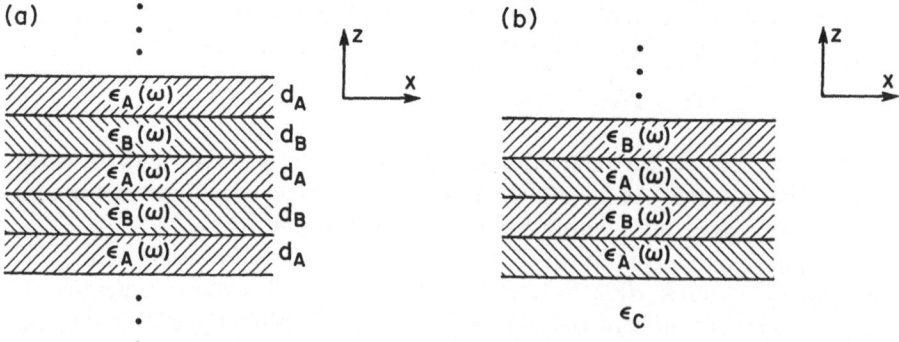

Fig. 2.2a, b. The two superlattice structures considered in the text. **a** Of infinite extent, consisting of alternating layers of material A and material B, and **b** the semi-infinite superlattice, with a film of material A as the outer film. In the lower half space $z < 0$, the dielectric constant is ε_c

Just as the surface waves on the two film surfaces interact, so do the waves on the various interfaces of the superlattice structure. The periodicity of the structure dictates that $\Phi(z)$, the function in (2.5) which describes the spatial variation of the potential normal to the interface, become a Bloch wave. We may write, very much as in electron energy band theory,

$$\Phi(z) = U(z)\exp(ik_\perp z) ,\tag{2.19}$$

where $U(z)$ has the periodicity of the superlattice.

There is a direct analogy with electron energy band theory, as follows. The wave localized on each interface may be viewed as a Wannier function in the tight-binding version of electron energy band theory. By virtue of their overlap, the waves link together to form a propagating Bloch state. In essence, we have the analog of a tight-binding band in a crystal with two atoms per unit cell. One interface in each unit cell has a B film above and an A film below, and the other the opposite ordering. For each choice of $k_\parallel$, we will thus find two collective mode *bands*, because the interface modes link together to form Bloch states which transport energy normal to the interfaces. The frequency of such a superlattice collective mode then depends on both $k_\parallel$ and $k_\perp$; $k_\perp$ lies within the first Brillouin zone of the superlattice, between $-\pi/d$ and $+\pi/d$.

Camley and *Mills* [2.5] have obtained the dispersion relation of these modes by a straightforward extension of the discussion given above for the isolated film. The result may be phrased in terms of a function $c(k_\parallel, k_\perp)$ which depends only on the geometrical structure of the superlattice:

$$c(k_\parallel, k_\perp) = \frac{\cosh(k_\parallel d_A)\cosh(k_\parallel d_B) - \cos(k_\perp d)}{\sinh(k_\parallel d_A)\sinh(k_\parallel d_B)} .\tag{2.20}$$

When $c(k_\parallel, k_\perp) > 1$, then the implicit dispersion relation of the superlattice collective modes may be written

$$\frac{\varepsilon_A(\omega)}{\varepsilon_B(\omega)} = -\exp(\pm\psi(k_\parallel, k_\perp)) ,\tag{2.21}$$

where we let $c(k_\parallel, k_\perp) = \cosh(\psi(k_\parallel, k_\perp))$. For an equivalent form of the dispersion relation see (3.41).

We remarked earlier that for each choice of $(k_\parallel, k_\perp)$, we obtain two solutions because each unit cell contains two inequalent interfaces. Each choice of sign in (2.21) yields a distinct frequency, upon noting the identity

$$\exp(\pm\psi(k_\parallel, k_\perp)) = c(k_\parallel, k_\perp) \pm (c^2(k_\parallel, k_\perp) - 1)^{1/2} .\tag{2.22}$$

We require $c(k_\parallel, k_\perp) > 1$ for a solution to exist, for any choice of $k_\parallel$ and $k_\perp$.

Then for fixed $k_\parallel$, and $k_\perp$ swept through the Brillouin zone, we obtain a band of collective excitations. In Fig. 2.3, we give an example of the collective mode bands [2.5] for a GaAs/AlAs superlattice (both materials undoped). It is the

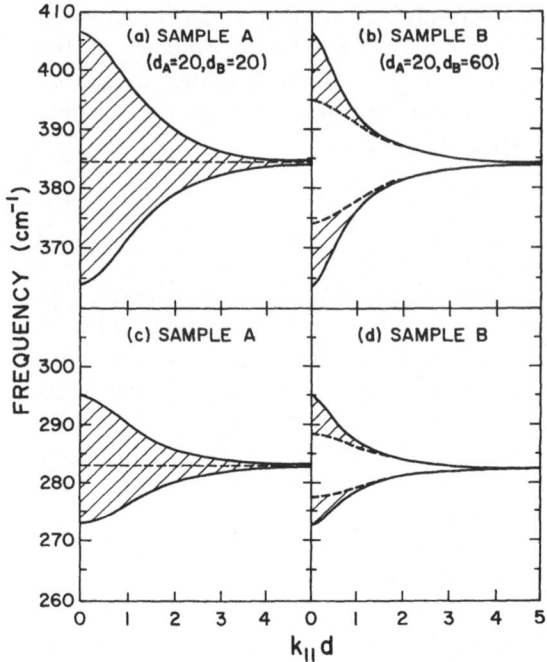

Fig. 2.3. Collective excitation bands in a GaAs/AlAs superlattice, for a and c $d_A = d_B = 20$ Å and b and d $d_A = 20$ Å and $d_B = 60$ Å. Material A is GaAs and material B is AlAs. One has two bands near the AlAs Restrahlen band (a and b) and near the GaAs Restrahlen band (c and d). With $d = d_A + d_B$, the solid lines are the dispersion curves for $k_\perp d = \pi$, and the dashed line $k_\perp d = 0$. The figure is reproduced from [2.14]

infrared active TO phonon that endows each material with a frequency dependent dielectric constant in the far infrared, driving the ratio $\varepsilon_A(\omega)/\varepsilon_B(\omega)$ negative in the vicinity of the Restrahlen band of each material.

As remarked earlier, the above discussion neglects the role of retardation. In a recent paper [2.7], *Szenics* et al. have explored the influence of retardation on the excitation spectrum of the superlattice. These authors find additional branches to the excitation spectrum, in the region $ck_\| \sim \omega$, which owe their existence to the presence of retardation. Quite clearly, retardation effects will be important in the discussion of experiments such as attenuated total reflection (ATR) studies of superlattice structures.

Quinn and collaborators have completed extensive theoretical studies of collective excitations in infinite superlattices [2.8–2.11], based on the following physical picture. They suppose the electrons to be confined within quantum wells localized at the interface; noting that the dielectric constant of GaAs and $Ga_xAl_{1-x}As$ are very similar, the background matrix is represented by a dielectric with dielectric constant ε. The electron response is treated microscopically within a formalism equivalent to the random phase approximation of many body theory. They include the influence of a magnetic field [2.8], and explore superlattices which contain alternating layers of electrons and holes [2.9] then they examine the role of inter-subband excitations [2.10], and finally introduce retardation effects [2.11].

An intriguing mode referred to as an acoustic plasmon is derived and discussed in these papers. The principal properties of these modes are obtained easily from the treatment given above. We may model a sequence of thin conducting sheets (the electrons in their quantum wells) by letting $\varepsilon_A(\omega) = \varepsilon - \omega_p^2/\omega^2$, and $\varepsilon_B = \varepsilon$, then allowing d_A to become very small. For this model (and with d_B arbitrary for the moment) the two branch dispersion relation assumes the form

$$\omega_\pm(k_\parallel, k_\perp) = \frac{\tilde{\omega}_p}{\sqrt{2}} \left[1 \pm \left(1 - \frac{1}{1 + c(k_\parallel, k_\perp)} \right) \right]^{1/2} \tag{2.23}$$

where $\tilde{\omega}_p = \omega_p/\sqrt{\varepsilon}$. If we now let d_A become very small,

$$c(k_\parallel, k_\perp) \cong \frac{\coth(k_\parallel d_B) - \operatorname{cosech}(k_\parallel d_B) \cot(k_\perp d_B)}{k_\parallel d_A \sinh(k_\parallel d_B)}. \tag{2.24}$$

In this limit, $c(k_\parallel, k_\perp) \gg 1$ always, hence to a good approximation we have for the mode whose potential has odd parity

$$\omega_-(k_\parallel, k_\perp) \cong \tilde{\omega}_p \left(1 - \frac{1}{4c(k_\parallel, k_\perp)} \right) \cong \tilde{\omega}_p \tag{2.25a}$$

and for the even parity mode

$$\omega_+(k_\parallel, k_\perp) = \tilde{\omega}_p / (2c(k_\parallel, k_\perp))^{1/2}. \tag{2.25b}$$

The mode ω_+ is the oscillation of the conducting layers normal to the interface. Depolarization fields render the frequency finite in the limit considered, and interactions between adjacent quantum wells are weak.

The acoustic plasmon mode emerges as the appropriate limit of (2.25b). If we assume also $k_\parallel d_B \ll 1$, a limit appropriate to modes excited in light scattering experiments discussed in Chap. 4 of this volume, we find the dispersion relation that emerges from the work of Quinn et al.:

$$\omega_+(k_\parallel, k_\perp) = \left(\frac{2\pi n_s e^2 d_B}{em^*} \right)^{1/2} \frac{k_\parallel}{(1 - \cos k_\perp d_B)^{1/2}}. \tag{2.26}$$

We have written the free carrier plasma frequency as $\omega_p^2 = 4\pi n e^2/m^*$, with n the volume number density of the free carriers in the thin layers, and have identified the combination $n d_A$ as the number of free carriers per unit area n_s in a given quantum well.

Other limits of (2.25b) are interesting to explore. If $k_\parallel d_B \gg 1$, then

$$\omega_+(k_\parallel, k_\perp) = \left(\frac{2\pi e^2 n_s k_\parallel}{m^* \varepsilon} \right)^{1/2} \tag{2.27}$$

which we now recognize as the dispersion relation of a plasmon in a two-dimensional electron gas [recall (2.18)]. The absence of a dependence on $k_\perp$ in

(2.27) shows that in this limit, there is no cross coupling between the thin layers; each supports its own two-dimensional plasmon.

If we let $k_\perp \to 0$ while $k_\parallel$ is finite [(2.26) follows upon holding $k_\perp$ finite while $k_\parallel$ becomes small], then

$$\omega_+(k_\parallel, 0) = (2\pi n_s e^2/m^*\varepsilon)^{1/2} (k_\parallel \coth(k_\parallel d_B/2))^{1/2} \qquad (2.28)$$

which, when $k_\parallel d_B \to 0$, approaches the finite value $(4\pi\langle n\rangle e^2/m^*\varepsilon)^{1/2}$, with $\langle n\rangle = n_s/d_B$ the *average* free carrier concentration in the superlattice structure.

Burstein et al. [2.12] have shown that for electrons in a quantum well localized at a single interface, a proper description of the long wavelength collective excitations including inter-subband excitations is obtained by endowing the thin conducting layer with an anisotropic dielectric tensor that is diagonal, with $\varepsilon_\perp(\omega)$ and $\varepsilon_\parallel(\omega)$ describing the response perpendicular to and parallel to the interface, respectively. One has $\varepsilon_\perp(\omega) = \varepsilon + \Omega_{01}^2/(\omega_{01}^2 - \omega^2)$ and $\varepsilon_\parallel = \varepsilon - \omega_p^2/\omega^2$, with ω_{01} the frequency of the inter-subband transition and Ω_{01}^2 a measure of its oscillator strength. It is straightforward to extend the dielectric theory description of collective excitation in superlattices to include such anisotropy in one layer of each unit cell, say layer A. When this is done, and we mimic quantum wells by letting d_A become small, one finds that (2.26) still describes the surface acoustic plasmon when $k_\parallel d_B \ll 1$. On the other hand, the frequency of the upper branch, which is now the collective mode branch to which the inter-subband excitations contribute importantly, approaches $(\omega_{01}^2 + \Omega_{01}^2/\varepsilon)^{1/2}$ in this limit. This is the frequency of the inter-subband transition stiffened by the surface depolarization field, as discussed first by *Chen* et al. [2.13].

Experimental studies of collective excitations in superlattices are reviewed in Chaps. 3 and 4 of this volume. We wish to comment on one issue here, however. *Cardona* and his colleagues [2.14] have provided light scattering data on GaAs/AlAs superlattices which contain clear structure with origin in the collective excitations examined here, and described by (2.21). In addition, they also observe [2.15] the standing wave LO phonon modes described by (2.10) and (2.11), which as we have seen, are modes confined to a single film in the superlattice structure. If $u(x)$ is the optical phonon amplitude associated with one of the modes, then within dielectric theory one easily shows that

$$u(x) = \frac{e^*\varepsilon_\infty}{M\Omega_p^2}\, \Phi_s \exp(i k_\parallel \cdot x_\parallel)\, [i k_\parallel \sin(k_z^{(n)}z) + k_n^{(n)}\hat{z}\cos(q_z^{(n)}z)]\,, \qquad (2.29)$$

where $k_z^{(n)} = n\pi/d$, e^* and M are the transverse effective charge and reduced mass in the unit cell, and all physical parameters refer to the film within which the mode is localized. In the light scattering experiment, the magnitude of $k_\parallel$ is roughly 10^5 cm^{-1}, and since the films in the structures examined were only a few tens of Angstroms thick, we have $k_z^{(n)} \gg |k_\parallel|$. Thus, to a good approximation,

(2.29) may be replaced by

$$u(x) \cong \frac{e^* \varepsilon_\infty}{M\Omega_p^2} \, \Phi_s k_z^{(n)} \hat{z} \exp(i\boldsymbol{k}_\parallel \cdot \boldsymbol{x}_\parallel) \cos(k_z^{(n)} z) \, . \tag{2.30}$$

The theory predicts that the displacement field in these modes will have an anti-node at the surface.

The data, when analyzed in view of the selection rules applicable to light scattering [2.15] suggest that $u(x) \sim \hat{z} \exp(i\boldsymbol{k}_\parallel \cdot \boldsymbol{x}_\parallel) \sin(k_z^{(n)} z)$, i.e. the displacement field has a node rather than an anti-node at each surface. Simply on the basis of a force constant model which ignores the presence of the macroscopic electric field, this is in fact the result that would be expected. There is a large mismatch between the optical phonon frequencies of AlAs and GaAs, and this will force the amplitude of the standing wave modes to a value close to zero at the interface. This is because, under these circumstances, a standing wave excitation couples inefficiently to atoms in the neighboring medium (Sect. 3.2.4).

What is required is a theory which explains why dielectric theory provides an adequate description of the interface modes, but fails for the standing wave LO phonons. This question has been addressed recently by *Huang* and *Zhu* [2.16], within a model which is simple in content, but which contains the essential physics, in the view of the present author. They consider a model of a superlattice made from films of material *A* and *B*, where *A* and *B* differ only in their transverse optical phonon frequency at zero wave vector. Both materials have the same dipole moment effective charge, high frequency dielectric constant, and each bulk material is endowed with bulk phonons with identical dispersion. For this model, Huang and Zhu derive an effective dynamical matrix which, when diagonalized produces interface modes very similar to those which emerge from dielectric theory, but they find the confined LO phonons have zero displacement at the surface. Throughout the film, the displacement field of the confined LO modes is very similar to that in (2.30), but within a quarter wavelength of either interface it rolls over to zero, rather than possess an anti node at the surface. In this picture, in a film many optical phonon wavelengths thick, dielectric theory describes the confined LO phonon everywhere, except near each surface.

Huang and Zhu note that in the long wavelength limit, all the standing waves described by (2.30) have the frequency ω_{LO}, independent of n. Then any arbitrary linear combination of these modes (modes of different n, with $\boldsymbol{k}_\parallel$ fixed) is also an eigenmode. In essence, the perturbation produced by the mismatch between materials *A* and *B* mixes these eigenvectors strongly, to produce a new eigenvector for which the displacement at the surface vanishes. The situation reminds one of degenerate perturbation theory in quantum mechanics, where even a weak perturbation can produce a new eigenfunction of well-defined character, formed as a linear combination of degenerate states. The interface collective modes, however, are non-degenerate; for a given $\boldsymbol{k}_\parallel$, there is only a single eigenvector associated with each frequency. These are then rather "rigid" states, and the character of the eigenvector provided by dielectric theory is thus not changed greatly by the perturbation generated by the mismatch between materials.

2.2.4 The Semi-Infinite Superlattice

In the previous subsection, we have seen how modes localized on the interfaces of a superlattice structure link together to form collective excitations of the structure as a whole. Schematically, we may view this phenomenon in the following way. Consider an interface at $z = z_n$, and denote the electrostatic potential associated with the mode on this interface by $\exp(i\mathbf{k}_\parallel \cdot \mathbf{x}_\parallel)\Phi_{\mathbf{k}_\parallel}(z - z_n)$. (Again, for simplicity, we ignore retardation effects.) Then the Bloch function which describes the collective excitation may be written, in a form more explicit than (2.19),

$$\phi_{\mathbf{k}_\parallel \mathbf{k}_\perp} = \exp(i\mathbf{k}_\parallel \cdot \mathbf{x}_\parallel) \left\{ \sum_n \exp(i\mathbf{k}_\perp z_n)\Phi_{\mathbf{k}_\parallel}(z - z_n) \right\}. \tag{2.31}$$

The expression in curly brackets in (2.31) is identical in form to a Bloch function in the theory of electron energy bands in one dimension, in the tight binding limit. There $\Phi_{\mathbf{k}_\parallel}(z - z_n)$ is the Wannier function localized at site z_n. As remarked earlier, the superlattice is formally equivalent to a one dimensional Kronig-

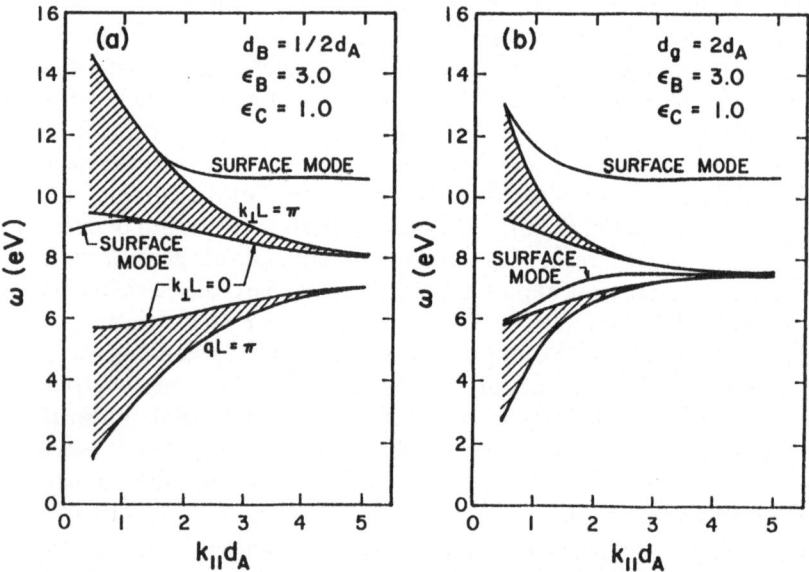

Fig. 2.4a, b. Dispersion relation of the superlattice surface modes for two values of the ratio d_A/d_B, **a** $d_A/d_B = 2$ and **b** $d_A/d_B = 1/2$. The calculations consider a superlattice formed from aluminum (outermost film, film A) and an insulator such as aluminum oxide. The shaded regions are the bulk collective excitations for the structure

Penney model. This discussion is a bit schematic; actually there are two inequivalent interfaces per unit cell and this should be denoted explicitly. But here we are interested only in the overall structure of the theory.

Let us suppose we have a semi-infinite superlattice. To be explicit, let the structure reside in the half space $z > 0$, the outermost film be of material A, and the lower half space be filled with material of dielectric constant ε_c. The two films are described by dielectric constants $\varepsilon_A(\omega)$ and $\varepsilon_B(\omega)$, as in our earlier discussion of collective excitations in the semi-infinite structure. The geometry is illustrated in Fig. 2.2.

For such a structure, a new mode may appear in the excitation spectrum, localized to the surface. Crudely speaking, such modes are described by replacing $k_\perp$ in (2.31) by $i\alpha_\perp$, so that the electrostatic potential within the superlattice is given by:

$$\phi^{(s)}_{k_\parallel k_\perp} = \exp(\mathbf{k}_\parallel \cdot \mathbf{x}_\parallel) \left\{ \sum_n \exp(-\alpha_\perp z_n) \Phi_{k_\parallel}(z - z_n) \right\} . \qquad (z > 0) . \qquad (2.32a)$$

The potential must be continuous at the surface, and thus extends into the material below the superlattice. There the potential has the form

$$\phi^{(s<)}_{k_\parallel k_\perp} = \exp(i\mathbf{k}_\parallel \cdot \mathbf{x}_\parallel) \exp(+k_\parallel z) \phi^{(s<)} , \qquad (2.32b)$$

since it must satisfy Laplace's equation and vanish as $z \to -\infty$.

The semi-infinite superlattice is an analog to the one-dimensional Kronig-Penney model; surface electronic states in the terminated form of this model were discussed many years ago by *Tamm* [2.17]. The surface excitations of the superlattice just described may thus be viewed as a direct analog of Tamm's electron states.

The first discussion of surface modes in a superlattice with the properties just outlined was given by *Camley* et al. [2.18], in their discussion of light scattering from spin waves in magnetic superlattices. We shall review this topic in Sect. 2.3. Shortly thereafter, *Camley* and *Mills* [2.19] discussed the collective excitations of the semi-infinite dielectric superlattice, modeled as just outlined. Subsequently, *Guiliani* and *Quinn* [2.20] presented a treatment of these modes, within the framework of their quantum well model of the semi-infinite GaAs based superlattice. The treatment of Guiliani and Quinn incorporates the influence of retardation.

As is often the case in these problems, a dispersion relation is found only in implicit form, and thus one must obtain a numerical solution for the dispersion relation of these modes in any particular instance. For the model described above, and in the present notation, *Camley* and *Mills* [2.5] find the implicit dispersion relation may be written as

$$2\cosh(k_\parallel d_A)\sinh(k_\parallel d_B)$$
$$+ \sinh(k_\parallel d_A)\left[P_2(\omega)\exp(k_\parallel d_B) - P_1(\omega)\exp(-k_\parallel d_B)\right] = 0 , \qquad (2.33)$$

where

$$P_1(\omega) = \frac{\varepsilon_A(\omega) - \varepsilon_B(\omega)\varepsilon_c}{\varepsilon_A(\omega)(\varepsilon_c - \varepsilon_B(\omega))} \tag{2.34a}$$

and

$$P_2(\omega) = \frac{\varepsilon_A^2(\omega) + \varepsilon_B(\omega)\varepsilon_c}{\varepsilon_A(\omega)(\varepsilon_c + \varepsilon_B(\omega))} . \tag{2.34b}$$

Given any solution of (2.33), one must check and be certain that the attenuation constant $\alpha_\perp$ introduced above is positive; otherwise, spurious solutions can emerge from such an implicit dispersion relation as (2.33).

Suppose $\varepsilon_A(\omega)$ is frequency dependent, while ε_B and ε_c are not. Then we call film A the *active* film. If film A is a semiconductor doped with free carriers or a metal, $\varepsilon_A(\omega)$ may be modeled as in (2.2). If we are describing an ionic material with an infrared active TO phonon, $\varepsilon_A(\omega)$ may be modeled as in (2.1). For these cases, in the previous subsection, we have seen that each choice of $k_\parallel$ yields two bands of collective excitations, as illustrated for a particular example in Fig. 2.3. In general, the surface modes lie outside these bands; we show two examples in Fig. 2.4, for a superlattice formed from aluminum metal, and a dielectric insulator such as its oxide. The figures are taken from [2.5], but rearranged to conform to the present notation. (In a recent paper [2.21], *Wallis* and co-workers have explored the dispersion relation of magnetoplasmon collective modes in the semi-infinite superlattice, with retardation included.)

For a structure such as that considered in Fig. 2.4, inelastic electron scattering offers a means of probing the collective excitations of the semi-infinite superlattice near its surface. This was proposed by *Camley* and *Mills* [2.5], who calculated the near specular electron energy loss cross section for such structures. *Streight* and *Mills* [2.22] have recently examined the influence of non-uniformity of the outermost layer on the electron loss spectrum to find, unfortunately, only a rather modest influence, unless the properties of the outer layer are modified very substantially. The dispersion curves of the structure are affected substantially by the modifications of the outer layer, but since the near-specular electron loss method averages over a rather wide range of wave vectors, the experimental spectrum is influenced only modestly.

Quinn and co-workers have carried out extensive numerical studies of the influence of surface modes on various experimental means of probing superlattices. Following a method introduced by *Katayama* and *Ando* [2.23] in their treatment of light scattering by collective excitations in the infinite superlattice, *Hawrylak* et al. [2.24] have discussed light scattering from the semi-infinite system, with collective modes formed from inter-subband excitations included. The same authors have explored the form of the near specular electron energy loss cross section for their model [2.25]. These two papers are preceded by a study of the influence of a surface on the collective excitations of a semi-infinite superlattice of quantum wells, each of which contains intra-subband as well as inter-subband excitations [2.26]. The excitation spectrum of superlattices with several inequivalent charge layers (quantum wells) per unit cell are also treated in this series of papers. Beginning with a discussion of the excitation spectrum of selected examples of such structures [2.27], *Mayanovic* et al. continue to examine collective excitations localized near the surface, along with the light scattering spectrum and electron energy loss spectrum, analyzed as before [2.28].

All of our discussion of surface effects assumes that the surface is strictly parallel to the interfaces between the superlattice. *Wu* et al. [2.29, 30] explore the behavior of magnetoplasmon modes near the surface of a semiconducting superlattice, again modeled as a collection of thin quantum wells, in which the surface is *normal* to the interfaces. They find a rich and interesting spectrum of surface modes. Finally, these authors explore the case where the surface is cut at an arbitrary angle θ with respect to the normal to the interfaces [2.31].

In configurations where the surface is nonparallel to the interfaces, the physics will prove intriguing. Quite clearly, there are now three classes of modes for such a structure that is semi-infinite. We have the bulk collective excitations, which are viewed as linear combinations of waves localized on the interfaces that are linked together to form Bloch waves through the overlap of their fields, as we have discussed. These may propagate up to and reflect off the surface, though since we no longer have translational symmetry, but rather periodic modulation of the structure parallel to the surface, in principle there may be several reflected waves (both propagating and evanescent) due to Umklapp processes. Formally, the reflection process reminds one of the Bragg scattering of a wave from a periodic structure. We then have surface modes, such as those discussed by *Wu* et al. [2.29–31], which necessarily lie outside the bulk bands. One would also expect surface resonances or leaky waves on the surface. These are surface waves which lie inside the bulk bands, and can leak their energy into the bulk by radiating bulk waves. This is possible if the coupling to the bulk waves is too weak for long lived surface resonances to occur. We would expect this to happen on intuitive grounds. A more complete theoretical analysis of such structures, which goes beyond the calculation of the dispersion relation of surface waves, will prove of great interest.

2.2.5 Finite Superlattices

While many theoretical analyses are performed on the infinitely extended superlattice or, as we have just seen, on superlattice structures that are semi-infinite in extent, in fact, all samples studied in the laboratory have finite number of unit cells. In a typical sample, there may be approximately fifty unit cells.

The collective mode structure of a finite superlattice may differ from that in the semi-infinite structures. Surface waves on the two surfaces may now couple and interact, to become a mode of the structure as a whole. Also, the bulk excitations no longer form a continuum for each value of the wave vector $k_{\parallel}$ parallel to the surface, but now $k_{\perp}$ will be quantized to assume values $k_{\perp} \cong m\pi/D$, with D the thickness of the sample, and m is an integer.

When the bulk excitations are explored by a technique such as light scattering, in general they produce a loss structure in the form of a band that extends over the continuum associated with the particular value of $k_{\parallel}$ probed by the scattering geometry. In the presence of the finite size quantization of $k_{\perp}$, these bands will acquire fine structure. Such fine structure has been observed in light scattering studies of spin waves in finite magnetic superlattices, as we shall see in Sect. 2.3.

A very nice discussion of the collective excitation spectrum of a finite dielectric superlattice has been given by *Johnson* et al. [2.32]. These authors explore the issues discussed above, and present instructive plots of the electrostatic potential associated with the various modes they consider. Finite size effects are incorporated also in the work of *Quinn* and co-workers [2.28, 33].

2.3 Spin Wave Excitations in Magnetic Superlattices

2.3.1 Introduction

In addition to the superlattices fabricated from semiconducting materials which were the primary topic of the previous section, high quality superlattice materials, in which one or more constituent is a magnetically ordered material, have also been fabricated. So far, these constituents are magnetic metals, either the transition metals Fe and Ni, or rare earth metals such as Gd or Dy.

Before we begin, we should remark that magnetic multilayers have been studied for many years. The early work concentrated on CoCr multilayers (Co is ferromagnetic, and Cr antiferromagnetic), and multilayers formed from the antiferromagnet FeMn and the ferromagnet Permalloy. These structures have properties that prove desirable in technological applications. Sample preparation techniques are such that the quality of the interfaces is very poor, compared to the semiconducting superlattices fabricated with MBE technology.

A new generation of magnetic superlattice structures, in which the interfaces are quite sharp on the microscopic scale, have appeared in recent years. These are the materials of interest in the present chapter, because they come so close to achieving perfect interfaces that quantitative contact between theory and experiment has been achieved in several cases. Sputtering techniques [2.34] have produced very high quality superlattices of such combinations as MoNi, WFe, and PdFe. These are structures in which ferromagnetics films alternate with films of nonmagnetic material. Film thicknesses in the various samples range from tens to hundreds of Angstroms. More recently, MBE techniques have been used to synthesize superlattices of Y and Gd, Y and Dy, and Dy with Gd, with interface quality as high as that achieved in the semiconducting superlattices [2.35]. One can expect other combinations to appear shortly. The phenomenology of magnetism in the rare earth metals (generally hexagonal) is very rich. Simple ferromagnetism occurs in Gd, spiral spin ordering in Dy, and while Y is a paramagnetic metal. Evidently it is close to an instability with respect to antiferromagnetism. As we shall see, in the Y-Gd structures, the Y couples the magnetic moments of adjacent Gd films over very long distances, to form a superlattice with a coherent magnetic ground state.

We begin with general remarks on the nature of spin waves in magnetically ordered materials. Direct analogies may be made between spin waves in magnetic materials and the phonons of (anisotropic) insulators, although, as we shall see, clear and distinct differences remain. We refer the reader to a review article on surface spin waves which further elaborates on these points [2.36]. Here we provide only a brief orientation to the reader who should also look at Chap. 7 of this volume.

When we examine the nature of long wavelength optical phonons in ionic crystals, two interactions between ions become important. We have long range electric fields of macroscopic character; these can be described by the dielectric theory approach of Sect. 2.2. There are also short range restoring forces between the ions that provide finite frequencies for those modes which fail to generate a macroscopic field. The TO phonons of cubic crystals are an example: the short range forces also provide the curvature in the dispersion relations as one moves away from zero wave vector. In general, the frequency of an optical phonon contains contributions from both sources. It is the long range fields that provide the additional stiffening to the LO phonon, so that $\omega_{LO} > \omega_{TO}$. In an anisotropic crystal such as quartz, the relative contribution of the short range and long range fields for various modes depends, at long wavelengths, on the direction between the wave vector and the optic axis [2.37].

The spin waves in magnetic media also receive contributions to their dispersion relation from long range fields, and short range couplings between the spins on the lattice. For the moment, we confine our attention to ferromagnets modeled within the framework of the Heisenberg model, a picture not strictly correct for the transition metal ferromagnets, but which can be regarded as a phenomenological description of them.

When a spin wave is excited, the long range fields have their origin in the magnetic dipole moments of the spins, which generate macroscopic magnetic fields with the frequency and wave vector of the spin wave. A rough measure of the strength of these fields is provided by their contribution to the frequency of the wave, which is on the order of $4\pi\gamma M_s$, with M_s the saturation magnetization, and γ the gyromagnetic ratio. This is a frequency which lies in the microwave regime (not the infrared) for nearly all materials. The direction of the magnetization plays the role of the optic axis in anisotropic dielectrics, in that the strength of the macroscopic dipole fields generated by the spin motion depends on the angle between the wave vector and the magnetization.

In general, an external Zeeman field of strength H_0 is present, and is applied parallel to the magnetization $\boldsymbol{M}_s$ (or rather, $\boldsymbol{M}_s$ aligns with the field). This contributes an amount γH_0 to the frequency of the wave. Typically one has $\gamma H_0 \sim 4\pi\gamma M_s$, so that the contribution from the Zeeman and magnetic dipole fields are comparable.

The short range interactions between the spins are the exchange interactions of quantum mechanical origin, written $-J_{12}S_1 \cdot S_2$ between two neighboring spins. If a spin wave is excited, and θ_{12} is the angle between spins 1 and 2, excitation of the spin wave increases the exchange energy over that in the fully aligned ground state by $J_{12}S^2\theta_{12}^2/2$ at long wavelengths, where $\theta_{12} \ll 1$. We have $\theta_{12} \sim ka_0$, with a_0 being the lattice constant and k the wave vector of the wave. Exchange couplings contribute a term proportional to k^2 to the spin wave frequency in the long wavelength limit. This is written γDk^2, where we see that D is on the order of $zJS^2a_0^2$, where z is the number of neighbors.

Spin waves in the ferromagnet are the analog of acoustical phonons in crystals, in that the exchange contribution to their frequency vanishes as $k \to 0$ (though the contribution from the Zeeman field and magnetic dipole couplings remain). With more than one spin per unit cell, however we have optical branches, for which the exchange contribution remains finite as $k \to 0$. The presence of anisotropy also has this consequence in antiferromagnets.

The exchange interactions are very strong, but their role is suppressed at long wavelengths in ferromagnets, as we have seen. When we consider short wavelength spin waves, where ka_0 is not small, in typical materials exchange dominates the Zeeman and dipolar couplings by orders of magnitude, so we may ignore the latter interactions, and consider only a lattice of exchange-coupled spins. An analogous statement cannot be made in the theory of phonons in ionic insulators, since everywhere in the Brillouin zone electrostatic (Coulomb) couplings are similar in magnitude to those provided by the short range interactions. Magnetic dipole fields are simply much weaker than electrostatic fields, and thus exert their influence strongly only at long wavelengths where, for reasons discussed above, the influence of exchange interactions is suppressed.

When all the ingredients are put together, the dispersion relation for spin waves within a ferromagnet at long wavelengths may be written

$$\omega(k) = \gamma[(H_0 + Dk^2)(H_0 + 4\pi M_s \sin^2\theta_k + Dk^2)]^{1/2} , \tag{2.35}$$

where for Fe, we have $4\pi M_s \cong 18$ kGauss, $\gamma \cong 1.8 \times 10^7$ rad/s, and $D = 2.5 \times 10^{-9}$ Gauss-cm^2.

At long wavelengths, we have seen in Sect. 2.2 that the theory of various excitations which generate a macroscopic electric field may be discussed simply and easily within the framework of electrostatic theory, when retardation is neglected. The imput is provided by specifying the nature of the frequency dependent dielectric constant (or for anisotropic materials the dielectric tensor) of the material or materials of interest.

In a very similar manner, such long wavelength spin waves in a ferromagnet or antiferromagnet, where the wave vector dependencies introduced by exchange may be ignored, are described by magnetostatic theory. Then we need the form of the magnetic susceptibility tensor as input. The macroscopic magnetic field $h^{(d)}$ generated by the precessing spins satisfies (with retardation neglected)

$$\nabla \times h^{(d)} = 0 , \tag{2.36}$$

which means we may write

$$h^{(d)} = -\nabla\phi_M , \tag{2.37}$$

where ϕ_M is the magnetic scalar potential. We also have the time varying component of the magnetic induction

$$b^{(d)} = h^{(d)} + 4\pi m , \tag{2.38}$$

where the total magnetization is written $M(x, t) = M_s + m(x, t)$ when a spin wave is excited. The magnetic induction satisfies

$$\nabla \cdot b^{(d)} = 0 , \tag{2.39}$$

and m is then related to $h^{(d)}$ by means of a frequency dependent susceptibility tensor

$$m_i = \sum_j \chi_{ij}(\omega) h_j^{(d)} . \tag{2.40}$$

For ferromagnets and antiferromagnets of the classical uniaxial type (MnF_2, FeF_2, for example), the susceptibility tensor elements are rather simple in form. For the ferromagnet, with magnetic field H_0 applied parallel to the z-axis, one has[2.37]

$$\chi_{xx}(\omega) = \chi_{yy}(\omega) = \frac{\gamma^2 M_s H_0}{(\gamma H_0)^2 - \omega^2} \quad \text{and} \tag{2.41a}$$

$$\chi_{xy}(\omega) = -\chi_{yx}(\omega) = +i \frac{\gamma M_s \omega}{(\gamma H_0)^2 - \omega^2} \tag{2.41b}$$

where all other elements of $\chi_{ij}(\omega)$ vanish. For the uniaxial antiferromagnet [2.37], again with magnetic field H_0 applied along the easy axis,

$$\chi_{xx}(\omega) = \chi_{yy}(\omega) = \frac{\Omega_s^2}{8\pi} \left[\frac{1}{\Omega_0^2 - (\omega + \gamma H_0)^2} + \frac{1}{\Omega_0^2 - (\omega - \gamma H_0)^2} \right] \tag{2.42a}$$

$$\chi_{xy}(\omega) = -\chi_{yx}(\omega) = \frac{\Omega_s^2}{8\pi} \left[\frac{1}{\Omega_0^2 - (\omega - \gamma H_0)^2} - \frac{1}{\Omega_0^2 - (\omega + \gamma H_0)^2} \right] , \tag{2.42b}$$

where Ω_0 is the antiferromagnetic resonance frequency in zero external field (which generally lies in the far infrared), and $\Omega_s^2 = 8\pi\gamma^2 M_s H_A$, where M_s is the magnetization of one sublattice, and H_A a parameter referred to as the anisotropy field. Parameters which characterize the classical antiferromagnetic materials have been tabulated by *Kittel* [2.38]. A generalization of (2.42) to the case where H_0 makes an arbitrary angle with respect to the easy axis has been recently given [2.39]. In general, the expressions are very complex, save for the case where H_0 is perpendicular to the easy axis.

Within the framework of this macroscopic theory, one may discuss surface spin waves on antiferromagnets or ferromagnets in the long wavelength limits, their counterparts in the thin film, and collective excitations in superlattices. The approach is very similar to that described in Sect. 2.2. Thus, in what follows, we omit derivations and concentrate on discussing the results and their implications.

2.3.2 The Semi-Infinite Ferromagnet and the Ferromagnetic Film

We should begin by discussing surface spin waves on the semi-infinite ferromagnet, since they have unique properties which are strikingly different in one regard from the surface plasmons discussed in Sect. 2.2. This occurs at wavelengths sufficiently long that the influence of exchange is unimportant. Thus, we ignore the influence of exchange in this discussion. The limit applies nicely to the surface waves studied in the light scattering (Brillouin) studies and in the experimental work on light scattering from spin wave excitations in the magnetic superlattices discussed below.

The most interesting geometry is that illustrated in Fig. 2.5a. We have a semi-infinite ferromagnet, with external magnetic field H_0 and magnetization M_s parallel to the surface, and to each other. Surface spin waves, described by a magnetic potential very similar to that encountered in the theory of surface plasmons, may propagate perpendicular to the magnetization, as shown. The modes have been known in the literature for many years, and are referred to as Damon-Eshbach modes, after the authors who discussed them many years ago [2.40]. For this propagation direction, their frequency is given by

$$\omega_s(k_\parallel) = \gamma(H_0 + 2\pi M_s) , \qquad (2.43)$$

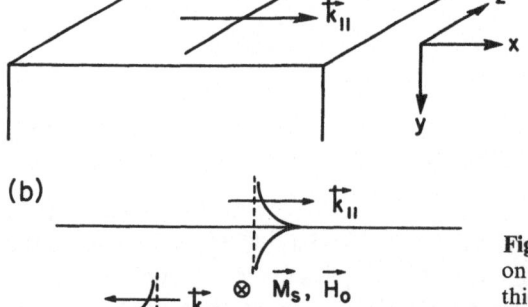

Fig. 2.5. Geometry of surface spin waves on (**a**) ferromagnetic surfaces, and (**b**) in thin ferromagnetic films. In (**b**) a cross section of the film is shown with both the Zeeman field H_0 and M_s parallel to its surfaces

which lies *above* that of the bulk spin waves, in the absence of exchange ((2.35) with $D=0$).

The most striking characteristic of the Damon-Eshbach spin waves is that they propagate in only *one* direction across the magnetization, from right to left. There are no solutions of the equations for the opposite sense of propagation. We may consider waves for which $k_{\parallel}$ makes an angle θ with the x axis in Fig. 2.5a. Then there exist solutions of the equation for angles θ which lie in the range $-\theta_c < \theta < +\theta_c$, where

$$\cos\theta_c = \left(\frac{H_0}{H_0 + 4\pi M_s}\right)^{1/2}. \tag{2.44}$$

As $\theta \to \theta_c$, the frequency of the surface wave drops to merge with the bulk spin wave band.

It is striking that the bulk spin wave dispersion relation is left unchanged as $k \to -k$, but the surface wave dispersion relation is so strikingly asymmetric. This rules out the breakdown of time reversal invariance by the presence of either H_0 or M_s. It may be shown [2.41] that the pseudo vector character of H_0 and M_s, combined with the fact with surface present, the xz plane is no longer a reflection plane, is responsible for this behavior.

One may observe the surface spin waves in light scattering (Brillouin) studies of excitations on magnetic surfaces. On a metallic ferromagnet, such as Fe or Ni, the optical skin depth is roughly 150 Å. The light scatters off the thermally excited spin waves whose wave vector $k_{\parallel}$ parallel to the surface is on the order of 10^5 cm^{-1}, by virtue of wave vector conservation. There are also features in the spectrum from bulk spin waves, which propagate up to and reflect off the surface.

The "one way" character of the Damon-Eshbach wave has the consequence that if the wave produces a line on the Stokes side of the Brillouin spectrum, it is absent on the anti-Stokes side. If the mode excited in the Stokes side has wave vector $k_{\parallel}$, the one in the anti-Stokes event is required to have wave vector $-k_{\parallel}$, and here no such wave exists. We illustrate this in Fig. 2.6, where we show Brillouin data on Fe reported by *Sandercock* and *Wettling* [2.42]. The prominent feature labeled SM is the surface spin wave line, and we see it on only the Stokes side. The broader and weaker structures are bulk spin waves; the shape of these structures is strongly influenced by exchange. The solid line is the result of a theoretical calculation, using the formalism developed by *Camley* and *Mills* [2.43]. The fit to the data is excellent; no adjustable parameters are required.

If the external field H_0 and magnetization M_s are normal to the surface, then no surface spin waves occur in the long wavelength limit with exchange ignored. *Rahman* and *Mills* have analyzed the dispersion relation of the Damon-Eshbach wave as H_0 is tilted out of the plane parallel to the surface [2.44], while *Camley* and *Mills* have explored the influence of exchange on these waves [2.43].

We next consider a thin film of thickness d, with M_s and H_0 both parallel to the surface. We now have two surface waves, one localized on each surface, as

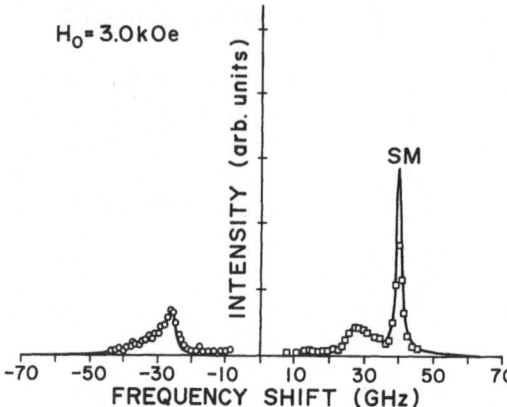

$H_0 = 3.0 \, kOe$

INTENSITY (arb. units)

SM

FREQUENCY SHIFT (GHz)

-70   -50   -30   -10   10   30   50   70

Fig. 2.6. Brillouin spectrum for scattering of light by spin waves on the surface of iron. The peak labeled SM is the surface magnon, and the weaker bands are produced by scattering from bulk spin waves which propagate up into the skin depth

illustrated in Fig. 2.5b. On the top surface, a wave runs from right to left as before, but on the bottom surface, the wave propagates in the opposite direction. The frequency of these waves is given by

$$\omega_s(k_\parallel) = \gamma \left[(H_0 + 2\pi M_s)^2 - (2M_0)^2 \exp(-2k_\parallel d) \right]^{1/2} , \qquad (2.45)$$

which of course reduces to (2.43) when $k_\parallel d \gg 1$. When $k_\parallel d \sim 1$, the frequency of the waves in the film lies *below* that appropriate to the Damon-Eshbach wave on the semi-infinite sample.

There is considerable theoretical literature on the properties of surface spin waves in the Heisenberg ferromagnet and antiferromagnet, in the limit where exchange contributions dominate the Zeeman and magnetic dipole contribution. It is these waves which enter the description of the thermodynamic properties of the magnet near its surface. The properties of such modes have been reviewed elsewhere [2.36], and we will not be greatly concerned with them in the present article.

2.3.3 Spin Waves in Superlattices with Ferromagnetic Films

As remarked earlier in this section, it is now possible to fabricate superlattices of high quality, in which one constituent is a ferromagnetic film. Such a superlattice structure is illustrated in Fig. 2.7a, where d_1 is the thickness of the ferromagnetic film, and d_2 that of the nonmagnetic material. We have also seen that an isolated ferromagnetic film supports surface spin waves. For each value of $k_\parallel$ we have one possible mode localized on the first surface (for example, the upper surface) if $k_\parallel$ describes propagation from left to right across the magnetization, and on the second (say the lower surface) if it describes prpagation from right to left.

In the superlattice, we may form collective excitations of the structure as a whole, very much as in our discussion of the collective modes in the dielectric superlattice. Each interface in the superlattice supports a mode, and when the spacing d_2 between adjacent ferromagnetic films is finite, the fields between

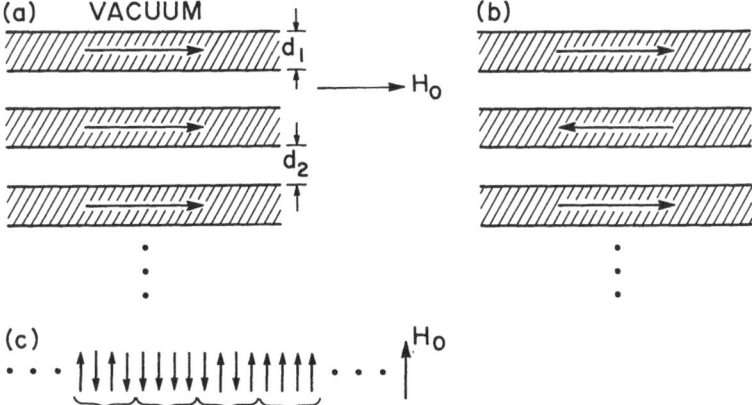

Fig. 2.7a–c. A sketch of various magnetic superlattice configurations discussed in the text. **(a)** A superlattice formed from ferromagnetic films, with a nonmagnetic medium between them, **(b)** a superlattice of ferromagnetic films, in which the inter-film coupling, mediated by the intervening nonmagnetic layer, is antiferromagnetic in character, and **(c)** a superlattice which consists of alternating layers of ferromagnetic and antiferromagnetic materials. We assume an external magnetic field of strength H_0 is applied as indicated

modes of neighboring interfaces overlap, with the result that we have Bloch states which may transport energy normal to the interface. The theory of such collective modes of the infinite and semi-infinite superlattice was discussed by *Camley* et al. [2.45]. These authors also developed the corresponding theory of Brillouin scattering, which has proved a successful means of studying spin waves in superlattices. We refer the reader to the article by Grimsditch (Chap. 7) for a detailed discussion of Brillouin scattering by spin waves in superlattices. About the same time as the paper by *Camley* et al. appeared, *Grünberg* and *Mika* [2.46] presented detailed numerical studies of spin waves in finite superlattices, along with light scattering data.

One difference between the magnetic superlattice and the dielectric superlattice is that for a given choice of $k_{\parallel}$, only one interface of each ferromagnetic film supports a surface mode, while in the dielectric superlattice each of the two interfaces of the "active" film supports a mode. Thus, in the present case, for each choice of $k_{\parallel}$ we have only one band of collective excitations in the infinitely extended superlattice, while in the dielectric case we had two. *Camley* et al. [2.45] have derived an analytic expression for the collective modes of the infinitely extended superlattice of ferromagnetic films, for propagation normal to the magnetization which, as illustrated in Fig. 2.7a is assumed to also lie parallel to the surface. Their dispersion relation reads (with exchange ignored).

$$\omega(k_{\parallel}, k_{\perp}) = \frac{\gamma}{[1 + \Delta(k_{\parallel}, k_{\perp})]^{1/2}}$$
$$\times \left\{ H_0(H_0 + 4\pi M_s) + 1/2\,[H_0^2 + (H_0 + 4\pi M_s)^2]\,\Delta(k_{\parallel}, k_{\perp}) \right\}^{1/2},$$

$$(2.46a)$$

where

$$\Delta(k_{\parallel}, k_{\perp}) = \frac{\sinh(k_{\parallel} d_1) \sinh(k_{\parallel} d_2)}{\cosh(k_{\parallel} d_1) \cosh(k_{\parallel} d_2) - \cos(k_{\perp} d)} .$$
(2.46b)

In (2.45b) $d = d_1 + d_2$ is the period of the superlattice. Notice that the above dispersion relation is an even function of $k_{\parallel}$, i.e., we have collective modes which propagate in either direction across the magnetization. For one sign of $k_{\parallel}$, these are composed of modes which are localized on the upper surface of each ferromagnet, while for the other sign they are localized on the lower surface.

If the superlattice is terminated, i.e., we have a semi-infinite superlattice as illustrated in Fig. 2.7a, we encounter the possibility of having collective spin waves localized near the surface, similar to the surface modes of the dielectric superlattice discussed in Sect. 2.2.4. The theory of these modes was discussed in [2.45]. The theoretical reuslt is quite remarkable. The frequency of the surface wave (in the absence of exchange) is given by *exactly* the same formula which applies to the semi-infinite ferromagnet! For propagation perpendicular to the magnetization, this is the result of (2.43), where in the superlattice M_s is the magnetization density in one ferromagnetic film (as opposed to the average magnetization density of the superlattice, $M_s d_1/(d_1 + d_2)$).

For propagation perpendicular to the magnetization, the attenuation constant α in the magnetic analog of (2.32a) is given by [2.45]:

$$\alpha = k_{\parallel} \frac{d_1 - d_2}{d_1 + d_2} .$$
(2.47)

We must have $\alpha > 0$ for the surface mode to exist. Thus, we require $d_1 > d_2$, i.e. the thickness of the ferromagnetic film must exceed that of the nonmagnetic "spacer" layer. As $d_1 \to d_2$ from above, $\alpha \to 0$, and it may be shown easily that the frequency of the surface wave merges with the top of the band of bulk collective modes described by (2.46). The surface mode is predicted to have "one way" propagation only, as for the semi-infinite superlattice.

The first detailed experimental studies of the collective spin wave modes of superlattice structures which incorporate ferromagnetic films are the Brillouin scattering studies reported by *Grimsditch* and his colleagues [2.47, 48]. These authors examined superlattices of Ni and Mo. To an excellent approximation the Mo films may be regarded as nonmagnetic spacers. For one sample, the Ni films were 300 Å thick and the Mo sample was 100 Å thick. The bulk spin wave bands and the surface wave were observed very much as predicted in [2.45] and the surface mode was absent as was also predicted for a sample with 100 Å thick Ni films, and 300 Å thick Mo films. As far as this writer knows, these experiments, and the data reported by *Grünberg* and *Mika* [2.46], are the first experimental studies of collective modes in superlattices which are coherent excitations of the structure as a whole, as opposed to excitations localized within particular films of the sample. Since this early work, numerous other experimental light scattering studies have been reported in magnetic superlattice structures of various sorts. *Hillebrands* et al. [2.49] review these, and also present new data that address the point discussed below.

Our discussion has so far been confined to the superlattice geometry illustrated in Fig. 2.7a, where the magnetization in each ferromagnetic films is parallel to the surface, and to an external Zeeman field applied to the sample, also parallel to the surfaces. Application of a strong magnetic field normal to the surface will rotate the film magnetizations normal to the surface. This produces dramatic changes in the spectrum of collective excitations of the superlattice. This case has been discussed by *Camley* and *Cottam* [2.50] who also explore the spectrum of collective excitations in a superlattice which incorporates antiferro- magnetic films as elements.

We have also assumed that the nominally nonmagnetic constituent of the superlattice is benign in its influence. This need not be the case, as illustrated by recent studies of Y − Gd superlattice [2.51]. While Y is a paramagnetic metal, and in this sense nominally non-magnetic, the magnetic configuration illustrated in Fig. 2.7a is produced only for a selected range of Y film thicknesses. One also generates, for some samples, the magnetic configuration in Fig. 2.7b in which the magnetic moments in the ferromagnetic Gd films alternate in sign as one moves through the superlattice. It is evident that the intervening Y films are providing a magnetic coupling between adjacent Gd films, whose sign alternates as one increases the Y film thickness. It is argued that the origin of this oscillatory coupling is spin polarization induced in the Y film by the neighboring ferromagnetic Gd films. The theoretical discussion presented in [2.51] show that there is indeed an oscillatory spin density induced in the Y film, with a wavelength that accounts quantitatively for the behavior found in the super- lattices.

If a magnetic field is applied normal to the plane which contains the magnetizations for an Y − Gd superlattice with the antiparallel ground state illustrated in Fig. 2.7b, then neutron data shows that the magnetizations in the film cant out of the plane perpendicular to the field [2.51]. The spin wave excitations of such a canted superlattice structure will prove fascinating. A theoretical description of these modes is a topic of current research [2.52].

Our theoretical discussion also assumes that the magnetic response of a ferromagnetic film in a superlattice may be described by the same physical picture employed in the analysis of bulk magnetic media. Actually, spins in the surface of a thin film experience strong anisotropy fields, absent in the bulk material, which are present in the surfaces as a consequence of its low symmetry [2.53]. Such anisotropy fields lead to dramatic shifts in the frequency of spin waves in a superlattice which contains ultra-thin ferromagnetic films, for which a substantial fraction of the spins reside in the surface. We refer the reader to [2.49] for a description of such shifts, and an analysis of Brillouin spectra on superlattices which are fabricated from such ultra thin films.

Brillouin scattering spectroscopy has proved to be a powerful probe of the collective excitations in magnetic superlattice structures. The classical method for study of spin waves in thin films is by ferromagnetic resonance. The sample is placed in a cavity with a fixed resonance frequency and a magnetic field is used to sweep the frequency through that of the cavity. Thus one, obtains information on the response of the system at only one, or possibly a small number, of particular frequencies. In a Brillouin scattering study of spin waves in magnetic films or multilayer structures such as superlattices, one obtains a complete spin wave spectrum for any value of the applied field H_0. Indeed, the strength of the applied field may be varied continuously over a wide range, and the frequency of each wave is tracked continuously as a function of H_0. Also, we have seen that the frequency depends on the direction and magnitude of $k_\parallel$; the former may be varied if desired by changing the scattering geometry. Thus, Brillouin spectroscopy provides a very large volume of data. One can use this information to extract accurate values for the magnetization in the ferromagnetic films in materials such as those discussed here, and the constants which characterize the surface anisotropy, if the superlattice contains very thin films. An excellent discussion of such analyses is given in [2.49].

2.3.4 Superlattices with Ferromagnetic and Antiferromagnetic Films; Theoretical Descriptions of Model Systems

The discussion in the previous subsection concerned superlattices in which one of the two films in each unit cell is ferromagnetic, while the second constituent is a nominally nonmagnetic material. One may envision fabrication of superlattices in which each film is a magnetically ordered material, possibly of a more complex character than the simple ferromagnetism just discussed. For example, Y −Dy superlattices have been fabricated [2.54]. Dy is an example of a spiral spin material. The crystal is hexagonal, and in the spiral spin state the magnetic moments lie in the basal plane and are arranged in the form of an incommensurate spiral parallel to the c axis. In the Y − Dy superlattice, the Y films provide magnetic coupling between adjacent Dy films, much as they do in the Y − Gd system, so that a coherent ground state results. One can also form a superlattice of Gd and Dy; in this case the ferromagnetic film intervenes between the films in which the spiral spin state exists.

Clearly, this is a potentially rich area offering the possibility of fabricating a diverse array of magnetically ordered structures. The simplest example of a superlattice in which two distinct kinds of magnetic order are produced is a structure in which antiferromagnetic order exists in one film, and ferromagnetic order in the second. No such superlattice has been generated in the laboratory yet, but theoretical studies [2.55, 56] suggest that their properties will be fascinating. In this section, we summarize in qualitative terms, the results of these analyses. First, however, a brief introduction to the properties of spin waves in antiferromagnets is in order [2.57].

In most antiferromagnets, the crystal structure has symmetry lower than cubic. For example, the much studied classical antiferromagnets FeF_2 and MnF_2 are body-centered tetragonal crystals. In these cases, there is an anisotropy energy which either favors alignment of the spins along the unique axis (called the easy axis), or within the plane normal to the easy axis (called the easy plane). The most common situation, found for example in MnF_2 and FeF_2, is for the unique direction to be an easy axis. The antiferromagnetic ground state then has half of the spins, say the A sublattice, directed in the $+z$ direction, and half, say the B sublattice, directed along $-z$, where the z axis coincides with the unique direction of the crystal. Phenomenologically [2.57], one introduces an effective internal magnetic field, the anisotropy field H_A, which pins the A spins along $+z$, and the B spins along $-z$.

Suppose the B sublattice is pinned in place, and the A sublattice is canted away from $+z$. The B spins will feel a torque from the A spins, due to the very strong quantum mechanical exchange couplings, $-JS_1 \cdot S_2$, which entered our introductory remarks on ferromagnets (in the antiferromagnet, however $J < 0$). The influence of exchange is represented as an effective magnetic field H_E, directed antiparallel to the magnetization of the A spin sublattice, and also antiparallel to that of the B spin sublattice.

We also have dipolar interactions in the antiferromagnet; $4\pi M_s$ serves as a measure of their strength, where M_s is the magnetization density of one sublattice. In general, $4\pi M_s$ is very small compared to H_E, and also to H_A in most cases, so we can ignore the influence of dipolar interactions here.

We have seen that, in the ferromagnet, exchange has only a modest effect on the spin waves, at long-wavelengths. This is not the case for the antiferromagnet, where one can show that at long wavelengths, the spin wave frequencies are given by [2.57]

$$\omega(k \to 0) = \gamma \left(\sqrt{H_A^2 + 2H_E H_A} \pm H_0 \right) , \tag{2.48}$$

where H_0 is the strength of an applied external magnetic field. The frequency in (2.48) generally lies in the far infrared (rather than the microwave range relevant to our earlier discussion of ferromagnets) because the exchange field H_E is very large.

With these remarks in hand, we are now ready to discuss the properties of a model superlattice which consists of alternating ferromagnetic and antiferromagnet films. The simplest geometry is one in which the antiferromagnet consists of sheets of ferromagnetically aligned spins, with the net moment in each sheet alternating in sign as one moves normal to the surface of the film. Such a geometry is realized in films of MnF_2 or FeF_2 with a (100) surface.

If the antiferromagnetic film consists of an even number of layers, and if for definiteness we assume ferromagnetic coupling between spins in adjacent films, then we have the ground state configuration shown in Fig. 2.7c. The main point is that the net moment of each *ferromagnetic* film alternates in sign as one moves down the superlattice. The length of the unit cell in the structure is $2(d_1 + d_2)$ and not $(d_1 + d_2)$, where d_1 and d_2 are the thicknesses of the ferromagnetic and antiferromagnetic films, respectively. The reader will recognize that the ferromagnetic films in the structure have a configuration quite identical to the staggered ground state of the Y – Gd superlattice [2.51], as discussed in the previous subsection. The spin density wave evidently induced in the Y films by the neighboring Gd films can be mimicked by induced antiferromagnetism, and this author believes that the model to be discussed below serves to describe, phenomenologically, the behavior expected in the Y – Gd system.

The ground state spin configuration illustrated in Fig. 2.7c is unstable with respect to application of an external magnetic field H_0 parallel to the easy axis. Half of the spins in the ferromagnetic films are oriented antiparallel to H_0, and wish to rotate parallel to it, but at low fields, are inhibited by the coupling to the spins in the antiferromagnet. Clearly, if H_0 is large enough, all the spins will align parallel to H_0, and one may inquire how this occurs.

These questions have been explored theoretically for the model just outlined [2.55]. The picture which emerges is the following. The ground state spin configuration shown in Fig. 2.7c is stable for a range of external fields $0 < H_0 < H_c^{(1)}$, where $H_c^{(1)}$ varies inversely with the number of spins in the ferromagnetic film, a result expected from the remarks in the previous paragraph. At $H_c^{(1)}$, there is a second order, magnetic field induced phase transition to a state of low symmetry, which is stable in a field regime $H_c^{(1)} < H_0 < H_c^{(2)}$. At $H_c^{(2)}$ we have another second order phase transition to a state with glide plane symmetry, referred to in [2.57] as the superlattice spin flop state. The spins become fully aligned along H_0 at a third field $H_c^{(3)}$; the lock-in to full alignment also has the character of a second order phase transition.

Thus, the spin system evolves from the ground state spin configuration of Fig. 2.7c to full alignment with the external magnetic field by passing through a sequence of second order phase transitions, referred to as spin reorientation transitions in [2.57].

Associated with each phase transition is a soft spin wave mode, which is a collective excitation of the structure as a whole. With increasing field, in a given phase, the theory shows that a mode is driven down to zero frequency, then stiffens with a further increase in field, after one passes into the next phase of the sequence of field induced phase transitions. Quite clearly, light scattering studies of the collective excitations in such structures will prove of very great interest. A field induced canted phase of the Y – Gd system, with the ground state spin configuration illustrated in Fig. 2.7c, has been reported in [2.51]. If we allow ourselves the liberty of modeling the system as a superlattice of alternating films of ferromagnets and antiferromagnets, mimicking the Gd-induced spin density wave in the Y films as an antiferromagnetic material, then the field-induced canted state is identified as the state with glide plane symmetry discussed by *Hinchey* and *Mills* [2.55]. Study of the collective spin wave excitations in this structure would be of great interest.

The above remarks apply specifically to the case where the antiferromagnet may be regarded as a sequence of sheets of ferromagnetically aligned spins, whose moments alternate in direction as one moves through the structure, normal to the interface. A theoretical description of a geometry in which the spin sheets contain an equal admixture of up and down spins has been proposed

[2.56]. One again finds a sequence of magnetic field induced spin re-orientation transitions, with soft spin waves, etc. We shall not discuss the details of the theoretical predictions here.

2.3.5 Concluding Remarks

We have seen that detailed experimental studies of collective excitations in magnetic superlattices have been performed by Brillouin scattering, and the spectra provide quantitative confirmation of the theoretical descriptions of these modes. We also have exciting new systems in hand, such as the $Y-Gd$ and $Y-DY$ superlattices. While their static magnetic properties have already been explored, there have been no experimental studies of collective excitations in these materials. Theoretical studies of related models suggest this to be a rich area for future research.

One difference between the magnetic superlattices discussed in this section and the semiconducting superlattices examined in Sect. 2.2 is that an external magnetic field of modest magnitude may influence the nature of the collective modes dramatically and alter the nature of the spin configuration by means of the field-induced spin re-orientation transitions discussed above. In this author's view, the study of superlattices which incorporate films of magnetically ordered materials will be a rich and fascinating area of future research, particularly since structures with interfaces of extraordinary quality may be fabricated in the laboratory.

2.4 Nonlinear Excitations in Superlattices; Gap Solitons

The elementary excitations of dielectric and magnetic superlattices described above are discussed within the framework of the classical theoretical framework of solid state physics. One exploits the translational invariance in the two directions parallel to the surface to endow the underlying wave fields with spatial variations proportional to $\exp(i\mathbf{k}_{\parallel} \cdot \mathbf{x}_{\parallel})$, and then, at least for the infinitely extended superlattice, we form Bloch waves in the direction normal to the interfaces. Thus, in a manner familiar from the theory of electrons, phonons, or spin waves, we have collective excitations that are wave-like in all three dimensions, and thus are collective excitations of the structure as a whole. We may also have states localized near the surface of the structure, as we have discussed, or near a defect within it. In the end, the spectrum of collective excitations controls the linear response of the superlattice to an external probe.

Recent theoretical work has explored the nonlinear response of a superlattice to an external probe [2.58, 59]. The model system is a dielectric superlattice, in which one (or possibly both) films within the unit cell exhibits a nonlinear response to an electromagnetic wave which propagates through the structure. So far, attention has been confined to the simple case of a plane-polarized electromagnetic wave propagating normal to the interfaces. The magnitude of the electric field then depends only on z, and in the nonlinear film the electric field obeys the wave equation

$$\frac{d^2 E}{dz^2} + \frac{\omega^2}{c^2}\, \varepsilon(1 + \lambda |E(z)|^2)\, E(z) = 0 \; , \tag{2.49}$$

where ε is the dielectric constant, and λ a nonlinear coefficient that may be positive or negative.

The theoretical studies show that under conditions described below, a wave incident on the structure may excite solitons. In the spectral regime where this occurs, soliton-mediated bistability is predicted [2.58, 59]. The solitons may be viewed as elementary excitations of the structure whose existence depends intimately on the presence of the nonlinearity. These are localized entities, in contrast to the extended Bloch waves of linear theory.

These results emerged within the framework of rather extensive numerical studies of the nonlinear optical response of model dielectric superlattices, in which one of the two films in the unit cell has the nonlinear response characteristics described in (2.49). The essential properties of the solitons follow from an approximate analytic treatment of a much simpler model. Consider a dielectric whose linear dielectric constant is periodically modulated:

$$\varepsilon(z) = \varepsilon + \Delta\varepsilon \cos(Gz) \; . \tag{2.50}$$

We shall assume that $\Delta\varepsilon \ll \varepsilon$, for simplicity. Plane polarized electromagnetic waves, which propagate parallel to the z direction obey the nonlinear wave equation,

$$\frac{d^2 E}{dz^2} + \frac{\omega^2}{c^2}\, [\varepsilon(z) + \varepsilon\lambda |E(z)|^2]\, E(z) = 0 \; . \tag{2.51}$$

First, consider wave propagation in the limit $\lambda = 0$. When $\Delta\varepsilon = 0$ also, we have electromagnetic waves which propagate in the uniform dielectric, with the well known linear dispersion relation $\omega = ck\varepsilon^{-1/2}$. When $\Delta\varepsilon \neq 0$, gaps in the dispersion relation open up, as in the theory of wave propagation in periodic media. Near $k = G/2$, the dispersion relation for the electromagnetic waves is given by (for $\Delta\varepsilon \ll \varepsilon$)

$$\omega_\pm^2(k) = \omega_G^2 \pm (\omega_G^2/2)\, [(\Delta\varepsilon/\varepsilon)^2 + (4\,\Delta k/G)^2]^{1/2} \; , \qquad \text{where} \tag{2.52}$$

$$\omega_G = cG/2\sqrt{\varepsilon} \qquad \text{and} \tag{2.53a}$$

$$\Delta k = k - G/2 \; . \tag{2.53b}$$

Thus, there is a forbidden gap in the frequency regime $\omega_- \lesseqgtr \omega \lesseqgtr \omega_+$, where

$$\omega_- = \omega_G (1 - \Delta\varepsilon/2\varepsilon)^{1/2} \quad \text{and} \tag{2.54a}$$

$$\omega_+ = \omega_G (1 + \Delta\varepsilon/2\varepsilon)^{1/2} . \tag{2.54b}$$

The solitons of interest emerge as solutions of the nonlinear wave equation, (2.51), for frequencies which lie within the forbidden gap. For this reason, we refer to them as gap solitons. One has such solutions for both algebraic signs of the nonlinear coupling constant λ; here we suppose $\lambda > 0$, to illustrate their form. The electric field in the soliton has the form [2.60]

$$E(z) = \mathscr{E}(z) \cos\left[(Gz/2) + \phi(z)\right] , \tag{2.55}$$

where $\phi(z)$ obeys one of the classical differential equations of nonlinear mathematics, the double sine-Gordon equation, whose (analytic) soliton solutions have been discussed in detail by *Campbell* et al. [2.61]. Explicitly, the phase $\phi(z)$ has been shown to obey [2.60]

$$\left(\frac{4}{G}\right)^2 \frac{d^2\phi}{dz^2} + \frac{\Delta\varepsilon}{\varepsilon} \left(\frac{\omega^2 - \omega_G^2}{\omega_G^2}\right) \sin(2\phi) + \left(\frac{\Delta\varepsilon}{2\varepsilon}\right)^2 \sin(4\phi) = 0 . \tag{2.56}$$

The phase angle $\phi(z)$ exhibits the spatial variation

$$\phi(z) = (2n+1)\frac{\pi}{2} + \tan^{-1}\left[\left(\frac{\omega_+^2 - \omega^2}{\omega_-^2 - \omega^2}\right)^{1/2} \tanh(z/d)\right] , \tag{2.57}$$

and for the amplitude $\mathscr{E}(z)$ we have

$$\mathscr{E}^2(z) = \frac{2(\omega_+^2 - \omega^2)}{3\lambda\omega_G^2} \frac{\mathrm{sech}^2(z/d)}{1 + ((\omega_+^2 - \omega^2)/(\omega^2 - \omega_-^2))\tanh^2(z/d)} , \tag{2.58}$$

where the size, d, of the soliton is given by:

$$d = \frac{4\omega_G^2}{G} (\omega^2 - \omega_-^2)^{-1/2} (\omega_+^2 - \omega^2)^{-1/2} . \tag{2.59}$$

Near either gap edge, the length d becomes very large. Near the upper band edge, to a very good approximation the soliton takes on the classical $\mathrm{sech}^{-2}(z/d)$ form, while near the lower gap edge, the core of the soliton is rather localized. (Equation (2.58) shows that in the region $z \ll d$, the soliton is localized to a region whose size is the order of $(2a_0/\pi)(\varepsilon/\Delta\varepsilon)$, where a_0 is the period associated with the spatial modulation of $\varepsilon(z)$. See also (2.51) of [2.60].)

Thus, for $\lambda > 0$, for frequencies near the upper band edge, we have a spatially extended soliton, whose envelope has, once again, the classical $\mathrm{sech}^{-2}(z/d)$ form. This is in agreement with the numerical studies of the model superlattice discussed in [2.58, 59]. For $\lambda < 0$, the analytic theory predicts a spatially extended soliton for frequencies in the vicinity of the lower band edge, $\omega \gtrsim \omega_-$, again in agreement with the full theory.

Chen and *Mills* [2.58, 59] have calculated the response of a finite superlattice to a normally incident plane-polarized electromagnetic wave. One film in each unit cell is endowed with a nonlinear film, as described above. If the superlattice is illuminated with a wave whose frequency ω lies in the stop gap of linear theory, $\omega_- \lesssim \omega \lesssim \omega_+$, at low powers where linear theory applies, the calculation shows that the transmissivity $|T|^2$, defined as the ratio of output to input power, is very small, as expected. The wave transmitted through the surface on the input side has an envelope which decays exponentially as one moves through the superlattice, also as expected from linear theory.

In the calculations, as the input power is increased, the transmissivity $|T|^2$ becomes a multi-valued function of the input power, as in the theory of bistability. For the example considered, a finite superlattice with twenty unit cells, at low powers $|T|^2 \cong 10^{-4}$. In the regime where bistability occurs, within the accuracy of the calculations the maximum value of $|T|^2$ is unity. Thus, with increasing power, at least in theory, the nonlinear superlattice becomes perfectly transparent to the incident radiation! The power in the incident beam must be such that the nonlinear film, when exposed to a field with magnitude equal to the incident field outside the superlattice, changes its index of refraction by roughly one half of one percent.

The gap solitons discussed above play a central role in the bistable behavior of the superlattice. At the point where $|T|^2 = 1$, numerical studies of the field intensity in the superlattice show a soliton-like object that is centered in the middle of the finite superlattice. The self-induced transparency is then due to a gap soliton excited by the incident electromagnetic field. This may be viewed as a resonance of the nonlinear structure which, when matched perfectly to the incident field, leads to a transmissivity of unity, very much like the shape resonances of (linear) quantum mechanics.

We have seen that in the infinitely extended superlattice gap solitons emerge as solutions of the wave equation which describes the nonlinear superlattice. The soliton is a localized elementary excitation of the nonlinear superlattice, in the sense that with increasing distance from its center, the amplitude of the field falls exponentially to zero [(2.58) for $|z| \gg d$]. If we now create two surfaces, the field associated with the gap soliton is finite at each surface. Then, necessarily, it acquires a finite lifetime because it radiates its energy into the vacuum on each side of the superlattice. It then has the character of a resonance level of the superlattice structure, similar to the virtual states of alloy theory, or to auto-ionizing levels of atoms. Since it can decay by radiating its energy into the vacuum, it can also be coupled to and thus excited by an incident electromagnetic wave. As remarked above, it is excitation of gap solitons by the incident electromagnetic wave that leads to the transmission resonance, skewed in shape by the nonlinearity in response, which is the origin of the bistability in the theoretical model.

The combination of periodicity and nonlinearity are essential for the existence of gap solitons. In the linear limit, we must have a structure which has gaps in the dispersion relation for wave propagation. Solitons can exist only for frequencies within these gaps, since it is only there that one can generate solutions of the wave equation which decay to zero exponentially as one moves away from the center of the soliton. Of course, the nonlinearity also plays an essential role in stabilizing the soliton structure.

The above discussion is based entirely on theoretical studies of model superlattices. It will be intriguing to see experimental studies of the optical response of such structures, under conditions where theory suggests solitons can be excited.

References

2.1   See the discussion in C. Kittel: *Introduction to Solid State Physics*, Fifth Ed. (Wiley, New York 1976) p. 191
2.2   See the theoretical discussion given by J.D. Dow, S.Y. Ren, K. Hess: Phys. Rev. B**25**, 6218 (1982)
2.3   D.L. Mills, E. Burstein: Rep. Prog. Phys. **37**, 817 (1974)
2.4   V.M. Agranovich, D.L. Mills (eds.): *Surface Polaritons*, (North-Holland, Amsterdam 1982)
2.5   R.E. Camley, D.L. Mills: Phys. Rev. B**29**, 1695 (1984)
2.6   A. Mooradian, A.L. McWhorter: In *Light Scattering Spectra of Solids*, ed. by G.B. Wright (Springer, New York 1969) p. 297
2.7   R. Szenics, R.F. Wallis, G.F. Giuliani, J.J. Quinn: Surface Sci. **166**, 45 (1986)
2.8   S. Das Sarma, J.J. Quinn: Phys. Rev. B**28**, 6144 (1983)
2.9   G. Qin, G.F. Giuliani, J.J. Quinn: Phys. Rev. B**28**, 6144 (1983)
2.10 A.C. Tselis, J.J. Quinn: Phys. Rev. B**29**, 3318 (1984)
2.11 A.C. Tselis, J.J. Quinn: Phys. Rev. B**29**, 2021 (1984)
2.12 E. Burstein, A. Pinczuk, D.L. Mills: Surface Sci. **98**, 451 (1980)
2.13 W.P. Chen, Y.J. Chen, E. Burstein: Surface Sci. **58**, 263 (1976)
2.14 A.K. Sood, J. Menéndez, M. Cardona, K. Ploog: Phys. Rev. Lett. **54**, 2115 (1985)
2.15 A.K. Sood, J. Menéndez, M. Cardona, K. Ploog: Phys. Rev. Lett. **54**, 2111 (1985)
2.16 Kun Huang, Bangfen Zhu: "On the Dielectric Continuum Model and Frohlich Interaction in Superlattices". To be published; also ibid Phys. Rev. B**38**, 2183 (1988) and T. Tsuchiya, H. Akera, T. Ando: Phys. Rev., in press
2.17 I. Tamm: Physik. Z. Sowjetunion **1**, 733 (1932)
2.18 R.E. Camley, T.S. Rahman, D.L. Mills: Phys. Rev. B**27**, 261 (1983)
2.19 R.E. Camley, D.L. Mills: Bull. Am. Phys. Soc. **28**, 408 (1983). The full length paper is [2.5]
2.20 G.F. Guiliani, J.J. Quinn: Phys. Rev. Lett. **51**, 919 (1983)
2.21 R.F. Wallis, R. Szenics, J.J. Quinn, G.F. Guiliani: Phys. Rev. B**36**, 1218 (1987)
2.22 S. Streight, D.L. Mills: Phys. Rev. B**35**, 6337 (1987)
2.23 S. Katayama, T. Ando: J. Phys. Soc. Japan **54**, 1615 (1985)
2.24 P. Hawrylak, J.W. Wu, J.J. Quinn: Phys. Rev. B**32**, 5169 (1985)
2.25 P. Hawrylak, J.W. Wu, J.J. Quinn: Phys. Rev. B**32**, 4272 (1985)
2.26 P. Hawrylak, J.W. Wu, J.J. Quinn: Phys. Rev. B**31**, 7855 (1985)
2.27 R. Mayanovic, G.F. Guiliani, J.J. Quinn: Phys. Rev. B**33**, 8390 (1980)
2.28 P. Hawrylak, G. Eliasson, J.J. Quinn: Phys. Rev. B**34**, 5368 (1986)
2.29 J. Wu, P. Hawrylak, J.J. Quinn: Phys. Rev. Lett. **55**, 87 (1985)

2.30 J. Wu, P. Hawrylak, G. Eliasson, J.J. Quinn: Phys. Rev. B **33**, 7091 (1986)
2.31 J. Wu, G. Eliasson, J.J. Quinn: Phys. Rev. B **35**, 860 (1987)
2.32 B.L. Johnson, J.T. Weiler, R.E. Camley: Phys. Rev. B **32**, 6544 (1985)
2.33 G. Eliasson, P. Hawrylak, J.J. Quinn: Phys. Rev. B **35**, 5569 (1987)
2.34 I. Schuler: In *Frontiers in Electronic Materials and Processing*, ed. by L.J. Brillson, (American Vacuum Society, Series 1; American Institute of Physics Conference Proceedings No. 138, 1986) p. 93
2.35 D.B. McWhan, C. Vettier: ibid, p. 80
2.36 D.L. Mills: "Surface Spin Waves in Magnetic Crystals", in *Surface Excitations*, ed. by V. Agranovich, R. Loudon (North-Holland, Amsterdam 1982) Chap. 3
2.37 A simple phenomenological discussion of such modes can be found in [2.3]
2.38 See the discussion in C. Kittel: *Introduction to Solid State Physics*, Fifth Ed. (Wiley, New York 1976) p. 522
2.39 N.S. Almeida, D.L. Mills: Phys. Rev. B (in press)
2.40 R. Damon, J. Eshbach: J. Phys. Chem. Solids **19**, 308 (1961)
2.41 R.Q. Scott, D.L. Mills: Phys. Rev. B **15**, 3545 (1977)
2.42 J. Sandercock, W. Wettling: J. Appl. Phys. **50**, 7784 (1979)
2.43 R.E. Camley, D.L. Mills: Phys. Rev. B **18**, 4821 (1978)
2.44 T.S. Rahman, D.L. Mills: J. Appl. Phys. **53**, 2084 (1982)
2.45 R.E. Camley, T.S. Rahman, D.L. Mills: Phys. Rev. B **27**, 261 (1983)
2.46 P. Grünberg, K. Mika: Phys. Rev. B **27**, 2955 (1983)
2.47 M. Grimsditch, M.R. Khan, I.K. Schuller: Phys. Rev. Lett. **51**, 498 (1983)
2.48 A. Kueny, M.R. Khan, I.K. Schuller, M. Grimsditch: Phys. Rev. B **29**, 2879 (1984)
2.49 B. Hillebrands, A. Boufelfel, C.M. Falco, P. Baumgart, G. Güntherodt, E. Zirngiebl, J.D. Thompson: Proceedings of the 32nd *Conference on Magnetism and Magnetic Materials*, J. Appl. Phys. to be published
2.50 R.E. Camley, M. Cottam: Phys. Rev. B **35**, 3608 (1986)
2.51 C.F. Majkrzak, J.W. Cable, J. Kwo, M. Hong, D.B. McWhan, Y. Yafet, J.V. Waszczak, C. Vettier: Phys. Rev. Lett. **56**, 2700 (1986)
2.52 N.S. Almeida, D.L. Mills (to be published)
2.53 A classic early description of these surface anisotropy fields and their influence on the ferromagnetic resonance spectrum of a thin film is found in G.T. Rado, J.R. Weertman: J. Phys. Chem. Solids **11**, 315 (1959)
2.54 M.B. Salamon, S. Sintra, J.J. Rhyne, J.E. Cunningham, R.W. Erwin, J. Borchers, C.P. Flynn: Phys. Rev. Lett. **36**, 259 (1986)
2.55 L.L. Hinchey, D.L. Mills: Phys. Rev. B **33**, 3329 (1986)
2.56 L.L. Hinchey, D.L. Mills: Phys. Rev. B **34**, 1689 (1986)
2.57 The basic concepts are discussed in C. Kittel: *Quantum Theory of Solids* (Wiley, New York 1963) Chap. 4
2.58 W. Chen, D.L. Mills: Phys. Rev. Lett. **58**, 160 (1987)
2.59 W. Chen, D.L. Mills: Phys. Rev. B **35**, 524 (1987)
2.60 D.L. Mills, S.E. Trullinger: Phys. Rev. B **36**, 947 (1987)
2.61 D.K. Campbell, M. Peyrard, P. Sodano: Physica **19**D, 165 (1986)

3. Raman Spectroscopy of Vibrations in Superlattices

Bernard Jusserand and Manuel Cardona

With 58 Figures

– Jo jamés haguera pogut creure experiència tal si de mos ulls no ho hagués vist. Ara no tinc res per impossible que los hòmens no sàpien fer. En especial tals sabers cauen en gents qui van molt per lo món; e prec-te, pare reverent, me faces gràcia de dir-me totes les coses necessàries per aquest fet del què hi és mester.

J. Martorell, M. J. de Galba: Tirant lo Blanch (N. Spindeler, València, 1490)

3.1 Introduction

As already discussed in Vols. I and II of this series [3.1, 2], light scattering spectroscopy of crystals is subject to rather stringent selection rules which arise from the conservation of wavevector k : k has to be conserved modulo a vector of the reciprocal lattice or simply conserved if we confine ourselves to the reduced Brillouin zone (BZ). The magnitudes of the wavevectors of the incident and scattered radiation, k_i and k_s ($2\pi n_{i,s}/\lambda_{i,s}$ where the ns are refractive indices and the λs wavelengths) are very small compared to that of a general vector in the BZ ($\sim \pi/a_0$, where a_0 is the lattice constant $\cong 5$ Å). Hence, in order to conserve k the elementary excitation created or annihilated (here a phonon) must have a wavevector of magnitude close to zero, i.e., near the center of the Brillouin zone (Γ-point). Thus, of the many existing excitations, for all ks within the reduced BZ (see Fig. 3.1 for phonons in GaAs) we can only investigate by means of (first order) light scattering the excitations for $k \cong 0$. The same restriction applies to most other optical spectroscopies: simple absorption and reflection (or ellipsometry) and non-linear spectroscopies such as hyper-Raman and four-wave mixing (CARS) [3.3] but not to inelastic neutron scattering [3.4]. The wavelength of thermal neutrons is on the order of 5 Å. Therefore, all points in the reduced BZ can be swept by simply changing the angle between the incident and the scattered beam and the crystal orientation. As is well known, the phonon dispersion relations of many crystals have been mapped in this manner [3.5]. Nevertheless, and in spite of the reduced amount of information it yields, light scattering spectroscopy is used frequently as it offers a number of advantages with respect to neutron spectroscopy:

i) The linewidth resolution and the accuracy in the frequency determination are one to two orders of magnitude better than in neutron spectroscopy.

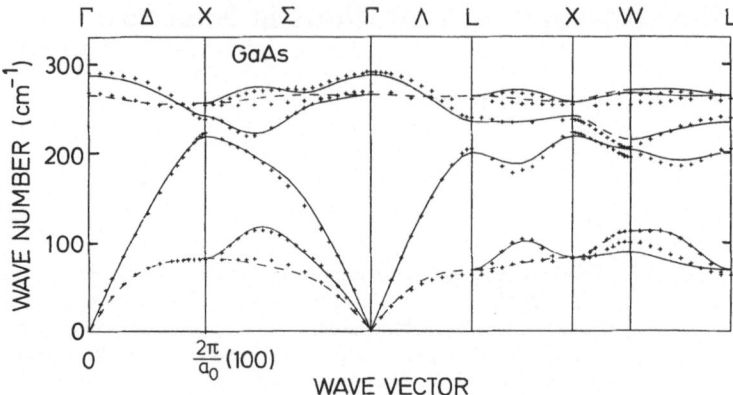

Fig. 3.1. Dispersion relations of phonons in GaAs measured at room temperature with neutron spectroscopy (*points*). The lines represent a theoretical fit. D. Strauch and B. Dorner, private communication

ii) Only microscopic amounts of material are needed for light scattering, as opposed to large crystals (several cm³) required for neutron spectroscopy. In fact, Raman microprobes using microscope optics have recently become commercially available [3.6]. The laser spot can be focused to ~ 1 μm and its penetration depth in a semiconducting sample is ~ 1000 Å. Hence, the technique is most suitable for the investigation of semiconductor microstructures.

iii) Light scattering spectrometers are extremely simple, inexpensive, and ecologically safe when compared with a neutron reactor spectrometer. They are a one-worker instrument and they are fast, especially when used with a multichannel detector [3.7, 8]. Not only spatial resolution but also time resolution (pico- and nanosecond) becomes possible [3.9]. Thus, they can be used for the characterization of semiconducting materials under industrial production conditions.

Several reviews of the applications of Raman spectroscopy to the investigation and characterization of bulk semiconducting materials have appeared [3.10–12]. Optical phonon branches are usually very flat at $k \cong 0$ (quadratic dispersion). Hence, no dispersion is seen in Raman scattering in bulk crystals and the phonon frequency for $k \equiv 0$ is obtained even if k differs somewhat from zero. For acoustic phonons, however, (Brillouin scattering) the dispersion relations are linear (see Fig. 3.1) and this linear dependence of ω on k (i.e., the speed of sound) can be measured in Brillouin spectroscopy [3.10, 13].

The severe restriction imposed upon optical spectroscopies by k-conservation is a direct consequence of the existence of a lattice of translations as symmetry elements. Thus, the thought arises that one may be able to circumvent that restriction by removing, in whole or in part, the translational symmetry operations. Several possibilities of doing so arise:

i) Making the materials amorphous or microcrystalline
ii) Introducing impurities or defects (e.g., mixed crystals such as $Al_xGa_{1-x}As$)
iii) Artificially fabricating a superlattice (e.g. $(AlAs)_7(GaAs)_7$). Such artificial crystals have a much larger lattice constant along the direction of growth than the corresponding single crystals: most of the bulk translation vectors cease to be symmetry operations. These subjects will be considered in detail in the present chapter.
iv) Observing second order (two phonon) spectra: one phonon destroys or lowers the translational invariance while the other scatters in the medium distorted by the first.

Examples of the lifting of selection rules in the cases (i) and (iv) have been given in [3.10, 14]. In these cases, broad bands which correspond to densities of phonon states weighted by a smoothly varying transition probability (matrix element), are observed. We illustrate this in Fig. 3.2 which compares the Raman and infrared spectra of amorphous silicon (a−Si) with spectral densities of phonons obtained both theoretically for a−Si and c−Si, and by neutron scattering for a−Si [3.15]. These spectra show four bands: TA, LA, LO, TO, as

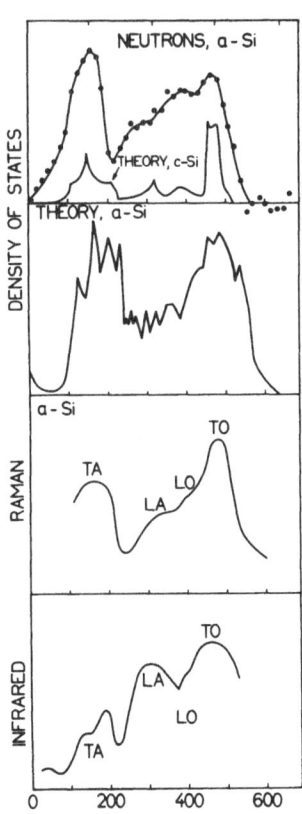

Fig. 3.2. Vibrational spectra of amorphous silicon (a-Si) as observed with infrared, Raman, and neutron scattering spectroscopies. They are compared with calculations of densities of phonons for a-Si and for c (crystalline)-Si [3.15]

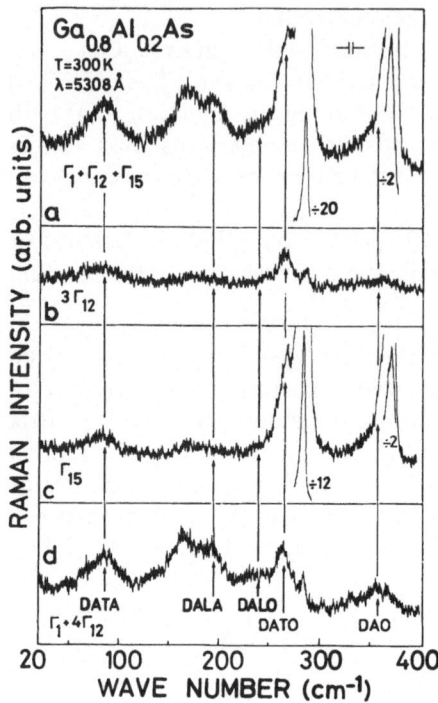

Fig. 3.3. Raman spectra of $Ga_{0.8}Al_{0.2}As$ in several scattering configurations at room temperature showing the disorder activated DATA, DALA, DATO, and DALO bands [3.16]

expected from the dispersion relations of $c - Si$ (roughly similar to Fig. 3.1 but with the vertical scale multiplied by ~ 1.7).

We show in Fig. 3.3 the Raman spectra of a bulk $Ga_{0.8}Al_{0.2}As$ crystal obtained [3.16] in four scattering configurations which correspond to different combinations of irreducible symmetry components A_1 (also called Γ_1), $E(\Gamma_{12})$, and $T_2(\Gamma_{15})$, see [3.10]. The four bands of phonons mentioned above are also observed in these spectra although they would be forbidden in the perfectly ordered single crystal constituents (GaAs, AlAs). They are thus activated by the chemical disorder of the Ga and Al atoms (disorder activated $\equiv DA$, DATA, DALA, DALO, and DATO bands). Since $Ga_xAl_{1-x}As$ is a constituent material of many superlattices, similar bands are also expected in the latter (see Fig. 15 of [3.42]).

This chapter is concerned with case (iii) above, i.e., with the reduction of the number of translational invariance operations produced by the formation of a superlattice or multiple quantum well MQW (MQW refers to the large period case in which the electronic states have zero dispersion along the superlattice direction z; see Chap. 4 of this volume). We discuss only superlattices made out of diamond or zincblende-like bulk constituents. Let us assume that the two constituents (e.g. GaAs, AlAs) have the same bulk lattice constants $2a_1 = 2a_2 = 2a$ and layer thicknesses $d_1 = n_1 a_1$, $d_2 = n_2 a_2$, the new translational period along z being $d = d_1 + d_2$ (Fig. 3.4). Because of the enhancement of the

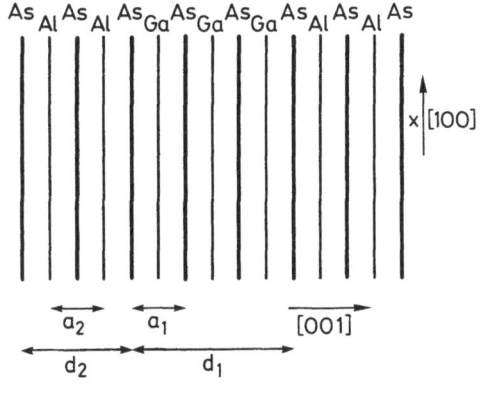

Fig. 3.4. Sketch of a $(GaAs)_{n_1}/(AlAs)_{n_2}$ superlattice with $n_1 = 3$, $n_2 = 2$. The growth direction is assumed to be $z = [001]$, since this is the most commonly investigated case

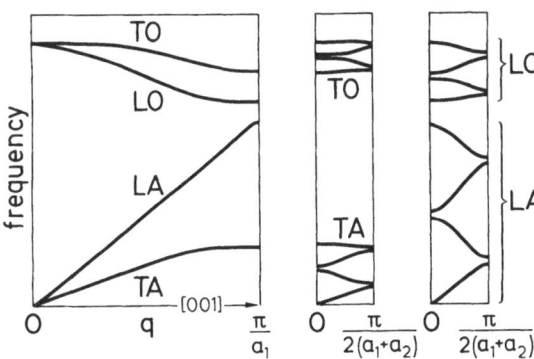

Fig. 3.5. Schematic representation of the folding to the Brillouin zone for $n_1 = n_2 = 2$. Note that the perturbation induced by the superlattice formation is small for the acoustic modes but can be large (e.g. $(GaAs)_2/(AlAs)_2$) for the optic ones

period along z the BZ must be folded into a smaller mini-BZ or superlattice BZ (SBZ) in order to stay within the reduced zone scheme. If this is not done we must, equivalently, accept non-conservation of k by reciprocal lattice vectors of the superlattice (umklapp processes). The usual (folded) reduced zone scheme, illustrated in Fig. 3.5, is more comfortable. After folding, new modes appear in the SBZ at $k = 0$ (6 longitudinal and 12 transverse in Fig. 3.5). These modes can now be Raman and optically (IR) active. Hence, superlattices are particularly interesting since they can increase the number of vibrational modes accessible to optical spectroscopies. The folding scheme is particularly useful if the phonon dispersion relations of the two materials are similar: one can fold the dispersion relations of one of the constituents (or the average of the two) and treat the difference with the real superlattice by perturbation theory. This scheme works rather well for the acoustic modes: the speeds of sound of most pairs of materials forming superlattices are indeed rather similar and, correspondingly, so are the TA and LA dispersion relations. We shall see that the main effect of the superlattice perturbation is to open "minigaps" in the folded dispersion relations at the center and edges of the SBZ. This treatment is similar to the nearly-free electron model in electronic band structures. Folded acoustic modes are discussed in Sect. 3.4.2.

The situation concerning the optical modes is quite different. The corresponding $\omega(k)$ branches of bulk materials such as GaAs and AlAs are well separated (because of the different cation masses) and rather flat (Fig. 3.1). It thus becomes meaningless to use "average" bands since a state with a given frequency for one material has no counterpart of the *same* frequency in the other, both bands being well separated. The corresponding vibrations in the superlattice are then *"confined"* to either one material or the other, decaying rapidly and exponentially beyond the interfaces. Observations of confined optical modes are discussed in Sect. 3.4.3. We note that folded modes were observed a long time ago by using the system of superlattices provided by the various polytypes of silicon carbide [3.17]. Confined acoustic modes can also be observed for single thin slabs in vacuum (or air) and for a superlattice of slabs using vacuum as a spacer. Such is the case of the *l*ongitudimal *a*coustic *m*odes (LAM) observed in folded polymer chains [3.18].

Under several simplifying assumptions the observation of folded acoustic and confined optic modes in superlattices yields information about the dispersion relations of the constituents. In some cases this information may not be available from other methods because of the lack of sufficiently large single crystals (e.g., AlAs). If independent information is available for the constituents the results for superlattices can be used to characterize their quality, in particular the details of the interfaces (see Sect. 3.4.4).

So far we have assumed that the bulk constituents have the same lattice constant. This is nearly the case for the GaAs-AlAs system (mismatch 0.1%). In many other cases the mismatch can be large (e.g. Ge − Si, 4%). Two possibilities then arise:

i) For thin layers (i.e., small periods) the lattice constants of both constituents remain matched at some average value determined by the substrate (pseudomorphic growth). The constituting layers are then under compressive or extensive strain depending on the details of the structure.

ii) For thicker layers (i.e., large periods) the individual layers relax their relative strain through the creation of misfit dislocations at the interfaces.

Even in the former case, some relaxation is expected to appear when the average superlattice parameter is not matched to that of the substrate. Above a given total thickness, some loss of strain along the axis of the structure then occurs, as recently reported [3.19]. Raman spectroscopy can be advantageously used to determine the strain in the various layers (see Sect. 3.4.4).

We have so far introduced the kinematics of light scattering in superlattices. The mechanisms leading to the scattering are also of interest since they yield detailed information about electron-phonon interaction. The simplest and oldest model is based on assuming bonds whose polarizability is modulated by the phonons [3.20]. Its implications for the case of superlattices are presented in Sect. 3.3.3. The dynamics of scattering by folded acoustic modes can be most simply discussed on the basis of the photo-elastic (elasto-optic) constants, i.e., the effect of strain on the refractive index (see Sect. 3.3.4). A more precise microscopic treatment is, however, required when the laser or scattered frequencies fall near strong electronic resonances such as those between confined valence and conduction levels in MQWs. In this case one introduces explicitly the electron-phonon interaction through deformation potential and Fröhlich mechanisms [3.10]. This is discussed in Sect. 3.5.1. In the bulk materials these mechanisms interfere with each other. In superlattices, however, they apply to phonons of different symmetries and the interference disappears.

We have so far implicitly confined our discussion to superlattice phonons which propagate along the axis of the superlattice. They are those which can lead to scattering in the conventional backscattering configuration. It is also possible, with some additional difficulties, to excite phonons which propagate perpendicular to the superlattice axis ($k_x \neq 0$). This is done by removing the substrate and using the forward scattering configuration (see Sect. 3.4.2 for acoustic phonons). Right angle scattering is also possible provided that one operates with a laser slightly below the lowest absorption edge and uses an appropriately clad superlattice, so as to transform it in an optical waveguide (see Sect. 3.4.3). Of particular interest are modes which decay exponentially (but not necessarily rapidly) around the interfaces and propagate along x or y. These have been labeled interface modes and are discussed in Sects. 3.2.4 and 3.4.3.

Several review articles have already been written on phonons in superlattices. We mention here the recent work of *Klein* [3.21], that of *Jusserand* and *Paquet* [3.22] and of *Cardona* [3.23]. The reader will also find a number of articles of interest in [3.24].

Interest in artificial semiconductor superlattices was kindled by the pioneering work of *Esaki* and *Tsu* [3.25]. The reliable experimental realization of such superlattices had to await the development of commercial molecular beam epitaxy (MBE) equipment. Most of the superlattices used for the work described here were grown by MBE. Some work has, however, been performed on superlattices prepared by the chemical vapor deposition technique involving metal-organic compounds (e.g. trimethyl gallium). A number of books, review articles, and conference proceedings devoted to the growth of superlattices, MQW, and heterojunctions have appeared. The interested reader should look at [3.26–28], other references given in these volumes, and the articles [3.29, 30] about MBE and [3.31a, 31b] about MOCVD. We should conclude by saying that all measurements reported here have been performed with superlattices grown along one of the cubic axes of the bulk constituents. Measurements on superlattices with other orientations (e.g. [111] or [110]) are highly desirable.

3.2 Lattice Dynamics of Superlattices

3.2.1 Survey

A number of models have been developed to describe the phonon dispersion curves of III–V cubic crystals [3.5]. They belong to different families such as shell models, bond charge models, etc. In these models, a large number of unknown parameters are introduced to fit the experimental dispersion curves. More recently, ab initio calculations based on the local density approximation have appeared [3.32]. They provide a good description of the phonon dispersion curves along high symmetry directions. As concerns the superlattices, a great amount of work using approaches of various degrees of complexity has been devoted to the derivation of the dispersion curves using only a knowledge of those of the bulk constituents. Several problems are encountered:

– the choice of the proper lattice dynamical model for the bulk constituents,
– the transferability of the bulk model parameters to the superlattice case, which is reasonable and straightforward for local quantities or very short range interactions but becomes difficult for long range forces,
– numerical difficulties arising from the size of the secular equations in the case of thick layer superlattices.

Lattice dynamical calculations for superlattices fall into two categories. The most general type are simple generalizations of the bulk calculations using force parameters between atoms and dynamic effective charges taken over from the bulk. To this category belongs the work of *Kanellis*[3.33], *Yip* and *Chang* [3.34], and *Richter* and *Strauch* [3.35, 35a]. These calculations yield the dispersion relations of the superlattice for any arbitrary direction of k, i.e., for finite k_x, k_y, and k_z.

The philosophy of the other, simpler group of calculations is based on the fact that most experiments (backscattering) measure only the dispersion relations for $k_z \neq 0$, $k_x = k_y = 0$. In the usual case of $k_z \parallel [001]$, one can calculate the dispersion relations by using force constants between $\{001\}$ planes (planar force constants). These planar force constants can be determined for the bulk materials either by fitting experimental dispersion relations or "ab initio" through total energy calculations [3.32, 36]. A schematic diagram of the planar nature of a zincblende-type superlattice grown along [001] is shown in Fig. 3.6. One should note that such a "planar" scheme is equivalent to the calculation of the vibrations of a one-dimensional chain with a basis (primitive cell) containing $2(n_1 + n_2)$ atoms. For vanishing in-plane wavevectors ($k_x = k_y = 0$), the longitudinal and transverse vibrations of the chain are not coupled, as they have different symmetry (see Sect. 3.3.2). Their frequencies are thus obtained from separate secular equations, a fact which simplifies the numerical analysis of the problem.

The most elaborate superlattice linear chain model [3.37] is based on ab initio local density calculations of the effective interplanar forces in GaAs [3.32]. In

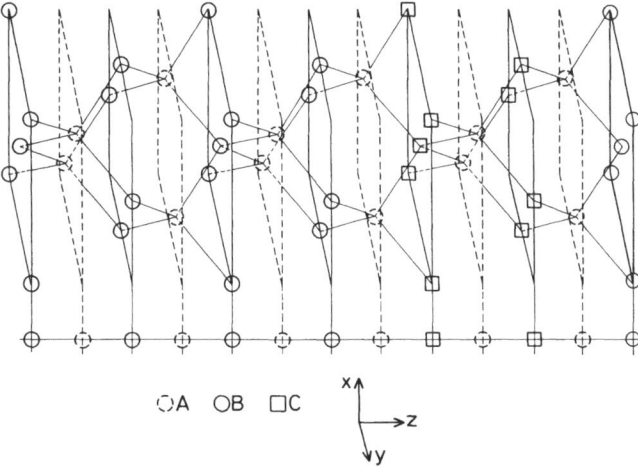

○A   ○B   □C

Fig. 3.6. Correspondence between the three-dimensional structure along the superlattice axis and the effective planar linear chain in the case of a $(AB)_4(AC)_2$ structure

this work both longitudinal and transverse dispersion curves are obtained by taking into account all significant interactions, including long range Coulomb forces, through a few short range effective forces. The problem of transferability of the bulk forces is circumvented in this description of GaAs/AlAs super-lattices: the difference between the vibrational properties of GaAs and AlAs is mainly attributed to the mass difference between gallium and aluminum. Some corrections in the Coulomb long range interactions are introduced to better describe the longitudinal-transverse splitting of AlAs. These corrections are obtained by introducing an additional term involving a Coulomb interaction between effective charges in a rigid ion model and calculating the extra charge needed to accurately describe the splitting. This somewhat artificial correction preserves the main advantage of the method: the modulated quantities (masses, effective charges) are local and thus directly transferable to the dynamical matrix of the superlattice. The only exception is the effective charge of the As interfacial atom which is different in GaAs and AlAs and must thus be interpolated. This model is particularly well adapted to obtaining realistic dispersion curves with a minimum of arbitrary assumptions. Other fitting procedures which describe the bulk dispersion curves with intermediate range forces, different in both compounds, indeed involve rather questionable interpolation schemes to generate the superlattice interactions.

On the other hand, the spatial extension of the interactions define the sensitivity of any atom in the structure to the existence of a modulation. In the model of [3.37] a large number of atoms is involved in each dynamical matrix and numerical diagonalization is needed to solve the problem, which becomes rather cumbersome for thick layer superlattices. Simpler resolution methods, independent of the thickness of the layers involved, and even providing analytical solutions in some cases, have been developed on the basis of simplified linear chain models. Such models, using a very small number of short range parameters, fitted to bulk dispersion curves, have been introduced in the early works on phonons in superlattices [3.38, 39]. The results, obtained by diagonalization of the corresponding dynamical matrices, appear to be rather good in spite of the rusticity of the models: nearest neighbor interactions often suffice, especially for longitudinal modes. Experimental constraints make the work of theoreticians easier as Raman *backscattering* selection rules for (001) faces only allow the observation of longitudinal modes (see Sect. 3.3.2). By reference to these early calculations, most of the experimental results have been successfully analyzed using the simplest imaginable model: a linear chain with only nearest neighbor short range interactions. We shall refer to it in the following as the "alternating linear chain model". All the parameters are then directly transferable from the bulk constituents to the superlattice, at least in the case where the two constituents AB (e.g. GaAs) and AC (e.g. AlAs) share a common atom A. The more general case (AB/CD) presents several specific aspects such as "wrong bonding" and "interface modes" which have been recently investigated in [3.40]. As pointed out in [3.41] the most interesting feature of alternating linear chain models consists in the reduced range of the interactions: only very few atoms are directly sensitive to the modulation. In the usual example of the GaAs/AlAs structure, only the As atoms located at each interface are affected: they are surrounded by gallium at one side and aluminum at the other. The eigendisplacement of all the atoms, except those at the interfaces, are basically known from bulk properties, and the solution of the dynamical problem thus reduces to the fulfillment of macroscopic boundary conditions at those interfaces. The method is independent of the thickness of the layers and provides, as first shown in [3.41], an analytical dispersion relation. This allows a qualitative analysis of the new vibrational features appearing in the superlattice, in particular the coexistence of:

- propagating modes for the energy ranges where allowed vibrations exist in both bulk constituents, and
- confined modes when the vibrational amplitude is evanescent in one of the constituents, since no propagating bulk modes are available at that frequency.

Since its introduction, this type of calculation has been refined somewhat by various authors to take into account, for instance, unequal nearest neighbor interactions [3.42] and to treat the case of AB/CD structures [3.43]. Some new mathematical methods have been introduced: *Albuquerque* et al. [3.44] have used a transfer matrix technique which allows to treat periodic structures with any given number of different layers per unit cell or even aperiodic structures. Green functions treatments of similar models have also been published [3.45].

The idea of taking advantage of the abrupt shape of the modulation profile and to reduce the problem to some interface boundary conditions, which was later applied in this context, is a usual one in the study of modulated structures with piecewise constant parameters. A classical example is the propagation of electromagnetic waves in layered compounds which gives rise to interference bands [3.46]. In the Kronig-Penney model, which handles the propagation of a free particle in a square wave potential [3.47, 48], some features appear, such as the confinement of the particles, which are formally reproduced in the alternating linear chain model. Moreover, the dispersion relations, obtained in both cases from interface boundary conditions, are very similar. In the context of lattice dynamics, the same ideas were first applied by *Colvard* et al. [3.49] to describe low frequency Raman lines in the elastic approximation. These authors made reference to an old calculation, by *Rytov*, of sound propagation in layered geologic systems [3.50]. Such a propagation is dominated, as in the case of electromagnetic waves, by interferences between sound waves transmitted or reflected at each interface between compounds of different acoustic impedances. Section 3.2.2 will be devoted to the elastic model, as this approximation is relevant to the analysis of numerous experimental results. We will first describe in detail the exact solution of the wave equation and analyze the dispersion relation in terms of band folding and gap opening. We will then compare the results with the more general analysis by Fourier transforms which gives better insight into the details of the eigenwaves. Section 3.2.3 will be devoted to the alternating linear chain model. We shall derive the dispersion relation in the simplest case and analyze the different vibrational behaviors of both dispersion curves and eigendisplacements. In the low frequency range the predictions of the elastic model are recovered, whereas the optical phonon range displays confinement effects which are best analyzed in terms of optical phonon quantum wells. In the case of strong confinement, the results are compared with various calculations relevant to isolated thin slabs. Some original features, such as interface modes appearing in AB/CD superlattices or surface modes in semi-infinite structures will also be briefly considered.

In Sect. 3.2.4 we shall consider the lattice dynamics for wavevectors off the superlattice axis and the connected problem of the superlattice effect on the long range Coulomb forces and the dielectric constant, which have been neglected up to this point. This subtle problem should be treated using three-dimensional lattice dynamical models in a manner similar to that used for isolated layers [3.51–53]. In spite of the scarcity of experimental results, this field has attracted much theoretical interest in the past few years in connection with two original observations of "slab modes" [3.54] and "interface modes" [3.55]. The former are confined optical phonons propagating in the layer planes whose transverse or longitudinal character is strongly affected by the dielectric modulation. The latter are optical modes, weakly localized at interfaces by the same dielectric modulation, in contrast to the "mechanical" interface modes previously considered which are strongly localized at interfaces by the short range modulation of some lattice dynamical parameter. From these two examples, the dominant role played by the long range forces in the understanding of the in-plane lattice dynamics clearly appears, a fact which gives rise to macroscopic analyses which neglect the microscopic nature of the problem. We will present these analyses and attempt to discuss their validity to describe the real three-dimensional lattice dynamics. Results of full three-dimensional calculations will be presented in Sect. 3.2.5.

3.2.2 Elastic Properties of Superlattices

We shall first consider the modulated structure corresponding to most of the experiments: the alternating stacking of two layers of different compounds A and B defined by their respective mass density ϱ_A and ϱ_B, elastic stiffness constant C_A and C_B (C_{11} is the stiffness constant relevant to the longitudinal acoustic modes propagating along the [001] direction in cubic crystals) and sound velocity v_A and v_B. There is no reference in this model to any crystalline order in the bulk constituents (i.e. it also applies to amorphous constituents) and a similar description is valid for longitudinal as well as transverse modes. The analysis we present in the following has been progressively refined in several papers [3.49, 56, 57] following improvement in the accuracy of the experimental results on samples of increasing quality. The equation for one-dimensional propagation of either longitudinal or transverse elastic waves along [001] reads:

$$\frac{\partial}{\partial t}\left[\varrho(z)\frac{\partial u}{\partial t}\right]=\frac{\partial}{\partial z}\left[C(z)\frac{\partial u}{\partial z}\right], \tag{3.1}$$

where $\varrho(z)$, $C(z)$, $u(z)$ are the local values at z of the density, elastic constant (different for longitudinal and transverse modes), and atomic displacement. This equation is also valid for propagation along [111] provided one uses the appropriate effective Cs.

For an alternating structure of two materials A and B (3.1) reduces to:

$$\varrho_{A,B}\frac{\partial^2 u_{A,B}}{\partial t^2}=C_{A,B}\frac{\partial^2 u_{A,B}}{\partial z^2} \tag{3.2}$$

in each layer.

At a given frequency ω, the sound waves in each layer are a linear combination of two plane waves of wavevectors $\pm k_A(\omega)$ for layer A and $\pm k_B(\omega)$ for layer B [$k_{A,B}(\omega)=\omega/v_{A,B}$]. As is well known for electromagnetic waves, and in order to take into account the reflection at each interface, we must write the displacement field in each layer as a linear combination of a forward ($+k_{A,B}$) and a backward ($-k_{A,B}$) elastic wave of unknown amplitudes. The solution of the problem then reduces to the determination of the four amplitudes of the four independent plane waves appearing for a given angular frequency ω.

Integrating the wave equation over an infinitesimal interval crossing any interface lying at z_i leads to the following boundary condition:

$$C_A\left.\frac{\partial u_A}{\partial z}\right|_{z_i}=C_B\left.\frac{\partial u_B}{\partial z}\right|_{z_i} \tag{3.3}$$

which expresses the continuity of the stress at each interface and holds together with the continuity of the atomic displacements:

$$u_A(z_i) = u_B(z_i) \ . \tag{3.4}$$

Imposing these two boundary conditions at each interface, and using Bloch's theorem to obtain the displacement field in the two unit cells on either side of the one at the origin, leads to a 4×4 secular determinant which provides the dispersion relation between the frequency ω and the superlattice wavevector $q = k_z$ [3.50]:

$$\cos(qd) = \cos\left(\frac{\omega d_A}{v_A}\right) \cos\left(\frac{\omega d_B}{v_B}\right)$$
$$- \frac{1}{2}\left(\frac{\varrho_B v_B}{\varrho_A v_A} + \frac{\varrho_A v_A}{\varrho_B v_B}\right) \sin\left(\frac{\omega d_A}{v_A}\right) \sin\left(\frac{\omega d_B}{v_B}\right), \tag{3.5}$$

where $d_{A,B}$ are the thicknesses of the A and B layers and $d = d_A + d_B$ is the period of the structure.

This relation has the same form as the well known Kronig-Penney dispersion relation for electrons in a square-wave potential [3.47] which reads (see Eq. (10.21) of [3.48]):

$$\cos(qd) = \cos(k_A d_A) \cos(k_B d_B)$$
$$- \frac{1}{2}\left(\frac{k_A}{k_B} + \frac{k_B}{k_A}\right) \sin(k_A d_A) \sin(k_B k_B) \ . \tag{3.6}$$

In both relations the physical details of the problem are contained in the coefficient of the second term which describes the nature and the amplitude of the modulation. The rest of the equation reflects the geometry of the structure. To better analyze the effect of this modulation on the dispersion, we write (3.5) in the equivalent form:

$$\cos(qd) = \cos\left[\omega\left(\frac{d_A}{v_A} + \frac{d_B}{v_B}\right)\right] - \left(\frac{\varepsilon^2}{2}\right) \sin\left(\omega \frac{d_A}{v_A}\right) \sin\left(\omega \frac{d_B}{v_B}\right), \tag{3.7}$$

where the parameter ε is given by:

$$\varepsilon = \frac{\varrho_B v_B - \varrho_A v_A}{(\varrho_B v_B \varrho_A v_A)^{1/2}} \ . \tag{3.8}$$

This parameter describes the acoustic modulation through the relative difference between the acoustic impedances $\varrho_i v_i$ of both bulk constituents. For available superlattices made of IV, III–V, or II–VI compounds, the acoustic modulation is usually small and $\varepsilon^2/2 \sim 10^{-2}$. This suggests to neglect, to a first approximation, the second term in (3.7) which is proportional to $\varepsilon^2/2$. In doing so, we only consider the "geometrical" contribution of the modulation which creates a new periodicity without modulating the physical quantities involved. Relation (3.7) then reduces to:

$$\cos(qd) = \cos\left[\omega\left(\frac{d_A}{v_A} + \frac{d_B}{v_B}\right)\right] \quad \text{or:} \tag{3.9}$$

$$qd = \pm\omega\left(\frac{d_A}{v_A} + \frac{d_B}{v_B}\right) + 2v\pi, \quad v = 0, \pm1, \pm2\ldots \tag{3.10}$$

which simply corresponds to the folding of an average elastic dispersion curve as shown in Fig. 3.7. The average sound velocity v follows from the averages of the inverse velocities weighted with the respective thickness d_A, d_B

$$v = \frac{v_A v_B}{(1-\alpha)v_B + \alpha v_A} \quad [\text{with } \alpha = d_B/(d_A + d_B)] \tag{3.11}$$

and reflects the inner structure of the supercell modulation. As will be emphasized later, this velocity is not the same as that of the average bulk compound. This velocity can be understood very simply in terms of transit time

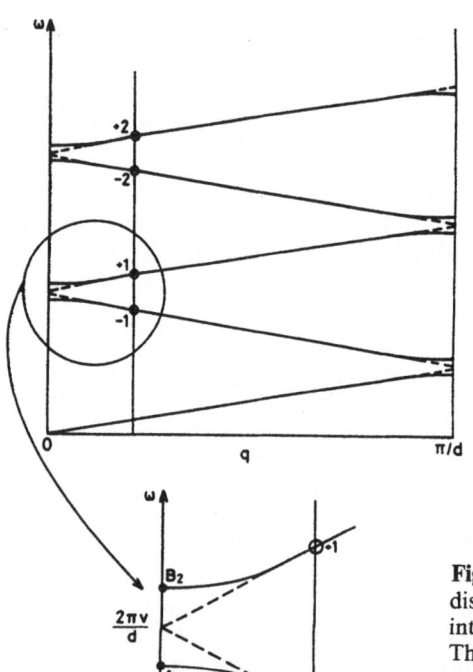

Fig. 3.7. Schematic diagram of a folded acoustic dispersion relation, neglecting (---) or taking into account (———) the acoustic modulation. The wavevector along the growth direction is represented by q. A_1 and B_2 indicate the symmetries found for these modes in gallium rich GaAs/AlAs superlattices

through the superlattice: in the absence of interface reflectivity ($\varepsilon = 0$), the transit time is the sum of the transit times through each layer, hence, inverse velocities weighted by path lengths must be added.

Within this approximation doubly degenerate modes appear at the zone center and zone edge of the SBZ (at $k=0$ and π/d) with the energies:

$$\Omega_v = \frac{v\pi v}{d},\tag{3.12}$$

where v takes even (odd) integer values at the zone center (edge). This degeneracy will be lifted when the finite magnitude of the modulation through the second term in (3.7) is taken into account. As well known from the theory of nearly free electrons in a periodic potential, the splitting for weak modulation is proportional to the amplitude of the modulation.

The limiting frequencies of these gaps at the zone center and edge correspond to eigendisplacements with equal amplitudes of the backward and forward propagating components. Thus, these modes do not transport energy. As shown recently [3.57], an exact analytic expression for the corresponding splittings can be obtained.

The splittings induced by the modulation at the zone center and edge can be also obtained with excellent accuracy by expanding the exact frequency ω to second order in $\Delta\Omega = \omega - \Omega$. The dispersion relation then reads:

$$\eta - \cos\left[(\omega + \Delta\Omega)\frac{d}{v}\right] = \frac{\varepsilon^2}{2}\sin\left[(\omega + \Delta\Omega)\frac{d_A}{v_A}\right]\sin\left[(\omega + \Delta\Omega)\frac{d_B}{v_B}\right],\tag{3.13}$$

where η equals $+1$ (-1) at the zone center (edge).

Keeping only the second order terms in $\Delta\Omega$ in the left hand side of (3.13) and the zeroth order terms in the right hand side, one obtains the shift of zone center and zone edge frequencies given by the same expression:

$$\Delta\Omega_v \cong \pm\varepsilon\frac{v}{d}\sin\left[\frac{v\pi}{2}\frac{(1-\alpha)v_B - \alpha v_A}{(1-\alpha)v_B + \alpha v_A}\right].\tag{3.14}$$

The band gap openings are therefore symmetrical relative to Ω and the average velocity of the two converging branches remains unchanged for this order of perturbation. The magnitude of the gap $2\Delta\Omega_v$, which is comparable for all v, is proportional to the modulation parameter ε and inversely proportional to the period d. It displays an oscillatory behavior as a function of α as illustrated in Fig. 3.8. Note that all zone center gaps vanish for the same value of α:

$$\alpha = \frac{v_B}{v_A + v_B}.\tag{3.15}$$

The eigendisplacement fields are piecewise sinusoidal functions. In each layer A,

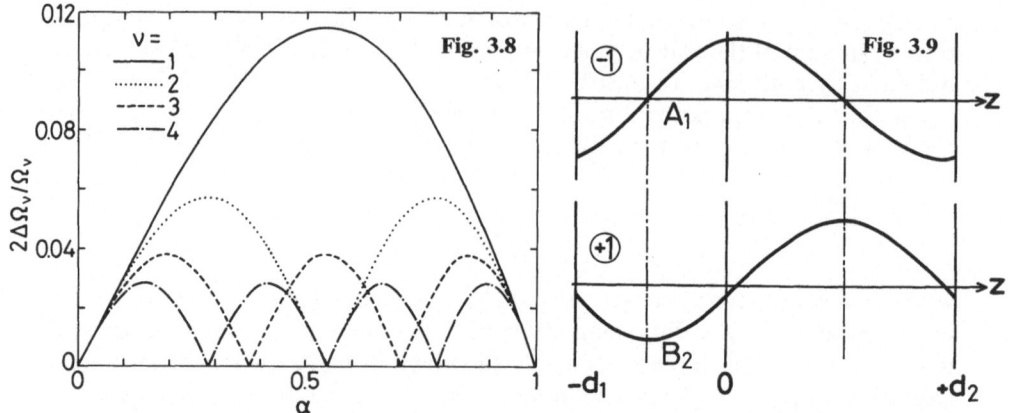

Fig. 3.8. Variation of the acoustic gaps at the lower zone center ($v=2,4$) and zone edge ($v=1,3$), normalized to the corresponding average frequency, as a function of the thickness ratio α in a GaAs/AlAs superlattice (the normalized gaps do not depend on the period) [3.60]

Fig. 3.9. Eigendisplacement for folded acoustic waves in a supercell with the two components of the lower zone center doublet (the symmetries are those found for Ga rich GaAs/AlAs) [3.22]

B, they are defined by the local wavevector $k_{A,B}$ which is slightly different in layer A and B. At the zone center, the two split modes correspond to very close local wavevectors but their displacement fields, represented in Fig. 3.9, display different symmetries: one is symmetric relative to each midlayer plane and the other antisymmetric. In Sect. 3.4.3 we shall give further details about the predictions of the elastic model and compare them with the results of various experiments. An outstanding feature is the extreme difficulty to obtain experimental evidence of the acoustic gaps. This is due to their small magnitude and to the fact that the backscattering wavevector usually lies beyond the range of significant coupling between folded zone center modes.

The same type of calculations has been recently performed [3.58] for superlattices with a period composed of three different layers. Very similar results were obtained, as should be the case for any piecewise constant profile. In such complex structures, a transfer matrix technique may, from the numerical point of view, be more convenient.

In the case of periodic structures with a smooth modulation, as for instance the erf function introduced in [3.59] to model the interdiffusion between neighboring layers of an abrupt structure, one cannot find an analytical solution of the elastic wave equation. Provided the modulation is not too strong, a perturbative approach can be applied and the wave equation approximately solved using a Fourier transform technique.

The Fourier method was first introduced by *Colvard* et al. [3.42] and applied to the square profile where a comparison with the exact solution is possible. Following [3.60] we present here a similar comparison but without the drastic approximations introduced in [3.42] to obtain analytical expressions. Such approximations give rise to less accurate predictions of the gap splittings. Let us first decompose the periodic quantities ϱ, C, and u_q in Fourier series:

$$u_q(z) = e^{iqz} \Sigma u_{n,q} e^{ingz}$$

$$C(z) = \Sigma C_n e^{ingz}$$

$$\varrho(z) = \Sigma \varrho_n e^{ingz} ,$$

where $g = 2\pi/d$ and u_q is the displacement field at frequency ω and wavevector q.
The wave equation then transforms into an infinite set of linear equations:

$$\sum_{n_1} [\omega^2 \varrho_{n-n_1} - (q+n_1 g)(q+ng) C_{n-n_1}] u_{n,g} = 0 \qquad (3.16)$$

whose secular determinant provides the dispersion relation. Usually, there is an infinite set of non-vanishing Fourier coefficients whose magnitude rapidly decreases with increasing Fourier order. Thus (3.16) can be truncated. Let us first neglect all the coefficients but the average ϱ_0 and C_0. The eigenfrequencies then reduce to:

$$\omega_v = (q+vg)(C_0/\varrho_0)^{1/2}$$

which just correspond to the folding of the dispersion relation of the average bulk compound (note that both ϱ and C must be averaged separately). This average dispersion curve does not coincide with the one obtained for zeroth order from the exact solution of the square profile where the inverse velocities were averaged (3.11). As a consequence, for such square profiles, accounting for the higher order Fourier component should not only induce a splitting between degenerate modes but also shift the whole set of eigenfrequencies. This feature cannot be obtained by only retaining the coupling between degenerate modes at the SBZ center and at the edge. The coupling between u_n and u_{-n} at the zone center or u_{n+1} and u_{-n-1} at the zone edge is of first order in the fluctuations of the material properties, as well known from the nearly-free-electron model. When one takes it into account, frequency gaps appear at the zone center proportional to $\sin(v\pi\alpha)$ [3.42] which differs from the oscillatory factor in (3.14). This difference, which is approximately proportional to squares of fluctuations in the material parameters (higher order perturbation theory), is due to non-degenerate couplings between branches of close Fourier order (n and $n\pm1$) which may be comparable with degenerate couplings between branches of distant Fourier order (n and $-n$). As concerns the square wave profile considered in [3.42] and here, the degenerate approximation is fairly good as illustrated on Fig. 3.10. The variation of the first and third zone center gaps is shown as a function of α, both taking into account either only degenerate couplings or a large set of couplings. In Fig. 3.11 we show for a given square profile the variation of the dominant Fourier components of the vibrational

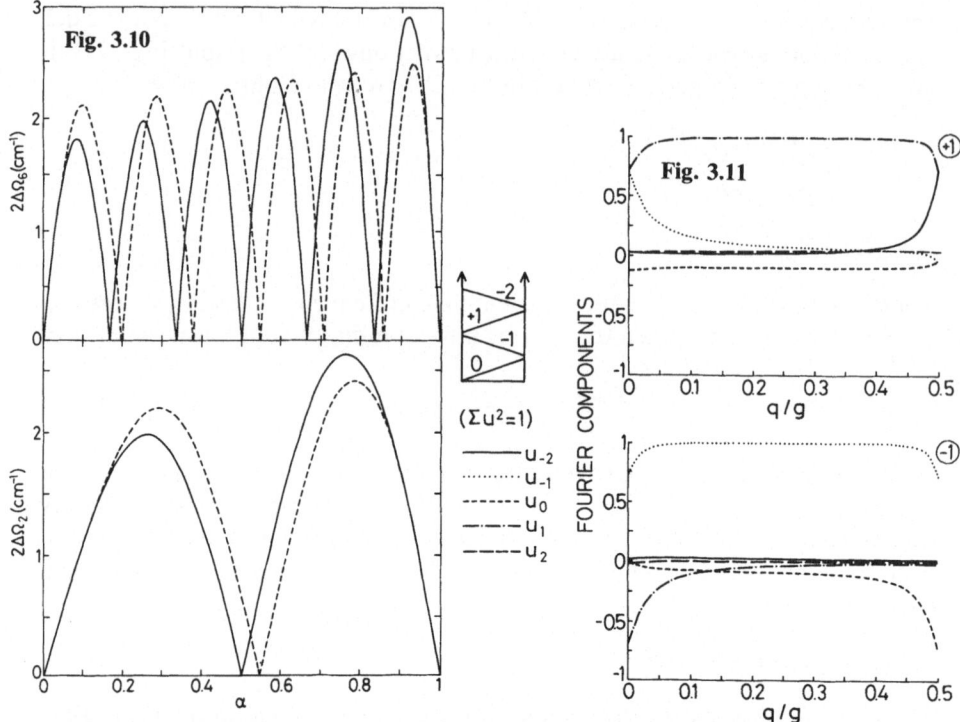

Fig. 3.10. Variation of the lower (*bottom*) and third (*top*) zone center gaps as a function of the thickness ratio in a GaAs/AlAs structure of period 40 Å, taking into account: (i) only the degenerate first order coupling (——) (ii) a large set of higher order couplings (---)

Fig. 3.11. Variation as a function of the wavevector q from mini-Brillouin zone edge ($q/g = 0.5$) of the dominant Fourier components of the eigendisplacements associated with the two lowest folded acoustic branches [3.60]

amplitudes of the lower branches as a function of the wavevector throughout the whole SBZ. Beside the single dominant component for intermediate q values, one clearly notices the effect of zone center and zone edge degenerate couplings. They induce pairs of Fourier components of equal or opposite magnitude, corresponding at the zone center to the symmetric and antisymmetric modes. One also notices the presence of additional components with a typical magnitude of 0.1 or less.

As outlined before, the Fourier transform analysis only becomes necessary for continuously varying (but well defined) profiles. Its application has been thus limited due to the lack of samples displaying such profiles. Moreover, the experimental observation of gap openings and other effects of the coupling between folded branches is extremely difficult due to their small magnitude and to limitations in the experimental conditions. However, as we will show later, the intensities of the folded acoustic Raman lines are much more sensitive to the details of the modulation and will allow us to obtain experimental confirmation of the main results of this section. The analysis of the intensities for interdiffused profiles [3.59] will in particular require the Fourier transform method.

3.2.3 The Alternating Linear Chain Model

We justified in Sect. 3.2.1 the description of the vibrational properties of superlattices along the [001] growth direction using linear chain models and discussed some more or less complex versions that one can find in the literature. The purpose of this subsection will not be to compare the validity of this description, which will be done in Sect. 3.4.2 for the GaAs − AlAs system, but to present the alternating linear chain model in its simplest version and to analyze the vibrational behavior predicted within its framework. They will not be qualitatively modified in more sophisticated models.

Let us consider the alternating linear chain AB/AC defined in Fig. 3.12 with its spring constants $K_{1,2}$ and the definition of the displacement $u_j^{(1,2)}$ and $v_j^{(1,2)}$ of atoms A or B/C in the AB or AC layers. Following the method used in the alternating elastic model, we will first look for the atoms which are not directly sensitive to the modulation. Due to the nearest neighbor range assumed for the interactions, the interface atoms A are the only ones perturbed: they are surrounded by B and C atoms. We can thus define the eigendisplacement at frequency ω of all atoms, except the A interface atoms, as a linear combination of corresponding eigendisplacements of the bulk:

$$u_j^{(1)} = \lambda_1 K_1 (1 - e^{-ik_1 a_1}) e^{ijk_1 a_1} + \mu_1 K_1 (1 + e^{ik_1 a_1}) e^{-ijk_1 a_1} \tag{3.17}$$

$$v_j^{(1)} = \lambda_1 (2K_1 - m_A \omega^2) e^{ijk_1 a_1} + \mu_1 (2K_1 - m_A \omega^2) e^{-ijk_1 a_1}$$

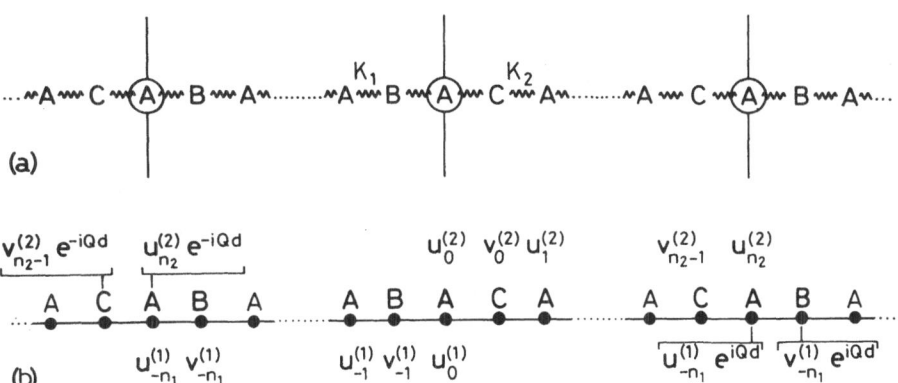

Fig. 3.12a, b. Description of the alternating linear chain model used in the text and definition of the eigendisplacements in the supercell. The A atoms are the boundaries between alternating chains

if the jth cell belongs to layer AB. Similar expressions are obtained for layer AC by replacing 1 by 2 in both expressions, a_1 and a_2 are the lattice spacings in AB and AC chains. Moreover, the interface atom displacement can be deduced from the dynamical equations of both of its neighbors and we obtain the first interface condition by identifying both expressions:

$$u_0^{(1)} = u_0^{(2)} = u_0 \qquad\qquad (3.18)$$

which reflects the identity of the interface atom displacement "seen" from AB and from AC. This displacement must also satisfy the dynamical equation of the interface atom itself:

$$-m_A \omega^2 = K_2 [v_0^{(2)} - u_0] + K_1 [v_{-1}^{(1)} - u_0] \qquad\qquad (3.19)$$

and this provides the second interface condition. Equation (3.19) can be simplified by drawing the analogy to the equation of motion of an A atom imbedded in AB (or in AC) to become:

$$K_2 [v_0^{(2)} - u_0] = K_1 [v_0^{(1)} - u_0] \,, \qquad\qquad (3.20)$$

where the $v_0^{(1)}$ displacement is a fictitious one obtained by extrapolating compound AB beyond the interface. For equal spring constants, the set of boundary conditions reduces to:

$$u_0^{(1)} = u_0^{(2)} \quad \text{and}$$
$$v_0^{(1)} = v_0^{(2)} \qquad\qquad (3.21a)$$

at the first interface ($z = 0$) and

$$u_{n_2}^{(2)} = u_{-n_1}^{(1)} e^{iqd} \quad \text{and}$$
$$v_{n_2}^{(2)} = v_{-n_1}^{(1)} e^{iqd} \qquad\qquad (3.21b)$$

at the other one.

In a model involving longer range force constants, the same method can be applied to define each of the unknown displacements progressively from the AB and AC layers. For equal spring constants in both constituents, one again obtains (3.21a, b) which can actually be written for any atomic site in the chain.

The dispersion relation found by solving the 4×4 secular determinant can thus be written (k_1 and k_2 are the wavevectors for frequency ω in the two bulk materials):

$$\cos(qd) = \cos(n_1 k_1 a_1) \cos(n_2 k_2 a_2)$$
$$-\eta \sin(n_1 k_1 a_1) \sin(n_2 k_2 a_2) \,, \qquad\qquad (3.22)$$

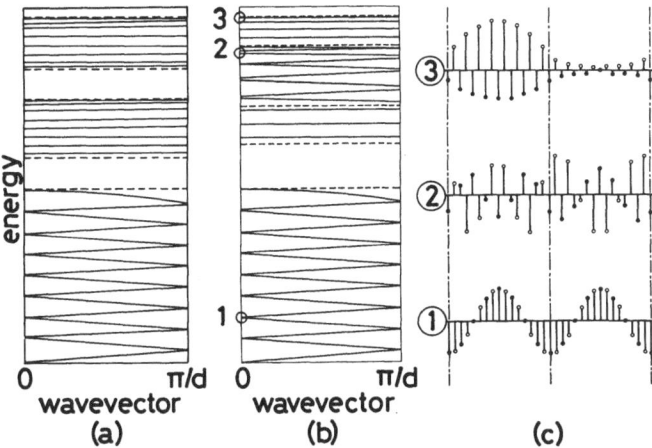

Fig. 3.13a–c. Dispersion relations for the LO phonons of a $(AB)_8/(AC)_8$ structure, with two different sets of parameters: $m_A = 70$ amu, $m_B = 50$ amu, $m_C = 25$ amu (**a**), or 40 amu (**b**), calculated with (3.22). **c** Three eigendisplacement sets in the supercell corresponding to the three zone center modes labeled by the same numbers of dispersion curve (**b**) [3.22]

where only the factor η, in this case given by:

$$\eta = \frac{1 - \cos(k_1 a_1)\cos(k_2 a_2)}{\sin(k_1 a_1)\sin(k_2 a_2)} \, ,$$

depends on the choice of the lattice dynamical model. One can find for instance in [3.42] the dispersion relation for a chain with modulated nearest neighbor interactions. The dispersion curves obtained with (3.22) for an alternating linear chain with $n_1 = n_2 = 8$ and two different choices of masses are shown in Fig. 3.13 together with some typical zone center eigendisplacements. For both cases, the common atom A is the heaviest one in both AB and AC. This choice allows a unified analysis of the acoustic and optic branches but prevents the emergence of such features as interface modes, confined acoustic modes, and mixed modes corresponding, respectively, to modes with evanescent character in both layers, with acoustic character in one and an evanescent in the other, and with acoustic character in one and optic in the other. These features have never been actually observed in AB/CD systems or Si/Ge superlattices where they have been predicted theoretically [3.40, 61]. We shall briefly describe their properties at the end of this subsection.

One easily notices the analogy of the dispersion relation (3.22) with (3.5). One indeed recovers (3.5) by taking the elastic limit for both AB and AC bulk compounds. The eigendisplacement labeled 1 in Fig. 3.13 corresponds to such a folded acoustic mode: the eigendisplacement of neighboring atoms is nearly the same. This analogy suggests to us to rewrite (3.22) in the following form:

$$\cos(qd) = \cos(\bar{k}d) - \frac{2\sin^2[(k_1 a_1 - k_2 a_2)/2]}{\sin(k_1 a_1)\sin(k_2 a_2)} \sin(n_1 k_1 a_1)\sin(n_2 k_2 a_2) \quad (3.23)$$

with the "average" $\bar{k}$ given by:

$$\bar{k} = \frac{n_1 k_1 a_1 + n_2 k_2 a_2}{n_1 a_1 + n_2 a_2} \tag{3.24}$$

and thus to separate the geometrical contribution from the physical ones. For acoustic frequencies, we could previously show the validity of this approach because $k_1 \cong k_2$. For optic frequencies, k_1 is usually very different from k_2. The energy splitting between the optic bands of two different III–V compounds is usually of the same order or even larger than the width of the bands. Thus, the perturbative approach is inadequate. At frequencies belonging to the optic band of AB, k_2 can even be complex when these frequencies fall out of the optic branch of AC. The corresponding displacements are then confined in the AB layers.

Two different behaviors actually appear corresponding to both sets of dispersion curves presented in Fig. 3.13. In the most frequent one, which is relevant to the GaAs/AlAs system, the two optic bands are well separated and all the superlattice modes constructed from these bands are confined either to the AB or to the AC layers (Fig. 3.13a). In the second case, which corresponds to the InAs/GaSb system and, with some approximation, to the GaAs/Ga$_{1-x}$Al$_x$As system, the two optic bands partially overlap and some optic modes display a confined character while some others propagate (Fig. 3.13b). Whereas the latter case must be treated exactly by solving (3.22), the former (strong confinement behavior) is suitable for an approximation "orthogonal" to the one used for elastic properties. One can treat the confined vibrations as *perfectly* confined, by making the imaginary part of the complex wavevector in the barrier (i.e. the layer where the mode is vanishing) infinitely large. One obtains, after some algebra [3.62]:

$$k_1 = \frac{\pi}{(n_1 + 1)a_1} m \qquad 1 \leq m \leq n_1 . \tag{3.25}$$

The modes are then simply standing waves of the $XAB \ldots ABX$ finite chain where the Xs are infinitely heavy atoms. This approximation appears to be extremely good when one compares the results with the numerical solution of (3.22). A similar comparison, illustrated in Fig. 3.14, has been performed in [3.63] using the results of Molinari's model [3.64]. Some small disagreement can be noticed for large values of m: the corresponding modes being less perfectly confined when calculated with (3.22). Even if for these m values the penetration depth becomes clearly larger than one interatomic distance, the perfect confinement approximation still provides an excellent estimate of the mode frequencies.

The idea of treating the optic phonons in GaAs/AlAs superlattices as perfectly confined is actually rather old and gave rise to numerous theoretical considerations in connection with the so-called slab modes. Already in [3.65] the confined frequencies were analyzed by analogy with the properties of a particle

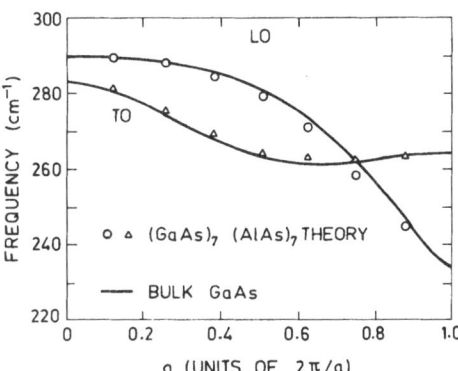

Fig. 3.14. Comparison between *calculated* GaAs-type TO and LO superlattice modes frequencies plotted as a function of $m\pi/a(n_1+1)$ and the corresponding reference bulk GaAs dispersion curves. From [3.63]

confined to a small box or a vibrating string with clamped ends, i.e., with the following relation:

$$k = \frac{\pi}{d_1} m , \qquad m = 1, 2, 3, \ldots \tag{3.26}$$

where d_1 is the thickness of the layer (size of the box or string). This relation has been used in several works [3.66a, b] and constitutes a reasonable approximation to the results of solving (3.22). However, it becomes questionable for small layer thicknesses where the difference between d and $(n_1+1)a$ [i.e. between (3.25) and (3.26)] is no longer small, as pointed out in [3.62]. In this case, the microscopic details of the thin vibrating layer and of the interface conditions have to be taken into account. The same conclusion was drawn for isolated thin GaAs layers by *Kanellis* et al. [3.53]. These authors treat in detail the lattice dynamics of isolated slabs using a rigid ion model with second neighbor short range forces and Coulomb long range forces evaluated by using a two-dimensional version of Ewald's summation method. They demonstrated that the optical modes of vanishing in-plane wavevector can almost be described as "small box modes". Some discrepancies, however, appear due to boundary conditions. In the isolated layer case, free displacement of the surface atoms has to be imposed instead of clamped neighboring atoms, as done for superlattices. This boundary condition also forces the transformation of some bulk modes into true microscopic surface modes, for which the displacement is maximum near the surface. These modes decay within a few lattice constants from the surface.

Let us now return to the alternating linear chain model in the case of overlapping optical bands. The corresponding dispersion curves are shown in Fig. 3.13b together with the displacement patterns of a confined and a propagating optic mode (Fig. 3.13c). The band ordering in this case is very similar to the one encountered in the study of electronic properties of superlattices, as already pointed out in [3.67]: in a finite energy range the modes are confined to potential wells and above the top of the potential barrier they become propagative. Thus, the notion of an "optic-phonon quantum well" is

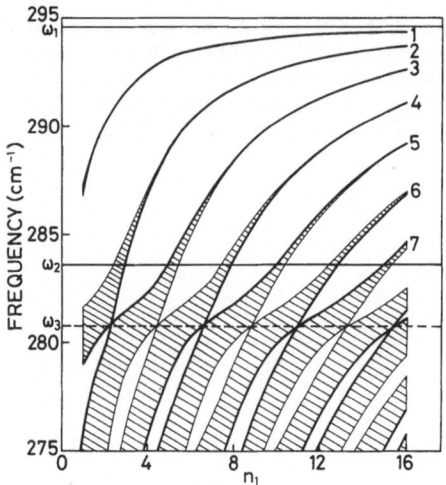

Fig. 3.15. Variation as a function of the well thickness n_1 of the highest allowed frequency bands calculated using an $(AB)_{n_1}/(AC)_5$ alternate linear chain. The bulk effective parameters are fitted to describe the GaAs-type optic modes in a GaAs/GaAlAs structure. From [3.67]

introduced. In Fig. 3.15 we show the upper optic mode frequencies of a typical alternating linear chain as a function of the well thickness n_1 for a fixed value of the barrier thickness n_2. The effect of confinement increases with decreasing n_1, a fact which is well known for electronic states, inducing a shift of the optical modes towards lower frequency. A single confined mode $(n=1)$ persists, whatever the width and the depth of the well, whereas the others progressively couple together across the barriers as their frequencies approach the top of the barrier, and become propagative beyond. Their dispersion curves then display minigaps and minibands. At the frequency ω_3, pairs of upper and lower band extrema cross without interacting since they are of different symmetries (different parity). For instance, the zone center modes are alternately symmetric (m even) and antisymmetric (m odd) relative to each midlayer plane.

As mentioned earlier, interface modes never appear in the AB/AC linear chain. On the other hand, they have been predicted for the AB/CD sequence [3.40] in the special case of InAs/GaSb structures. This pair of compounds is favorable to the emergence of such modes since one can demonstrate that they easily appear when the two optic bands strongly overlap. If such conditions are verified, the interface vibration which corresponds to a "wrong bond", to be specific a GaAs bond in GaSb/InAs structures, is fairly similar to both the gallium local mode in InAs and an arsenic local mode in GaSb. It thus becomes a common local mode and can be strongly localized at the interface. Another type of localized mode, namely a surface mode appearing in semi-infinite structures, has been described within different simple theories and, in particular, within the alternating diatomic linear chain model [3.68]. Surface modes are predicted in the gaps of the dispersion relations of bulk constituents of the infinite superlattice. Whereas the surface optic modes are localized near the surface in a manner similar to that of the interface modes near interfaces, low frequency surface modes corresponding to acoustic minigaps lie very close to the band edges and are weakly bound to the surface. The propagation properties of such surface elastic waves in the layer plane have been extensively studied using continuous three-dimensional models [3.68] which give evidence of the layering effect on the various types of surface acoustic waves.

Except in the last paragraph, we have only considered vibrations propagating along the superlattice axis because they are the only ones involved in almost all of the published experimental results and also because they are by far the easiest to model. One-dimensional models are indeed unable to account for propagation perpendicular to the superlattice axis. Three-dimensional lattice dynamics is thus unavoidable if one wants to tackle the problem of in-plane superlattice vibrations. Three-dimensional microscopic calculation have been performed rather recently for small period superlattices of GaAs − AlAs [3.35]. The few experimental results involving in-plane vibrations have been analyzed using macroscopic treatments of the layering effect on the electromagnetic field associated with the polar optical vibrations. They will be the topic of the next subsection.

3.2.4 The Effect of Layering on the Macroscopic Field: Interface Modes

In polar crystals, the zone center longitudinal-transverse (LO − TO) splitting is well known to arise from the association of a macroscopic electric field with the zone center LO mode [3.69]. Whereas a correct treatment, taking into account the polaritonic effect, yields zone center properties consistent with symmetry considerations (see Fig. 2.10 of [3.10]), the retardation effect in the long range Coulomb forces for finite k is generally neglected, however, an approximation which leads to a good description of the optic dispersion curves down to very small finite values of k, ($k > 2\pi/\lambda$, where λ is the *reststrahlen* wavelength in the medium). This was illustrated for superlattices in the early work of *Tsu* and *Jha* [3.38] who calculated the polaritonic effect on vibrations along the growth axis of a GaAs/AlAs structure. Since the experimental results are mainly obtained by Raman backscattering, which involves phonons with a large wavevector as compared with those in the region of strong polaritonic interaction, further theoretical work in this field has systematically neglected the retardation effect. Calculations involving retardation have been recently presented in [3.70] for the case of free carrier plasmons which is susceptible to generalization to polar phonons [3.70a].

The first reference to the dielectric modulation in superlattices and its effects on Raman spectra was made by *Merlin* et al. [3.71] to explain the emergence near resonance of an additional peak between TO and LO modes. They assigned this peak to an optical mode of an "effective medium" propagating in the plane of the layers, activated in Raman backscattering by an unidentified relaxation of the selection rules. To describe the frequency of the mode, they derived the anisotropic superlattice dielectric constant from those of the bulk constituents by imposing interface continuity conditions for the electric field (continuity of

E_x, E_y) and displacement (continuity of εE_z). The axial and in-plane components of the dielectric constant of the "effective medium" then read:

$$\varepsilon_z = \frac{\varepsilon_1 \varepsilon_2 (d_1 + d_2)}{\varepsilon_2 d_1 + \varepsilon_1 d_2} = \langle \varepsilon^{-1} \rangle^{-1} , \tag{3.27}$$

$$\varepsilon_{x,y} = \frac{\varepsilon_1 d_1 + \varepsilon_2 d_2}{d_1 + d_2} = \langle \varepsilon \rangle .$$

Within this model the zeros of ε_z and the poles of $\varepsilon_{x,y}$ correspond, respectively, to longitudinal and transverse modes propagating along the superlattice axis; these are basically the confined modes discussed in Sect. 3.2.3. Their frequencies coincide with the bulk values. This is a specific feature of the macroscopic approaches which neglect dispersion. Thus, their validity becomes questionable for small layer thicknesses. Additional propagating modes appear when:

$$d\langle \varepsilon \rangle = d_1 \varepsilon_1 + d_2 \varepsilon_2 = 0$$
$$d\langle \varepsilon^{-1} \rangle = d_1/\varepsilon_1 + d_2/\varepsilon_2 = 0 . \tag{3.28}$$

These obviously correspond to LO and TO modes of an effective average medium, propagating in the layer plane.

The corresponding frequencies lie between the TO and LO bulk frequencies. For $d_1 = d_2$, these two solutions become degenerate and satisfy the following equation:

$$\varepsilon_1 = -\varepsilon_2 \tag{3.29}$$

which is the standard equation for modes localized at the interface between semi-infinite materials in the absence of retardation [3.71].

Merlin et al. [3.71] assigned an additional peak observed in the Raman spectra of GaAs – GaAlAs superlattices to a solution of (3.28) and described the corresponding mode as the long wavelength limit of bulk optic modes propagating in the plane of the superlattice layers. They also noticed the connection with modes localized at and propagating along the interface between semi-infinite media. This point of view has been shown to be basically correct [3.55].

As mentioned before, the bulk optic modes in GaAs/AlAs superlattices are strongly confined in one type of layer and their frequencies can be described using a single layer model. In the macroscopic approach, where the dispersion is neglected, the small differences due to details at the interface [e.g., (3.25)] disappear. With this caveat, it is justified to use a single-slab electrostatic model [3.72] to analyze the bulk-confined optic modes whereas the more sophisticated models dealing with the complete layered dielectric medium [3.73] allow analysis of the interface modes. We follow the simple treatment given in [3.21] to describe the basic features of these models.

We proceed with the detailed mathematical treatment of the dielectric model which leads to interface and confined modes. In this model, the properties of the bulk materials 1 and 2 are solely represented by the frequency-dependent dielectric constant:

$$\varepsilon_{1,2}(\omega) = \varepsilon_{1,2}^{\infty} \frac{\omega^2 - \omega_{L1,2}^2}{\omega^2 - \omega_{T1,2}^2} , \tag{3.30}$$

where, in principle, ω_L and ω_T may be made to depend on the wave vector k so as to represent the dispersion of the longitudinal $[\varepsilon_{1,2}(\omega_L, k) = 0]$ and transverse $[\varepsilon_{1,2}(\omega_T, k) = \infty]$ frequencies. We shall neglect this dependence unless otherwise specified.

In the absence of retardation, the electric field (displacement) in each medium can be represented as the sum of a curl-free and a solenoidal part [3.21], i.e., derived from a scalar ϕ and a vector potential V:

$$E = -\nabla\phi + \frac{1}{\varepsilon}\nabla \times V . \tag{3.31}$$

This field must fulfill the equation:

$$\varepsilon\nabla \cdot E = 0 \tag{3.32}$$

which is automatically satisfied by the solenoidal part $(\nabla \times V)$. If the term in the scalar potential ϕ is not to vanish it must fulfill Poisson's equation:

$$\varepsilon\nabla^2\phi = 0 , \tag{3.33}$$

which is also automatically satisfied for $\varepsilon(\omega) = 0$. According to (3.30) this condition leads to longitudinal modes which can only have non-vanishing fields in one of the two media since, in general, $\omega_{L1} \neq \omega_{L2}$. These modes are of longitudinal character, confined either to medium 1 or 2. Another possible solution of (3.32) is obtained for $\nabla^2\phi = 0$. As we shall see, this leads to interface modes which propagate along the axis of the superlattice. Concerning the solenoidal term $\nabla \times V$ the continuity of the x, y-components of E and the z-component of D requires that $\varepsilon_1 = \infty$ if E is not to vanish in medium 1. This leads, according to (3.30), to transverse modes localized in medium 1. Both localized LO and TO modes so obtained can be made to depend on the k vector defined by (3.25, 26) simply by using k-dependent ω_L and ω_T frequencies in (3.30).

Before treating the interface modes ($\nabla\phi_{1,2} = 0$ plus appropriate boundary conditions), we want to discuss some particular properties of the LO modes found through the dielectric formalism. Fourier expansion of ϕ_1 leads to the two families of components:

$$\phi_1(x, z) = \phi_0 e^{ikx} \cos(qz) , \tag{3.34a}$$

$$\phi_2(x, z) = \phi_0 e^{ikx} \sin(qz) , \tag{3.34b}$$

where the dependence on y has been neglected, a rather inconsequential fact. In order to obtain localized modes we must assume that $E_2 = 0$ and impose the corresponding boundary conditions for E_x and D_z at the interface. For the $\phi_1(x, z)$ of (3.34a) this leads to (assuming that the origin of coordinates is at the center of layer 1):

continuity of E_x: $ik\phi_0 e^{ikx} \cos(qd_1/2) = 0$,   (3.35)

continuity of D_z: $\varepsilon_1 q\phi_0 e^{ikx} \sin(qd_1/2) = 0$,   (3.36)

and similar equations for the case of (3.34b). Equation (3.36) leads to $\varepsilon_1 = 0$, i.e., to the frequencies of bulk longitudinal modes. While (3.35, 36), representing the boundary conditions for the electric field, are automatically fulfilled for $\varepsilon_1 = 0$ and $k = 0$, one must also keep in mind that the superlattice is a mechanical system and thus the mechanical displacement of the atoms, u, must also be continuous across the boundary. This mechanical displacement is proportional to the electric polarization it generates which, in turn, is proportional to the electric field associated with the vibration. Hence the continuity of u_x is equivalent to that of E_x which we have already imposed in (3.35); that of u_z leads to an equation similar to (3.36) but without the factor ε:

$$q\phi_0 e^{ikx} \sin(qd_1/2) = 0 .$$ (3.37)

Except for the cases $q = 0$ or $k = 0$ it is impossible to simultaneously fulfill (3.35) and (3.37). For $k = 0$ we find from (3.35)

$$q = \frac{\pi}{d_1} m \quad m = 2, 4, 6, \ldots$$ (3.38a)

an analogous reasoning for (3.34b) leads to, in the case of $k = 0$:

$$q = \frac{\pi}{d_1} m \quad m = 1, 3, 5, \ldots$$ (3.38b)

Equations (3.38a, b) are equivalent to (3.26) which was derived for one-dimensional models that imply $k = 0$. Note that it is reasonable to obtain (3.26) and not (3.25) since the dielectric model is a continuum model in which the boundary conditions are imposed *exactly* at the interface.

The result of (3.38) can also be applied to the case $k \neq 0$ provided k is small compared with the qs of (3.38a, b). Such is the case which applies to phonons excited in optical experiments provided d_1 is not too large. In this case the boundary condition in (3.35) is more stringent than that in (3.37) and the latter can be neglected: the results obtained for $k = 0$ remain approximately valid. Ultimately of course, an additional weak field pattern will have to be developed around the interface in order to strictly fulfill (3.37). For most purposes, however, this additional field can be disregarded. For a recent discussion of the somewhat confusing simultaneous role of mechanical and electrostatic boundary conditions see [3.73a].

After illustrating the origin of confined modes with the electrostatic continuum model we discuss the interface modes. They arise by replacing in (3.34a,b) the pure imaginary values of q, ie., by making for ϕ the ansatz:

$$\phi_{1,2} = \phi_0 e^{ikx} e^{\pm Qz} , \tag{3.39}$$

where the origin is chosen to be at an interface. The potential of (3.39) fulfills Laplace's equation (3.33) provided $k = Q$. For a simple interface between semi-infinite media 1 (left) and 2 (right) the sign in front of Q in (3.39) must be chosen so as to avoid unphysical divergences for $z \to -\infty$ (medium 1) and $z \to +\infty$ (medium 2). The boundary condition for D_z then leads to (3.29). In the case of a single interface between semi-infinite media it is easy to modify (3.29) so as to include retardation [3.74].

For a periodic array of interfaces, such as that found in a superlattice, linear combinations of both types of solutions in (3.39) are acceptable. We impose electrostatic boundary conditions at each interface and use Bloch's theorem in order to relate the potential ϕ_0 in slab 0 to that in an equivalent slab (n). For this purpose we reintroduce a wavevector component q along z and write:

$$\phi_0(x, z) = \phi_0(x, z) e^{iqnd} . \tag{3.40}$$

As shown in [3.73] this procedure leads to the secular equation:

$$\cos(qd) = \cosh(kd_1)\cosh(kd_2) + \frac{1}{2}\left[\frac{\varepsilon_1}{\varepsilon_2} + \frac{\varepsilon_2}{\varepsilon_1}\right]\sinh(kd_1)\sinh(kd_2) \tag{3.41}$$

[note that (2.20, 21) are equivalent to (3.41)] where q is the axial superlattice wavevector. [Note also the similarity with other equations derived so far, e.g. (3.6, 22)]. The numerical solutions of (3.41) have been studied in detail in [3.73] and compared with experimental results for GaAs/AlAs superlattices in [3.55]. The results are illustrated in Fig. 3.16. For each value of k and q, four different frequencies are obtained which form two different energy bands in the range of the optical modes of each constituent. Their frequencies lie between the TO and LO bulk frequencies.

Some limiting cases of (3.41) can be further treated analytically. Let us first consider the limit of vanishing superlattice wavevector q. The dispersion relation then reduces to the two bands

$$-\frac{\varepsilon_1(\omega)}{\varepsilon_2(\omega)} = \begin{cases} \tanh\left(\dfrac{kd_1}{2}\right)\coth\left(\dfrac{kd_2}{2}\right) \\[2ex] \tanh\left(\dfrac{kd_2}{2}\right)\coth\left(\dfrac{kd_1}{2}\right) . \end{cases} \tag{3.42}$$

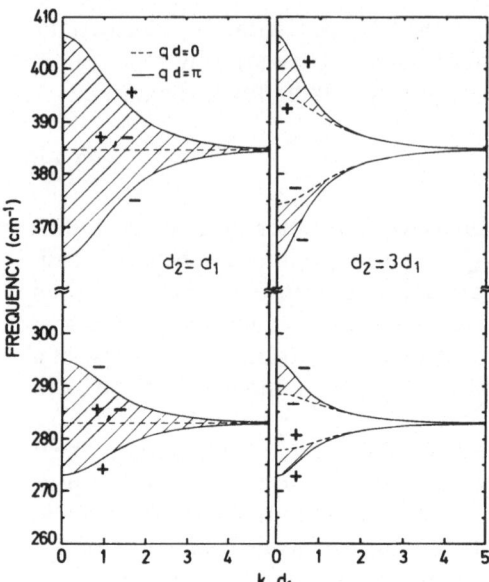

Fig. 3.16. Allowed frequency bands corresponding to the electrostatic interface modes in GaAs/AlAs structures with two different thickness ratios. The *plusses* and *minusses* indicate the parity of the corresponding electrostatic potential relative to the center of the GaAs layers. The *hatched bands* represent the axial dispersion (vs. q) for a given value of the in-plane wavevector k. From [3.55]

For interface bands corresponding both to compound 1 and compound 2, the two solutions of (3.42) lead to the limiting dashed lines of Fig. 3.16 which evidently coalesce for $d_1 = d_2$. In the latter case the interface frequencies are then obtained by solving:

$$\varepsilon_1(\omega) = -\varepsilon_2(\omega)$$

which is again equivalent to the secular equation for the interface between semi-infinite media (3.29).

In this limit ($q = 0, k \to 0$), the superlattice interface modes can be understood in terms of an interaction between equidistant single interface modes, which leads to the formation of collective modes symmetric and antisymmetric with respect to the bisector plane of an individual layer. When $d_1 = d_2$, both components appear to be degenerate, as is the case for zone-center folded acoustic modes for some values of α. Similarly, the symmetric mode belongs either to the upper or the lower branch depending of the relative magnitude of d_1 and d_2.

The limit for vanishing in-plane wavevector k of the $q = 0$ dispersion relation coincides with the early result of *Merlin* et al. [3.71]. In this limit, both solutions only coalesce if both thicknesses are equal. Otherwise, we obtain two frequencies corresponding to (3.28). These modes are then a limiting case of the interface solutions but display a special behavior: the decay length around the interface becomes infinite. These modes are thus not really localized at the interfaces even if their dependence on the dielectric properties of both bulk constituents is a consequence of their "interfacial character". Note also that for $k \to \infty$ (3.41) leads to $\varepsilon_1(\omega) = -\varepsilon_2(\omega)$ regardless of the values of d_1 and d_2, i.e., the condition for interface modes of a semi-infinite slab: for an infinite wavelength along the slab, and infinitesimal decay length, the vibrations of one interface are independent of those of the others. If the background dielectric constants of both constituents are similar and the corresponding bulk frequencies very different we obtain the approximate frequency at the zone edge $(q = \pi/d)$:

$$\omega_{1,2}^2 = \frac{(\omega_{T1,2})^2 + (\omega_{L1,2})^2}{2} . \tag{3.43}$$

The electrostatic analysis just given has a broad range of applicability. It can be easily generalized to other longitudinal excitations such as plasmons which have been recently observed by Raman backscattering in infinite and finite layered structures [3.75]. To apply this theory to the optic phonons of real superlattices one must examine the validity of the approximations involved.

As concerns bulk modes, the electrostatic approximation predicts, for $k \ll \pi m/d_1$, modes near the LO frequency of the bulk which, accordingly to (3.35, 36), are mainly polarized with the field perpendicular to the layer plane $(E_z \gg E_x)$. Likewise, one finds transverse-like modes for E parallel to either x or y. This fact explains the results of *Zucker* et al. [3.54].

For interface modes, the validity is more questionable: the calculated dispersion of interface modes overlaps with allowed bulk modes. This overlap does not take place in the non-dispersive macroscopic model where the optic band is assumed to be perfectly flat. Moreover, the mechanical boundary conditions have not been included [3.73a]. Thus, there is a need for a microscopic treatment of the superlattice vibrations of the type to be discussed in the next subsection [3.35]. Such calculations are only easy to perform for small period superlattices due to the large dimension of the secular equation for larger periods. Work performed for larger period superlattices with simplified lattice dynamical modes, even if not perfectly realistic from the quantitative point of view, should certainly bring new insight to the understanding of the vibrational properties of polar superlattices.

In the case of isolated slabs, microscopic calculations have been performed by several groups. We present here some details relevant to superlattices. The calculations by *Kanellis* et al. for GaAs slabs [3.53] are unfortunately restricted to axial modes. Thus, we must consider older publications devoted to thin slabs of NaCl structure such as those of *Tong* and *Maradudin* [3.51] or *Jones* and *Fuchs* [3.52]. The latter authors, in particular, analyze the in-plane dispersion curves in order to identify the two Fuchs-Kliever surface modes [3.72]. In Fig. 3.17 the predictions of Fuchs and Kliever for the in-plane dispersion together with the nondispersive bulk TO and LO modes are illustrated. Although axial dispersion is precluded by the model, the general trends of the surface modes are rather similar to those of interface modes in superlattices. Depending on the zone center curvature and the width of the real optic bands, these surface modes should or should not cross the whole set of bulk-like

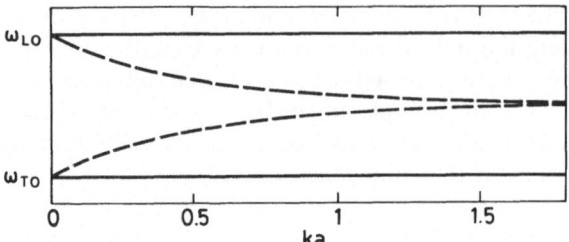

Fig. 3.17. Schematic representation of the in-plane dispersion of the bulk LO and TO modes and of both surface modes as predicted for an isolated slab in the electrostatic approximation. From [3.72]

confined modes before reaching a forbidden energy range. For instance, in the NaCl case, the LO band is very broad and decreases with k, so that no real upper surface mode appears. In the overlap range, couplings between surface and bulk modes of the same parity take place so that some bulk modes display partial surface-like character. Similar conclusions should hold for GaAs/AlAs super-lattices where interface modes and bulk quantized ones lie in the same energy range. Due to the axial dispersion of the interface modes, their crossing with the quantized bulk modes will depend on k and the whole dispersion pattern should be very complex [3.76]. The interpretation of some experimental results [3.55] in terms of interface modes and wavevector relaxation will be discussed in Sect. 3.4.

Finally, the presence of a free surface for semi-infinite superlattices, already considered from the mechanical point of view, also imposes special boundary conditions on the electric fields, giving rise to surface electrostatic modes of the whole superlattice. These modes have been observed recently by Electron Energy Loss Spectroscopy (EELS) [3.77]. To analyze their frequencies, the authors of [3.77] introduce an original formulation of the dispersion relation of interface and surface modes which, in particular, allows the study of non-periodic stackings of layers.

3.2.5 Full Three-Dimensional Microscopic Calculations for Superlattices

a) Symmetry Considerations

We start this subsection with some considerations of the symmetry properties of zinc-blende-type superlattices (e.g. $(GaAs)_{n_1} - (AlAs)_{n_2}$) grown along [001]. We first note that the primitive cell of the superlattice is composed of n_1 (n_2) primitive cells of constituent 1 (2). Two space groups are possible for such superlattices depending on whether $n_1 + n_2$ is even or odd. In the even case the translation lattice of the superlattice is primitive tetragonal, with space group $P\bar{4}m2$ (D_{2d}^5 in Schönfliess notation) while in the odd case it is body centered tetragonal, with space group $I\bar{4}m2$ (D_{2d}^9) [3.78]. The point group is in both cases $\bar{4}m2$ (D_{2d}). We give here for future reference, the character table of this group [3.79]:

	E	C_2	$2S_4$	$2C_2'$	$2\sigma_{\mathrm{d}}$	
A_1	1	1	1	1	1	$x^2+y^2,\ z^2$
A_2	1	1	1	-1	-1	$(x^2-y^2)z$
B_1	1	1	-1	1	-1	x^2-y^2
B_2	1	1	-1	-1	1	$xy,\ z$
E	2	-2	0	0	0	$xz, yz;\ x, y$

$$(3.44)$$

where C_2 and C_2' represent two-fold rotations, the former about [001], the latter about either [100] or [010]. The symmetry elements $2\sigma_{\mathrm{d}}$ are the two (100) and (010) reflection planes, while $2S_4$ are improper fourfold rotations about [001]. The last column in (3.44) gives the simplest combinations of the coordinates which belong to the corresponding representation: they are useful for figuring out optical (Raman, IR) selection rules: transverse phonons propagating along the superlattice axis have E-symmetry, their longitudinal counterparts either A_1 or B_2 symmetries (3.55, 56).

We show in Fig. 3.18 the Brillouin zone of the superlattice with $n_1 = n_2 = 1$. It is isomorphic to that for other values of n_1, n_2 provided $n_1 + n_2$ is even: one simply has to change the $\Gamma - Z$ length to ΓX divided by $n_1 + n_2$. For $n_1 + n_2$ odd we must use the Brillouin zone of the body centered tetragonal lattice, which can be found in standard textbooks [3.80].

b) Generation of the Dynamical Matrix from that of the Bulk Constituents

As we have seen, the primitive cell of the superlattice, SPC, is composed of $n_1 + n_2 = N_0$ primitive cells (PC) of the original bulk crystals. Its volume V_s is thus

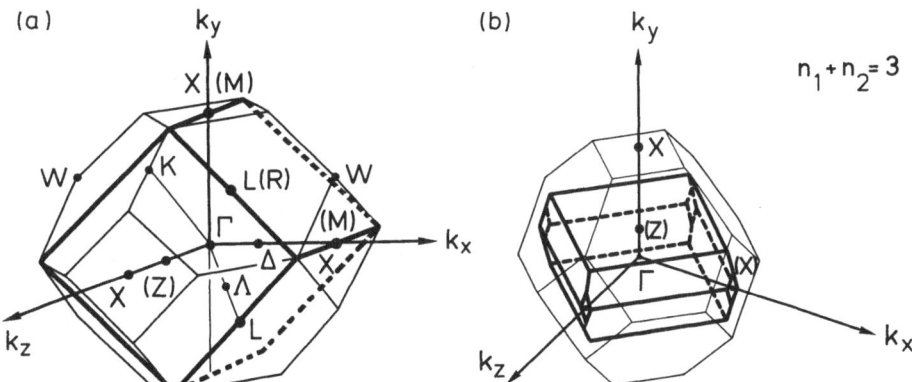

Fig. 3.18. (a) Brillouin zone (BZ) of a [001] (GaAs)$_1$(AlAs)$_1$ superlattice (*thick line*) inscribed in the BZ of the bulk crystal (*thin line*). The special symmetry points are given outside (*inside*) parentheses for the bulk (superlattice). (b) Same as Fig. 3.18a but for a superlattice with $n_1 = n_2 = 3$ (BZ of body centered tetragonal lattice)

equal to $N_0 V_0$, where V_0 is the volume of the bulk PC. The folded Brillouin zone of the superlattice (SBZ) has a volume equal to $(2\pi)^3/N_0 V_0$, i.e., that of the bulk crystal divided by N_0. Hence, if we describe a bulk crystal using only the translational symmetry of the superlattice, each band of the original BZ gives rise to N_0 folded bands. They are obtained by solving the corresponding secular equation for the dynamical matrix of the bulk:

$$|\omega^2 \delta_{\alpha\beta} - D^0_{\alpha\beta}(\kappa\kappa'|\boldsymbol{b}_m + \boldsymbol{k})| = 0 , \tag{3.45}$$

where α, β represent cartesian components, κ the two atoms of the bulk PC, $\boldsymbol{b}_m$ the N_0 reciprocal lattice vectors of the superlattice involved in folding the bulk BC, and $\boldsymbol{k}$ a vector of the SBZ. $\underset{\approx}{D}^0$ can be used to generate the dynamical matrix of the superlattice $\underset{\approx}{D}^s$ in the following way. We first construct a block diagonal $\underset{\approx}{D}^b$ of dimensionality N_0-times that of D^0 (i.e., $6(n_1 + n_2)$) [3.33]:

$$D^b_{\alpha\beta}(\kappa\kappa'; mm'\boldsymbol{k}) = D^0_{\alpha\beta}(\kappa\kappa'|\boldsymbol{b}_m + \boldsymbol{k})\delta_{mm'} . \tag{3.46}$$

The matrix $\underset{\approx}{D}^b$ must be related to that of the bulk material in the standard superlattice representation $\underset{\approx}{D}^s$ (i.e., using instead of two atoms κ, all atoms of the SPC) through a unitary transformation since it gives the bands of the superlattice in the SBZ.

$$\underset{\approx}{D}^s = \underset{\approx}{G} \cdot \underset{\approx}{D}^b \cdot \underset{\approx}{G}^{-1} . \tag{3.47}$$

Kanellis [3.33] finds for the matrix elements of $\underset{\approx}{G}$:

$$G_{\alpha\beta}(\kappa\kappa'mm') = \sqrt{N_0} \exp[i\boldsymbol{b}_m \cdot \boldsymbol{r}(\kappa, s)]\delta_{\alpha\beta}\delta_{\kappa\kappa'} , \tag{3.48}$$

where s labels the bulk PCs within the SPC.

Once the matrix $\underset{\approx}{D}^s$ for the bulk in the superlattice representation is found with (3.47) it becomes easy to introduce the superlattice modulation, i.e., to change atomic masses and interatomic force constants. The former is the easiest to vary. This can be achieved following the prescription [3.33]

$$D^s_{\alpha\beta}(\bar{\kappa}, \bar{\kappa}'...) \rightarrow \left(\frac{M_\kappa M_{\kappa'}}{M_{\bar{\kappa}} M_{\bar{\kappa}'}}\right)^{1/2} D^s_{\alpha\beta}(\kappa, \kappa'; ...) , \tag{3.49}$$

i.e. by by renormalizing the matrix elements of $\underset{\approx}{D}^s$ with the square root of the new masses of atoms κ and κ' (which now become $\bar{\kappa}$ and $\bar{\kappa}'$).

Kanellis has applied this technique to $(GaAs)_1 (AlAs)_1$ superlattices grown both along [001] and [111]. The calculations were performed with an 11-parameter Born-von Karman rigid ion model which included first and second nearest neighbors interaction and an effective ionic charge [3.81]. Unfortunately, in [3.33] only results for $\boldsymbol{k}$ along [001] and no information about "interface" modes (propagating perpendicular to [001]) is given. This information should be obtainable within this formalism.

A somewhat similar calculation has been performed for $n_1 = n_2 = 1$ and $n_1 = n_2 = 2$) in [3.82]. In this work a rigid ion model with *central forces* up to third neighbors was used (four force constants). Starting values for these force constants were taken from the bulk. They were then adjusted and modulated so as to fit observed values in these superlattices. Calculations were also only performed for $k \cong 0$ along [001].

c) Off-Axis Propagation

The most complete three-dimensional calculations have been performed for $[GaAs]_{n_1}[AlAs]_{n_2}$ superlattices by *Richter* and *Strauch* [3.35, 35a]. They include calculations for $n_1 = n_2 = 1, 2, 3, 4, 5$ and also for $(n_1, n_2) = (10, 6), (6, 10),$ $(5, 4), (1, 7), (2, 6),$ and $(3, 5)$. The thesis in [3.35] contains not only the dispersion relations for some of these superlattices for k along and perpendicular to [001] but also a large amount of information concerning eigenvectors, in particular comparison with interface modes (optical phonons) and with partly confined *acoustic* modes which also propagate perpendicular to [001] (related to the Lamb modes and Love modes of single slabs [3.83]). The authors of [3.35] chose as a lattice dynamical model the Valence-Overlap Shell Model (VOSM) [3.84] with 10 parameters fitted to neutron data for GaAs. The same model was used for the force constants involving Al instead of Ga (i.e., only the masses were changed).

Figure 3.19 shows the results obtained by Richter and Strauch for $n_1 = n_2 = 1,$ 2, 3, 4 along various high symmetry directions of the SBZ (Fig. 3.18) parallel and perpendicular to the growth axis. A number of interesting trends can be recognized in these figures. For k along [001] ($\Gamma \to Z$ line) one sees the confined modes predicted by (3.25). Examples for a $n_1 = n_2 = 5$ superlattice can be seen in Fig. 3.20. Figure 3.20b shows clearly the correctness of (3.25) and the inaccuracy of (3.26): The Ga atoms immediately outside the interface stand still while the As atoms at the interface move.

In Fig. 3.21 we show some of the corresponding eigenvectors calculated in [3.35] for modes propagating along [100] (i.e. $k_z \neq 0$). In the optical region we find the characteristics of the interface modes discussed in Sect. 3.2.4 (both components of the displacement u_z and u_x are given). Note, however, that the origin of the exponential decay for the AlAs-like modes is shifted from the interface to the nearest Al atoms, no doubt an effect of the mechanical boundary conditions. Only the antisymmetric (B_2-like for $k_x \to 0$) interface modes are shown. Note also that the acoustic-like Love and Lamb modes are predominantely localized in the GaAs.

The interface optical modes can also be recognized in Fig. 3.19 for k vectors perpendicular to [001], e.g., along the ΓX and ΓM directions (see modes labeled "I"). The dispersion relations along these directions are indeed remindful of those calculated for the interface modes with the electrostatic model for $q = k_z = 0$ (Fig. 3.16) and $d_1 = d_2$. Contrary to the predictions of the electrostatic model, however, the odd and even I modes are slightly split for $k_{x,y} \to 0$. We also note that the dependence on the direction of k for $k \to 0$, associated with the

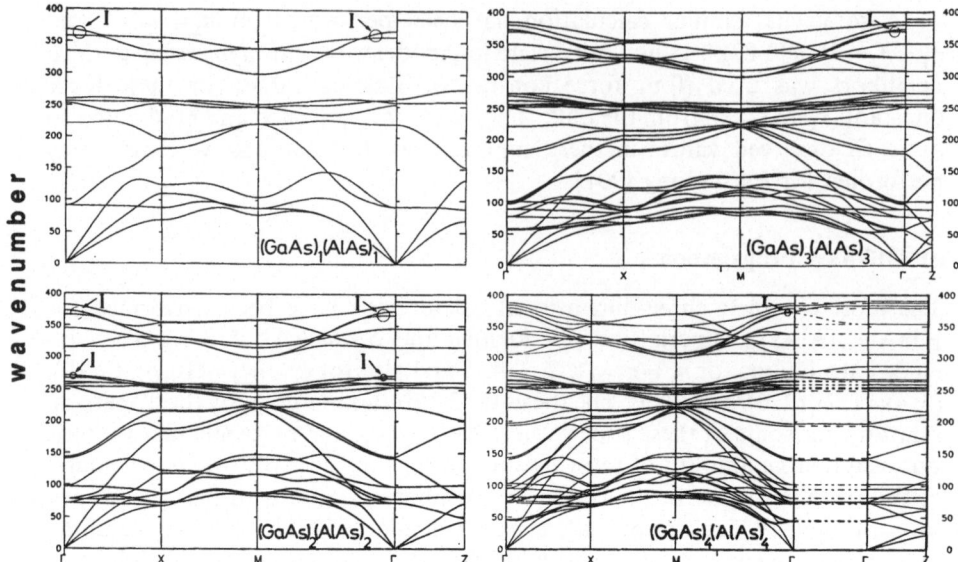

Fig. 3.19. Dispersion relations of phonons in $(GaAs)_{n_1}(AlAs_{n_2})$ superlattice for $n_1 = n_2 = 1, 2, 3, 4$ calculated by *Richter* and *Strauch* ([3.35] and private communication). "*I*" indicates interface modes propagating parallel to the layer planes. For $n_1 = n_2 = 4$ we show the angular dispersion ($\Gamma ... \Gamma$) of interface modes for $k \simeq 0$

a) Folded LA Modes, $\vec{k} \parallel [001] \rightarrow 0$

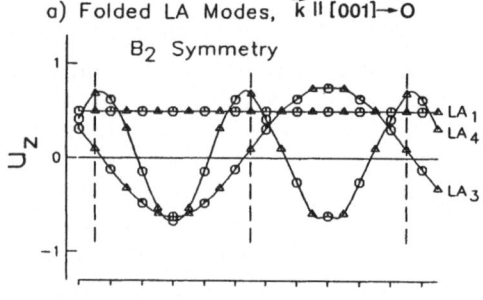

b) Confined LO Modes, $k_x = 0, k_z \rightarrow 0$

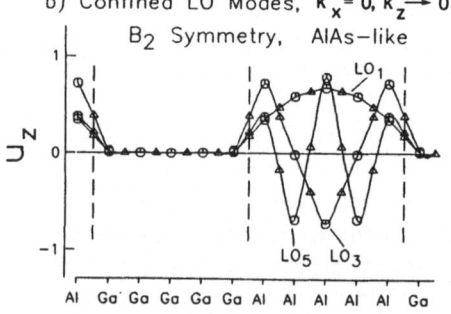

Fig. 3.20. Displacement pattern u_z of folded (acoustic) and confined (optical) modes of B_2 symmetry (3.44) for a $(GaAs)_5(AlAs)_5$ superlattice. The triangles represent u_z for the As atoms, with the sign reversed for the optical modes, while the circles represent u_z for Ga or Al. From [3.35]

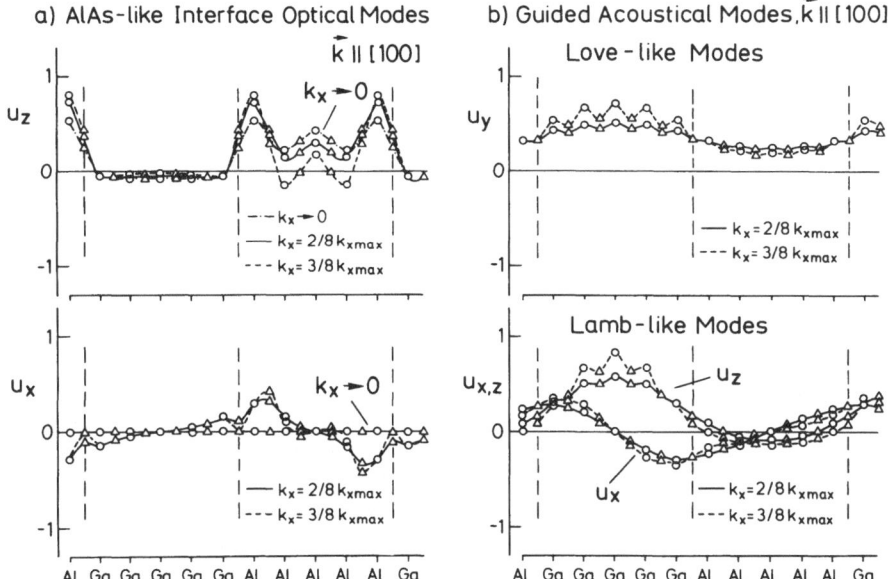

Fig. 3.21. Eigenvectors for modes propagating along [100] in a $(GaAs)_5(AlAs)_5$ superlattice grown along [001]. The optical modes are AlAs-like interface modes while the acoustical ones correspond to Love and Lamb modes of a slab. For the optical modes the signs of the displacements of cations and anions have been reversed. From [3.35]

appearance of interface modes, is clearly observable for the first confined LO and TO modes ($m = 1$ in (3.25)). For m even no such dependence occurs while for m odd > 1 the calculated dependence is small. This also agrees qualitatively with the electrostatic model: for $q = k_z \gtrsim \pi/d$ and k_x small the frequencies of the interface modes tend to the LO and TO frequencies found for $k = 0$ and $q = k_z \neq 0$.

Another interesting observation that can be made in Fig. 3.19 is the variation of the LO − TO splitting for $\mathbf{k} \parallel [001] \to 0$ (ΓZ direction near $k = 0$) with $n_1 = n_2$.

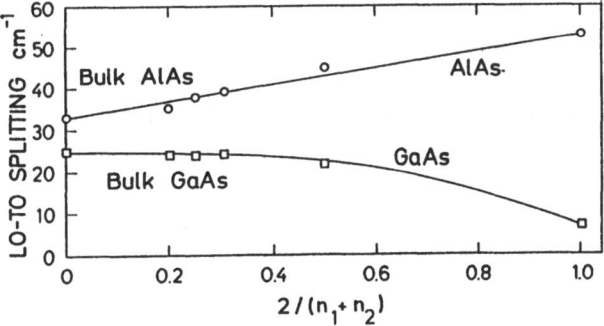

Fig. 3.22. LO-TO splitting of the GaAs- and AlAs-like modes of $(GaAs)_{n_1}-(AlAs)_{n_2}$ superlattice vs. $2/n_1 + n_2$. From the calculations of [3.35]

As shown in Fig. 3.22 this splitting becomes smaller for GaAs and larger for AlAs as the period of the superlattice decreases. This striking effect seems to have been qualitatively confirmed by experiments [3.82]. It is related to the fact that the LO — TO splitting of bulk AlAs increases with increasing k while that of GaAs decreases (the LO and TO bands actually cross with increasing k_x, see Fig. 3.1), a difference which is simply due to the fact that the mass of Al is much smaller than that of Ga: At the X point for the LO modes only Ga moves while at Γ both Ga and As move. Thus, the effect of replacing Ga by Al is larger at X than at L.

3.2.6 Conclusions

In this section we have presented the different approaches which have been used to describe the lattice dynamics of superlattices: elastic, linear chain, electrostatic models, and full three-dimensional calculations. These approaches provide a rather good description of the axial vibrations in periodically layered polar structures. The primary result is the coexistence of propagating modes (mainly acoustic) and modes strongly confined in one type of layers (mainly optic). Although a complete three-dimensional description is only available for small period superlattices, an approximate macroscopic description of the large period case has been developed to explain the limited experimental information which exists about out-of-axis superlattice vibrations. The concepts of interface optical modes and Lamb and Love acoustical modes have been introduced.

3.3 The Light Scattering Activity of Superlattice Vibrations

3.3.1 Experimental Methods

The usual methods to study lattice vibrations involve one of three types of spectroscopies:

- far-infrared spectroscopy (reflectivity, transmission)
- inelastic neutron scattering
- light scattering (Brillouin or Raman spectroscopy)

Actually, almost all experimental results available for superlattices have been obtained by means of light scattering. Before describing the mechanisms involved in the case of superlattices, we shall briefly compare the different techniques mentioned above.

Far-infrared spectroscopy has been little used to study vibrations of III–V compounds since the development of high power visible lasers for Raman spectroscopy [3.84a]. Besides technical difficulties and only a moderate frequency resolution, another drawback of this technique is the limited possibility to vary the experimental conditions. For instance, the incident frequency is fixed for each vibration probed. In contrast, visible light scattering is a non-linear process involving one (or more) phonons and two photons. One can thus vary the incident photon frequency and probe the variation of the scattering efficiency. This gives information on electron-phonon interaction [3.10]. Nevertheless, because far-infrared spectroscopy and light scattering involve different selection rules, they can provide complementary information. However, for the non-centrosymmetric III–V compounds and their superlattices all vibrations at the BZ center are Raman active for some polarization configurations and far-infrared spectroscopy may seem superfluous. Due to limitations in the available polarization configurations and sample orientations, however, some Raman active modes cannot be observed. For instance, GaAs/AlAs samples have only been available till recently for [001] oriented substrates. Backscattering experiments on [001] faces are not sensitive to transverse vibrations. Therefore, studies in the far-infrared would be very useful [3.84a]. The infrared reflectance work of [3.85] was performed for GaAs/Al$_{0.35}$Ga$_{0.65}$As superlattices of rather large period. It was interpreted on the basis of effective medium theory and yielded no microscopic information on superlattice vibrations.

As concerns neutron scattering, its main advantage lies in the neutron wavelength, which is on the order of a few Å instead of a few μm as for photons. As a consequence, with neutrons one can probe the whole Brillouin zone of crystals with lattice constants of a few Å, while, as discussed in Sect. 3.1, photons can only probe near the zone center. The application of inelastic neutron scattering is, however, limited by the very small magnitude of the scattering cross sections and neutron fluxes available. Large samples are therefore needed, a fact which hinders the application of neutron scattering to epitaxial layers and superlattices. Also, the frequency resolution of neutron scattering is never better than a few cm^{-1}, i.e., worse than for light scattering and clearly inadequate for the study of superlattice vibrations. Moreover, as we will show in Sect. 3.4, the superlattice mini-zone may be very small (comparable with a typical photon wavevector). For all these reasons, and also for its relative experimental simplicity, light scattering has been and remains the preferred technique to probe superlattice vibrations.

3.3.2 Some General Properties of Light Scattering

In the classical description of the light scattering process, the incident light at frequency ω and the lattice vibrations at frequency Ω cannot efficiently interact directly: their interaction takes place through the electronic susceptibility [3.10]. In the adiabatic or quasi-static approximation, the electronic susceptibility is modulated at the frequency Ω by the atomic displacements. The radiation, emitted by the electronic polarization induced by the incident light, is also modulated by the phonon and contains oscillating components at frequencies $\omega_i \pm \Omega$. In a light scattering experiment, the frequency shift between incident and scattered light provides the phonon frequencies, whereas their relative intensities are related to the scattering efficiency. This quantity reflects two multiplicative contributions: the average squared amplitude (actually the matrix element) of the atomic displacements and the squared magnitude of the derivative of the electronic susceptibility relative to the phonon displacement. In the case of thermal vibrations, the first contribution is proportional to $n(\omega)$ for anti-Stokes ($\omega_s = \omega_i + \Omega$) and to $n(\omega) + 1$ for Stokes scattering ($\omega_s = \omega_i - \Omega$), where $n(\omega)$ is the Bose-Einstein population factor at energy ω and temperature T. The analysis of the second contribution, involving the so-called Raman susceptibility, will be the main topic of this section. We shall leave resonance effects, i.e., the strong enhancement in Raman susceptibility in the neighborhood of electronic interband energy gaps, for Sect. 3.5.

In crystalline solids, the analysis of the Raman activity is usually done in two steps. One first derives by symmetry considerations the non-vanishing independent components of the Raman tensor. No reference is made to the microscopic details of the mechanism involved. In the second step one calculates the Raman activity in the framework of a model for the modulated susceptibility. Before developing the case of superlattices of zincblende-type bulk compounds, we must recall the selection rule, mentioned in Sect. 3.1, which expresses the wavevector conservation during the scattering process:

$$k_s = k_i \pm q , \qquad (3.50)$$

where k_s, k_i, and q are wavevectors inside the sample of the scattered and incident light and the phonon involved, respectively and $-(+)$ applies to Stokes (anti-Stokes) scattering. The vectors of (3.50) are confined to the first (mini)-Brillouin zone (SBZ).

Due to the very large value of the photon frequency of visible light (~ 20.000 cm^{-1}) as compared to that of a typical phonon (~ 200 cm^{-1}), the difference between k_i and k_s is often neglected, and one obtains:

$$k_i = \frac{2\pi n_i}{\lambda_i} \cong k_s = \frac{2\pi n_s}{\lambda_s} , \qquad (3.51)$$

where n_i (n_s) is the refractive index of the sample at the incident λ_i (or scattered λ_s) wavelength: $n_i \cong n_s \cong n$. Depending on the experimental configuration, the magnitude of the allowed phonon wavevector lies between ~ 0 (forward scattering) and $4\pi n/\lambda_i$ (backscattering). Its maximum value is much smaller (by a factor of 100) than the typical Brillouin zone extension of a bulk III–V compound and one usually considers that light scattering experiments probe only zone center excitations. Raman selection rules are thus determined using the zone center (Γ) symmetry. Taking into account the finite value of q will relax some selection rules: forbidden lines become allowed and their intensity increases when q moves away from zone center. This feature is important for the acoustic modes: at the zone center they correspond to a rigid displacement of the whole crystal which evidently induces no Raman scattering (translational invariance). To treat this q-dependent scattering process in the vicinity of $q = 0$, one introduces a new term in the modulated susceptibility which describes the effect of the strain on the susceptibility $(\partial\chi/\partial e)$. The resulting phenomenon is called Brillouin scattering. This term actually does not give rise to a q-dependent scattering intensity since the Bose population factor diverges for $\omega\to 0$ like q^{-1}. Taking also into account the proportionality of the phonon amplitude to $q^{-1/2}$ and that of the strain to q, we find a scattering efficiency independent of q.

Another case of q-dependent processes is encountered for the LO component of IR-active phonons near resonance: intra-band Fröhlich terms, which give a vanishing contribution at the zone center, become large (see Sect. 3.5 and [3.10]).

To determine the dipole (q-independent) allowed modes at the zone center, one must find the irreducible representation of the crystal point group to which they belong, as well as those of the components of the susceptibility tensor which is usually assumed to be symmetric (see, however, Sect. 3.5.3b). A mode will be allowed if its representation coincides with one of those of the susceptibility tensor [3.10]. For zincblende-type crystals (point group T_d) the representation which expresses the total symmetry of the atomic displacements Γ_M can be decomposed into irreducible components as:

$$\Gamma_M = 2T_2 , \tag{3.52}$$

where T_2 corresponds to the triply degenerate zone center acoustic and optic modes, respectively [3.10]. The corresponding representation of the *symmetric* susceptibility tensor reads:

$$\Gamma_s = A_1 + E + T_2 \tag{3.53}$$

and contains T_2. The zone center optic mode is thus allowed (note that A_1 corresponds to the trace of the tensor, E to the diagonal components less the trace and T_2 to the off-diagonal components). The related scattering efficiency is then proportional to:

$$|\boldsymbol{e}_s \cdot \boldsymbol{R}_{T_2} \cdot \boldsymbol{e}_i|^2 , \tag{3.54}$$

where $\boldsymbol{e}_s$ and $\boldsymbol{e}_i$ are the polarization vectors of the scattered and incident light. $\boldsymbol{R}_{T_2}$ is the Raman tensor associated with the vibration proportional to T_2 components (off-diagonal) of the susceptibility tensor. It is easy to show that for the backscattering configuration on a [001] face of a T_d crystal the longitudinal optical mode is allowed. This mode distinguishes itself from the transverse one because of the concomitant field which slightly alters the allowed Raman tensor. This alteration is determined by the Faust-Henry coefficient of zincblende materials [3.10, p. 58].

Let us now consider superlattices grown along the [001] axis. The superperiodicity induces (see Sect. 3.2.5a)

1) a lowering of the crystal symmetry from cubic (T_d) to tetragonal (D_{2d}).
2) an expansion of the primitive cell from 2 to $2(n_1 + n_2)$ atoms.

The lowering of the symmetry comes from two different features: the superperiodicity in the axial direction, which then becomes inequivalent to the two orthogonal ones, and the elastic deformation of the constituent compounds which appears in order to accommodate any lattice mismatch between them. The latter effect is often neglected (a good approximation for $(GaAs)_{n_1} - (AlAs)_{n_2}$). Some experimental results on strained systems will be presented in Sect. 3.4.4. While the point group of (001) oriented superlattices is always D_{2d} (tetragonal), their space group depends on the details of the structure (3.78). For instance, AB/AC structures are simple tetragonal or body centered tetragonal depending on the relative parity of n_1 and n_2. For AB/C superlattices where C is a group IV element (Ge, Si), the situation should be even more complex since n_2 can be half-integer, a fact which induces a doubling of the superlattice unit cell.

The representations of the mechanical vibrations at Γ are (see character table (3.44) and [Ref. 3.10, p. 50] for the method to derive these representations):

$$\Gamma_M = 2(B_2 + E) + (n_1 + n_2 - 1)(A_1 + B_2 + 2E) , \tag{3.55}$$

including the $k = 0$ acoustic phonons. Those of the susceptibility tensor (3.44):

$$\Gamma_S = 2A_1 + B_1 + B_2 + E . \tag{3.56}$$

Thus all modes are again Raman active. In (3.55) we have divided the mechanical representations into two sets. The first one corresponds to cubic representations at Γ which are split by the tetragonal distortion. The second one contains modes which come from out-of-zone-center points in the cubic Brillouin zone folded to the zone center, due to the increase in the size of the primitive cell. The emergence of a Raman active A_1 representation is a specific consequence of this increase. The intensity of all modes activated by supercell formation vanishes with vanishing amplitude of the modulation. In the backscattering configuration along the (001) direction, the E modes (transverse) are forbidden whereas the A_1 and B_2 modes (longitudinal) are allowed in $z(x, x)\bar{z}$ and $z(x, y)\bar{z}$ configurations, respectively. The E and B_2 modes thus follow the same selection rules as the cubic modes from which they arise, while the A_1 modes appear in a cubic *Raman inactive* configuration (remember, however, that in this configuration LO modes become Raman active near resonance due to intraband Fröhlich interaction).

Before discussing the microscopic models used to describe the modulated susceptibility, we shall reconsider the symmetry of the superlattice eigenmodes as predicted from the one-dimensional models described in Sect. 3.2. In the AB/AC case, zone center eigenmodes correspond to symmetric or antisymmetric displacement amplitudes of the rigid layer planes (relative to each midlayer plane). The symmetric *amplitudes* correspond to the B_2 modes and the antisymmetric to the A_1 modes, a fact which seems surprising since the A_1 representation is totally symmetric. The displacement amplitudes, however, are lengths of vectors whose direction is antisymmetric.

3.3.3 The Bond Polarizability Model

One of the most popular models to calculate the polarizability and its derivatives with respect to the atomic displacement is the so-called bond-polarizability model, due originally to Wolkenstein [3.20], which describes the polarizability of the whole crystal as the sum of the independent contributions of each bond. One then obtains a local description of these quantities. Such models have been extensively used for group IV [3.86] and III–V bulk crystals [3.87], various layered crystals [3.88–90] and introduced for GaAs/AlAs superlattices by *Barker* et al. [3.39]. Its predictions were considered in more detail and compared with experiments in [3.91]. In a bond-related set of axes, the polarizability tensor simply reads:

$$\begin{pmatrix} \alpha_{\parallel} & 0 & 0 \\ 0 & \alpha_{\perp} & 0 \\ 0 & 0 & \alpha_{\perp} \end{pmatrix}, \tag{3.57}$$

where the bond direction has been taken to be the x axis. The rest of the calculation is simply a transformation of all tensors related to each bond in the unit cell to crystal-related axes. While differentiating the crystal polarizability tensor relative to the atomic displacements, one has to take into account (i) the variation of the individual bond polarizabilities and (ii) the variation of the bond orientation which makes itself felt through the coordinate transformation. One usually assumes that the bond polarizability (3.57) only depends on bond lengths. In zinc-blende-type superlattices the differential polarizability tensor only involves the following parameters of each of the constituent bonds:

$$\alpha_{xx} = \frac{2}{\sqrt{3}} \frac{d\alpha_0}{dl}\bigg|_{l_0} - \frac{4}{l_0\sqrt{3}} \alpha_2$$

$$\alpha_{xy} = \frac{2}{\sqrt{3}} \frac{d\alpha_2}{dl}\bigg|_{l_0} - \frac{4}{l_0\sqrt{3}} \alpha_2 \tag{3.58}$$

$$\alpha_{zz} = \frac{2}{\sqrt{3}} \frac{d\alpha_0}{dl}\bigg|_{l_0} + \frac{8}{l_0\sqrt{3}} \alpha_2 \qquad \text{where:}$$

$$\alpha_0 = \tfrac{1}{3}(\alpha_{\parallel} + 2\alpha_{\perp})$$

$$\alpha_2 = \tfrac{1}{3}(\alpha_{\parallel} - \alpha_{\perp}) \tag{3.59}$$

and l_0 is the bond length.

For a bulk crystal and the zone center LO modes propagation along [001], one obtains [3.86]:

$$\Delta\chi \propto 2 \begin{pmatrix} 0 & \alpha_{xy} & 0 \\ \alpha_{xy} & 0 & 0 \\ 0 & 0 & 0 \end{pmatrix} \delta u_x \tag{3.60}$$

which, when replaced into (3.54) reproduces the well-known selection rules [3.10]. δu is the bond vector variation: its *magnitude* is the same for all bonds in the crystal in the case of this phonon.

For AB/AC superlattices, two different bonds appear and $\Delta\chi$ is obtained through a summation over all the bonds in the supercell:

$$\Delta\chi = \begin{pmatrix} A_{xx} & A_{xy} & 0 \\ A_{xy} & A_{xx} & 0 \\ 0 & 0 & A_{zz} \end{pmatrix} \quad \text{where:} \tag{3.61}$$

$$A_{xx,zz} = \alpha^{(1)}_{xx,zz} \sum_{i=-n_1}^{-1} (u_i^{(1)} - u_{i+1}^{(1)}) + \alpha^{(2)}_{xx,zz} \sum_{i=0}^{n_2-1} (u_i^{(2)} - u_{i+1}^{(2)})$$

$$A_{xy} = \alpha^{(1)}_{xy} \sum_{i=-n_1}^{-1} (u_i^{(1)} + u_{i+1}^{(1)} - 2v_i^{(1)}) \tag{3.62}$$

$$+ \alpha^{(2)}_{xy} \sum_{i=0}^{n_2-1} (u_i^{(2)} + u_{i+1}^{(2)} - 2v_i^{(2)}) \ .$$

The indices 1, 2 specify whether the bond belongs to compound 1 or to 2. In this manner one recovers the previously derived selection rules as A_{xx} vanishes for symmetric modes (B_2) and A_{xy} for antisymmetric ones (A_1). The activity A_{xx} of the A_1 modes can be considerably simplified and reads:

$$A_{xx}(As) = 2u_1(\alpha^{(1)}_{xx} - \alpha^{(2)}_{xx}) \ , \tag{3.63}$$

where u_1 is the displacement of the interface atom. This result can be understood by noticing that the A_1 representation is not compatible with a zone center representation of phonons in the bulk materials. In a model where the cubic symmetry is locally preserved almost everywhere, contributions to the Raman activity of these modes can only come from the non-cubic regions, namely the interfaces, and is proportional to the local polarizability modulation.

In contrast, the $A_{xy}(S)$ terms are nonzero even for a non-modulated polarizability. As we pointed out previously, due to the large optic frequency modulation, the superlattice optic modes are not well described in terms of folding. As a consequence, the successive confined modes, and not only the first one, have significant intensities. They decrease with increasing order m, independently of the modulation of their interaction with radiation. For large layer thicknesses it is easy to show that their intensity decreases with mode order m (m odd) like m^{-2}. We illustrate these features in Fig. 3.23. The intensity variation of the upper optic modes confined to AB for the alternate linear chain also used in Fig. 3.15 is shown as a function of the AB layer thickness. The bond polarizability has been assumed to be unmodulated (the same in AB as in AC). These parameters are not well known for III–V bulk compounds, especially in the visible where resonances occur, and one can only expect to obtain a

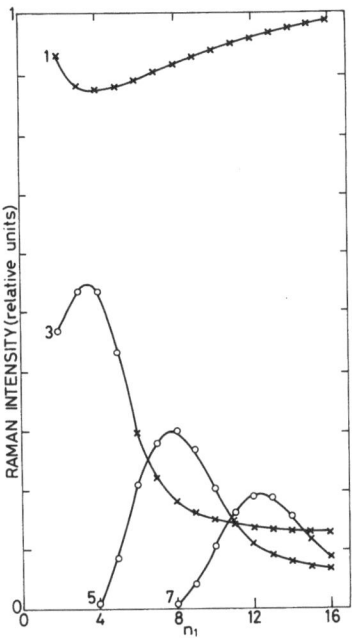

Fig. 3.23. Variation as a function of the well thickness n_1 (in monolayers) of the Raman intensity of the highest B_2 modes whose frequencies are shown in Fig. 3.15, calculated using the bond polarizability model. The results are shown by circles or crosses depending on whether the modes are propagative or confined [3.91]

qualitative description when using that assumption, as will be illustrated in Sect. 3.2.4. Moreover, this model completely fails in reproducing the light scattering intensities of the acoustic modes. It predicts very weak folded acoustic lines due to the weak deviation of the folded modes from those of the bulk, which also reduces the intensity of the acoustic Raman lines (the true Brillouin line excepted).

Actually a large modulation of the bond polarizability may allow to reproduce the typical folded acoustic intensity as suggested in [3.39]. We have already mentioned that a macroscopic description of the scattering efficiencies by folded acoustic modes becomes possible by using the derivatives of the susceptibility with respect to the strain, i.e., the so-called photoelastic constants. These photoelastic constants can, in principle, be related to the differential bond polarizabilities α_{xx}, α_{xy}, and α_{zz} [3.86, 92]. However, in semiconductors it is not possible to describe both the Raman tensor and the photoelastic constants with the same bond polarizabilities [3.86], the problem being the existence of electronic resonances in the visible region (note that such a description is possible for insulators such as diamond [3.93]). Hence, we decouple the photoelastic constants from the bond polarizabilities and give below the treatment of the scattering efficiencies for folded acoustic modes on the basis of photoelastic constants.

We should mention at this point a bond polarizability calculation that appeared recently [3.94] which includes the effect of disorder in the $Ga_{1-x}Al_x$ layers.

3.3.4 The Photoelastic Model

The polarizability modulation per SPC [3.42, 95] due to the photoelastic coupling associated with an acoustic mode reads:

$$\Delta\chi = \int_0^d \Pi(z)e_i^q(z)dz \tag{3.64}$$

where $e_i^q(z) = d(u_i^q(z))/dz$ is the strain related to the mode of wavevector q in branch i and $\Pi(z)$ is the relevant profile of the photoelastic coefficient along the superlattice axis z. For longitudinal modes propagating along [001], i in (3.64) represents the xz or yz cartesian components and thus $\Pi(z)$ is related to the p_{12} component of the photoelastic tensor p through:

$$4\pi\Pi(z) = -\frac{1}{\varepsilon^2}\,p_{12}\ , \tag{3.65}$$

where ε should be the dielectric constant of the laser or scattered frequency (assumed to be the same). The values of p_{12} and its spectral distribution have been determined for many diamond and zincblende-type semiconductors below the absorption edge [3.96]. In this region $\Pi(z)$ is real. Actually most measurements for superlattices are performed in the region where at least one of the bulk constituents absorbs and thus the corresponding ε and p_{12} become complex. In this region little information is available for p_{12} [3.97]. Hence, at the present time, the information about the photoelastic tensor p of the bulk constituents required to interpret the intensities of scattering by folded acoustic modes is not available. Even if this information were available the effects of the superlattice on the electronic properties would considerably modify it. For a discussion see Sect. 3.5.4 and [3.10].

In spite of this lack of precise information about the tensor p it is possible to obtain expressions for the efficiency for scattering through folded acoustic modes using the components of p in both media as parameters.

It is useful to discuss this problem with the Fourier formalism of Sect. 3.2.2 [3.42] which can be applied not only to the case of the sharp interfaces but also to diffuse ones of arbitrary one-dimensional profile. We expand the quantities $\Pi(z)$ and $u_i^q(z)$ in Fourier series:

$$\Pi(z) = \sum_n{}' P_n e^{ingz}$$

$$u_i^q(z) = e^{iqz} \sum_n u_{i,n}^q e^{ingz}\ . \tag{3.66}$$

One then obtains:

$$\Delta\chi = \sum_n P_{-n} u_{i,n}^q(ng + q)\ , \tag{3.67}$$

where we have omitted the common multiplicative factors of (3.65) in $\Pi(z)$.

The scattering efficiency for each mode is proportional to $|\Delta\chi|^2$. The eigendisplacements $u_l^q(z)$ have already been determined by solving the wave equation (see Sect. 3.2.2) and different approaches have been used to describe the photoelastic profile. The simplest one, performed by *Babiker* et al. [3.98] crudely assumes an unmodulated photoelastic coefficient P_0. These authors take into account the acoustic modulation obtained from a Green's function treatment of the wave equation. The resulting efficiencies are simply proportional to $|P_0|^2$ and can be compared without any knowledge of the photoelastic properties. As we pointed out before, the acoustic modulation in the GaAs/AlAs system is small, and correspondingly, these authors predict very small intensities which are strongly decreasing with increasing folding order. They arise, in this model, exclusively from the coupling of the folded modes with the Brillouin mode of the bulk. The predictions of this model for the intensity variation of the line ± 1 and ± 2 as a function of the relative thickness α of the two layers constituting the supercell are illustrated in Fig. 3.24. The model of [3.98] also includes the effect of optical absorption, which induces some uncertainty in the wavevector of the phonons involved and thus some broadening of the corresponding Raman lines. Due to this broadening, and to the very small magnitude of the folded lines as compared to the Brillouin line, strong coupling between the different components takes place, giving rise to complex line shapes. All these features have not yet been observed because on all the available systems, the contribution of the photoelastic modulation appears to be dominant.

This contribution was first introduced by *Klein* et al. [3.95] in a complementary model in which any effect of the acoustic modulation is neglected. The folded eigendisplacements are then plane waves propagating at the average velocity of the superlattice, a correct approximation far from zone center and edge. The calculated intensities only depend on the non-zero Fourier components of $\Pi(z)$. In the usual case of the periodic stacking of abrupt layers of two

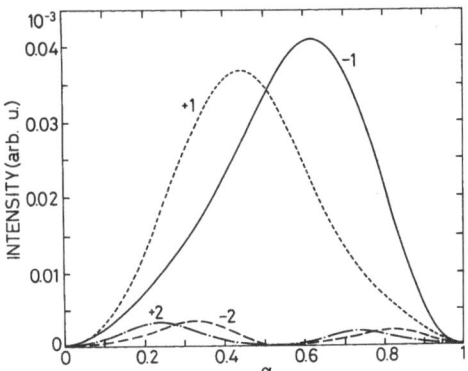

Fig. 3.24. Variation, as a function of the thickness ratio $\alpha = d_2/(d_1 + d_2)$ in a GaAs/AlAs structure, of the room temperature intensity of the four lowest folded branches at $q/g = 0.06$ (g is the length of the first reciprocal lattice vector of superlattice), calculated neglecting the photoelastic modulation, as done in [3.98], The line labelling was introduced in Fig. 3.7

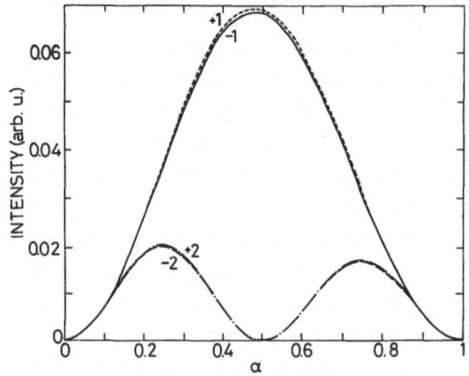
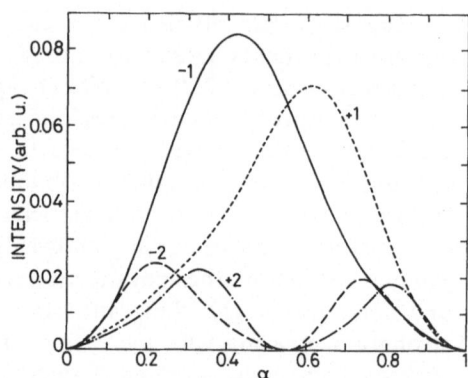

Fig. 3.25. Same as Fig. 3.24 but neglecting the acoustic modulation, and considering only the photoelastic one, as done in [3.95]

Fig. 3.26. Same as Fig. 3.24 but taking into account both acoustic and photoelastic modulations and assuming $P(\text{AlAs})/P(\text{GaAs}) = 0.1$. From [3.60]

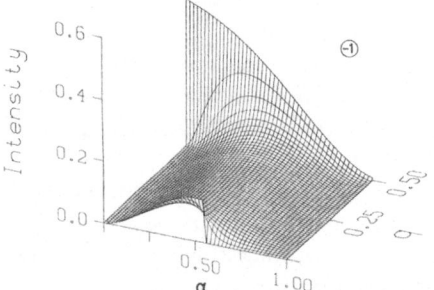
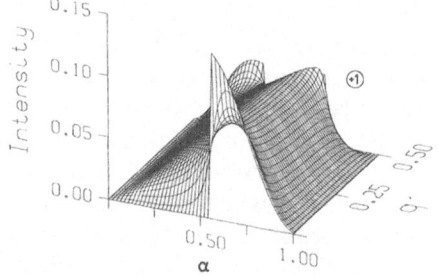

Fig. 3.27. Calculated variations of the intensity of the two lowest folded ($v = \pm 1$) branches at room temperature as a function of the thickness ratio α and of the wavevector q normalized to g, calculated assuming $P(\text{AlAs})/P(\text{GaAs}) = 0.1$. From [3.60]

different bulk compounds, all these components are simply obtained from the Fourier components of the compositional profile, multiplied by the difference between the (complex) photoelastic constants of the bulk constituents. Again one can compare the *relative* intensities without any knowledge of the photoelastic parameters. The results of the model are illustrated in Fig. 3.25 for the same phonons as Fig. 3.24. In the present approximation, the intensities of lines $\pm n$ are equal and almost independent of wavevector. These two features are the main deficiencies of a model which otherwise provides a good qualitative description of the experimental results.

In [3.60] a numerical calculation of the intensity of the folded lines, taking into account both acoustic and photoelastic modulations was performed. With a reasonably large set of Fourier components, one can determine the scattered intensities to a good approximation. We show in Fig. 3.26 the results for the same phonons as Figs. 3.24, 25 obtained under this assumption for a wavevector

q typical of those involved in a backscattering experiment. As a matter of fact, the intensities of lines $+n$ and $-n$ are now different and this difference strongly depends on the scattering wavevector. This effect can be understood from the results shown in Fig. 3.27. For an intermediate wavevector, the intensities of lines $+n$ and $-n$ are similar and reflect the nth Fourier component of the profile. At the zone center, due to the strong mixing of the two modes, one is allowed and the other forbidden. For a typical backscattering wavevector, the intensity difference reflects the progressive softening of the zone center selection rules. Furthermore, due to the couplings between the "Brillouin" acoustic and the folded branches the calculation becomes sensitive to the relative magnitude of the photoelastic coefficients of the bulk constituents. The line intensities are then the result of the interference between the acoustic and the photoelastic modulations. Such models can be easily applied to any periodic profile with a piecewise constant modulation. In the case of a smooth modulation, however, a detailed knowledge of the photoelastic profile is needed: a linear approximation to the dependence of the photoelastic coefficient on the composition is often inadequate [3.99].

We mention that an attempt to incorporate in the intensity model superlattice effects on the optical properties, i.e., to take into account the modulation of the complex refractive index along the z axis, has been recently presented by *He* et al. [3.100].

3.3.5 Conclusions

We have presented in this section two types of models introduced to describe the light scattering activity of superlattice vibrations. The bond polarizability model refers to the local deformation of bonds induced by the (optic) vibrations while the photoelastic one involves the macroscopic strain associated with the folded acoustic modes. Although the two models may be quantitatively related to each other, this connection leads to an inaccurate description of Raman and Brillouin scattering in the corresponding bulk materials. In the first model, the amplitudes of the B_2 optic-like modes ensuing from the cell multiplication are strongly modulated and their scattering intensity is significant, even in the absence of modulation of their coupling with radiation. On the other hand, the acoustic modulation being extremely small, the observation of the folded lines requires the modulation of the photoelastic process.

We must emphasize that these models are only useful far from electronic resonances, which will be considered in Sect. 3.5.

3.4 Selected Light Scattering Results for Vibrations in Superlattices

3.4.1 Introduction and Historical Aspects

The trends in experimental work on phonons in superlattices have been influenced by the evolution of the type of available structures and their crystalline quality. Until recently, the dominant systems were based on GaAs and AlAs bulk compounds. In the late 1970s, following the ideas of *Esaki* and *Tsu* [3.25], structures including a very few molecular layers of pure GaAs or AlAs as superperiod were considered. Due to the immaturity of the Molecular Beam Epitaxy (MBE) technique, these samples did not clearly exhibit the expected properties, including those concerning phonons [3.101], and were often found to be very close to random $Ga_{1-x}Al_xAs$ alloys. Consequently, the emergence of the concept of multi-quantum-wells, and the exploding interest in the remarkable electronic properties of these structures [3.102], less demanding on growth quality, slowed down the interest on real superlattice effects and the experimental work on phonons. One must mention some resonant Raman scattering investigations from this period, mainly devoted to elucidation of confinement effects on electronic states [3.103] or some controversial reports on additional lines in the optical phonon range [3.71, 104]. One fundamental result was obtained by *Colvard* et al. [3.49]: the first observation by light scattering of folded acoustic phonons, a result that one can consider as a milestone of the experimental work we describe in this section.

Since 1983, increasing interest has been devoted to the vibrations of superlattices, leading to a rather detailed understanding of these properties, at least for propagation along the growth axis. Again, the main stages concern results on GaAs/GaAlAs structures but new systems appeared involving other III–V, II–VI compounds, Si and Ge, and more recently, amorphous semiconductors. Even if these systems display a new additional feature, i.e., they involve layers strained because of lattice constant mismatch, their vibrational properties are not qualitatively different from those of the GaAs/GaAlAs system. Light scattering studies of strain effects have thus found their main interest in sample characterization. Among these systems, that based on bulk Si and Ge and their alloys has attracted the most attention and some interesting results have been obtained in both the acoustic and optic phonon ranges. Related results have also been reported for the folded acoustic modes of systems based on the periodic stacking of amorphous semiconducting layers. In the following we will illustrate all these results in three subsections. As already emphasized, the folded acoustic and the optic vibrations display very different properties and will be considered separately in Sects. 3.4.2, 3. Section 3.4.4 will focus on the results related to material characterization, mainly including strain and interdiffusion effects.

3.4.2 Experimental Results for Folded Acoustic Phonons

As we mentioned in the introduction, the first clear evidence from light scattering of a superlattice effect on vibrations was reported in 1980 by *Colvard* et al. [3.49]. These authors observed (see Fig. 3.28) a doublet around 60 cm^{-1} for a superlattice consisting of alternating 13.6 Å GaAs and 11.4 Å AlAs layers and attributed it to the lower near-zone-center folded longitudinal acoustic doublet. Before them, *Narayanamurti* et al. [3.105] had reported the observation of a dip in the acoustic transmission spectrum through a superlattice and attributed it to the lowest zone edge gap opening due to the superperiodicity. Recently, acoustic transmission has again attracted some interest [3.106, 107] and provided some results complementary to light scattering.

Following this first observation, obtained under resonance conditions (laser line close to strong electronic absorption), many reports have been devoted to these lines in:

- GaAs/AlAs and GaAs/Ga$_{1-x}$Al$_x$As structures [3.41, 42, 56, 59, 60, 65, 78, 82, 95, 100, 108–119]
- other III–V compounds: GaSb/AlSb [3.120–123], InGaAs/InP [3.124], InAs/GaSb [3.124], InGaAs/GaAs [3.124, 125]
- II–VI compounds: CdTe/CdMnTe [3.126, 127]
- Si/Ge and Si/Si$_x$Ge$_{1-x}$ layers [3.128–131]
- amorphous semiconductors:
  a–Si: H/a.SiN$_x$:H [3.57, 132–133], a.Si:H/a.Ge:H [3.134].

Thus, these modes have been observed for a wide variety of systems, with periods ranging from 10 Å [3.111, 112] to 500 Å (for instance, in [3.100]). Several doublets are often recorded even out of resonance (17 doublets are identified in [3.130]). In Fig. 3.29, we present a few Raman spectra of folded acoustic phonons obtained for several typical superlattices. Recently, folded acoustic lines have been investigated in samples whose period is composed of more than two different layers [3.58, 60], large and complex supercells [3.135], built up according to Fibonacci sequences [3.136] and even on aperiodic

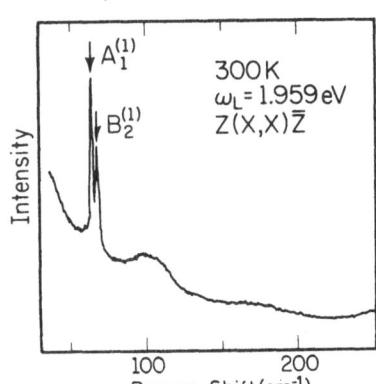

Fig. 3.28. First reported observation of a folded acoustic doublet in a GaAs/AlAs superlattice [3.49]

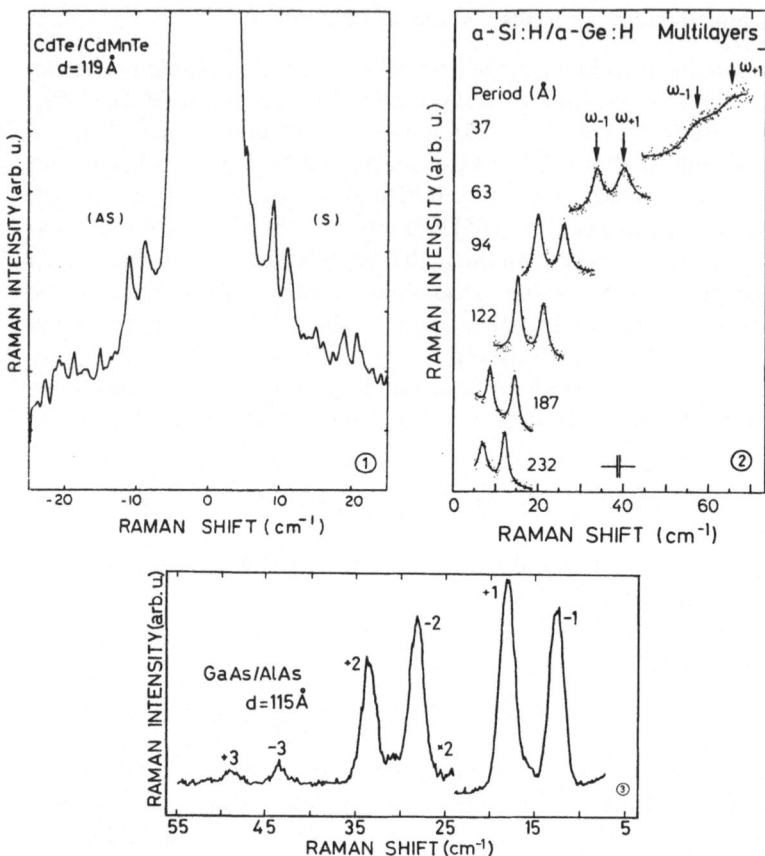

Fig. 3.29. Some typical Raman spectra exhibiting folded acoustic lines on: (*1*) a CdTe/CdMnTe superlattice [3.126]; (*2*) amorphous Si/Ge multilayers [3.134]; (*3*) a GaAs/AlAs superlattice [3.60]

structures [3.137, 138], a topic which will be considered in Chap. 5 of this book. One must also note that the folded acoustic modes are similar to those observed for SiC polytypes [3.17] and for graphite intercalated compounds [3.139] as a consequence of the periodic introduction of dopants in these crystals. Though less extensively studied experimentally, these systems exhibit essentially the same features as those described in this section.

Among these numerous reports, some only mention the observation of folded lines or roughly compare their frequencies with the period of the sample. Nevertheless, two particular topics have attracted increasing attention in the past few years:

- the quantitative understanding of the line frequencies, and in particular, the doublet splittings
- the stringent Raman selection rules and the quantitative description of the measured intensities.

These topics will be considered in the two following sections. The line profiles have attracted less attention [3.116], the main result being that they are extremely narrow, usually narrower than the resolution of classical Raman set-ups. A detailed investigation of these profiles has been undertaken recently using a Brillouin-Raman coupled apparatus [3.140].

a) The Raman Frequencies of Folded Phonon Branches

In [3.49] the folded LA line frequencies were already successfully analyzed by using the elastic model. The parameters one needs for this comparison are bulk properties such as densities and sound velocities, and the structural parameters of the superlattice. There are limitations to checking the validity of a model of the GaAs/AlAs structures. The first is the lack of knowledge of the sound velocities of AlAs [3.140a], the second the uncertainties in the determination of the superlattice parameters. As explained in Sect. 3.2, the acoustic dispersion curves are obtained first by folding an average dispersion curve and then by opening energy gaps at the zone center and boundaries. As long as the acoustic modulation is small, an assumption, which appears to be valid in all systems considered so far, the average frequency of a given near-zone-center doublet reads [see (3.11, 12)]:

$$\Omega_v = v\,\frac{\pi}{d}\left(\frac{\alpha}{v_B}+\frac{1-\alpha}{v_A}\right)^{-1}; \quad \alpha=\frac{d_B}{d}, \quad v=0,1,2,\ldots \tag{3.68}$$

and does not depend on the wavevector.

Again due to the relatively small difference between v_A and v_B, this quantity essentially varies as $1/d$, nearly independently of α. Such variation has been checked by different groups [3.95, 141] by using values of the period estimated from the growth conditions or, even better, values determined by x-ray diffraction [3.142]. This method provides an excellent determination of d from the distance between superlattice satellites which accompany the bulk reflexes. Moreover, taking into account the actual value of α in the samples provides a moderately accurate determination of the sound velocity in bulk AlAs which is found to be $\sim 5.7 \times 10^5$ cm/s (for longitudinal waves). This value corresponds to almost the same elastic constants as in GaAs: therefore, the acoustic modulation must essentially come from the density modulation.

From the determination of the doublet splittings one expects complementary information and more insight on the physics of the problem. These quantities are indeed proportional to the relevant modulation. The approximate relation for the splitting $\Delta\Omega_v$ (3.14) was obtained in Sect. 3.2. It shows that the gaps depend strongly on the details of the supercell structure, i.e., in the simplest case, on the ratio α. The experimental results are in apparent disagreement with these features. The doublet splitting seems to be independent of sample details and doublet order. It amounts to about 5 cm^{-1} for the usual Raman backscattering experiments performed at 5145 Å, a much larger value than predicted from the elastic model at $k=0$ using the sample parameters deduced from the average frequency measurements. As mentioned in [3.49, 109] and studied in detail in [3.110], this result can be understood by taking into account the finite wavevector of the phonon involved in the light scattering process. This quantity is usually neglected in Raman studies of bulk materials because it is much smaller than the size of the Brillouin zone. Also, optic branches are often weakly dispersive around the zone center. Nevertheless, Brillouin scattering experiments are well known to be sensitive to this parameter since the frequency of the true acoustic branch is linear near $k=0$ [3.13].

A similar effect has to be taken into account for superlattices. Due to the weak modulation, the wavevector range of interaction between a pair of converging acoustic folded branches is very small, and these "optic" modes, with quadratic dispersion at the zone center, rapidly transform into "acoustic"-like modes in the sense that their dispersion becomes linear. From a comparison between both modulation and dispersion contributions to the doublet splittings, one concludes that the second one is dominant in usual backscattering conditions on GaAs/AlAs structures for periods larger than 20 Å, i.e., for almost all available samples. This unusual feature in Raman scattering has both negative and positive consequences on the amount of information to be deduced from the line frequencies. On the negative side, Raman backscattering is nearly unable to probe the physical nature of the modulation and is also only weakly sensitive to the inner structure of the supercell, though being an excellent probe of its periodicity. On the positive side, as emphasized in [3.110], Raman backscattering does provide dispersion curves as neutron scattering does for bulk compounds. By varying the incident wavelength, one can change the k_z since it is inversely proportional to λ when neglecting the refractive index variation. This will change the doublet splitting without changing its average frequency. Such dispersion measurements have been performed by several groups on the GaAs/AlAs [3.95, 100, 110], GaSb/AlSb [3.121–123], Si/Ge [3.129], and on amorphous structures [3.132]. They are able to provide large portions of the dispersion curves. Representative results are shown in Fig. 3.30.

Different ways of circumventing the difficulty involved in probing very close to the zone center by Raman backscattering and to obtain information on the gaps have been presented. The most natural one is to extrapolate to the zone center the frequencies obtained at various wavelengths. As already pointed out in [3.95] and apparent in Fig. 3.30, this extrapolation provides no evidence for a zone center gap in the GaAs/AlAs system. For a. Si:H/a. SiN$_x$:H, evidence of a zone center gap has been obtained [3.133] thanks to the lower refractive index of these compounds, which allows one to get closer to the zone center. The second way is based on the possibility of adjusting both the period and the incident wavelength in order to probe the region close to the SBZ edge. This method has been first successfully applied by *Brügger* et al. to Si/Ge [3.129]. In this system, thanks to the large average sound velocity, the lowest zone edge doublet is accessible to a Raman set-up (at ~ 5 cm^{-1}), which is not the case for the GaAs/AlAs system (~ 2.5 cm^{-1}). The agreement between measured and calculated splittings near the zone edge is good, as illustrated in Fig. 3.31. More

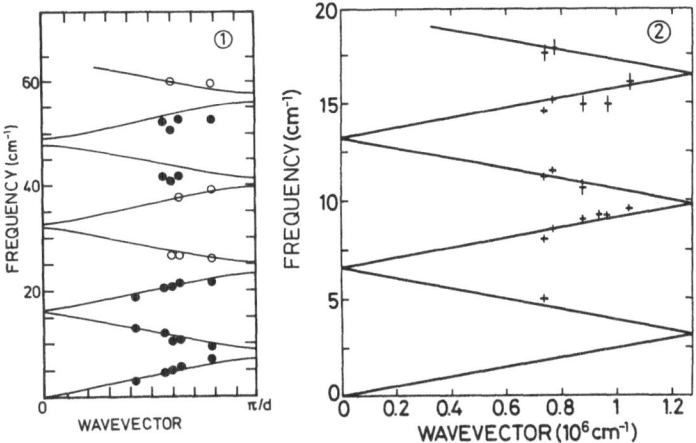

Fig. 3.30. Folded acoustic dispersion curves, calculated using the alternating elastic model, compared with the Raman frequencies measured at different incident wavelengths for: (1) a Si/SiGe superlattice ($d = 163$ Å) [3.129]; (2) a GaAs/GaAlAs superlattice ($d = 257$ Å) [3.110]

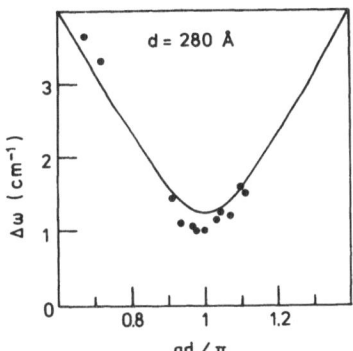

Fig. 3.31. Frequency splitting of the lowest folded acoustic doublet recorded on a Si/SiGe structure at different incident wavelengths, compared with calculation around the zone boundary [3.129]

recently, similar ideas have been applied to GaAs/AlAs large period superlattice [3.100] using a Brillouin-Raman apparatus which allows recording of Raman spectra down to 1 cm^{-1} [3.140] and also to amorphous structures [3.132] using a Brillouin (Fabry-Pérot) spectrometer.

The first reported attempt to obtain information on the folded acoustic mode coupling actually involved a forward scattering configuration [3.56]. This configuration allows one to probe the dispersion curves very close to the zone center and is thus perfectly adapted to this problem. Serious technical difficulties, related to the opacity of the involved compounds to usual ion laser frequencies have, however, restricted its application. In order to collect the scattered light on the other side of the sample, one must first remove the opaque substrate from a small area of the sample by using selective etching. Moreover, the parameters of the superlattices must be chosen so that they become sufficiently transparent. Figure 3.32 shows both forward and backward Raman spectra obtained in this way for one of the two different samples considered in

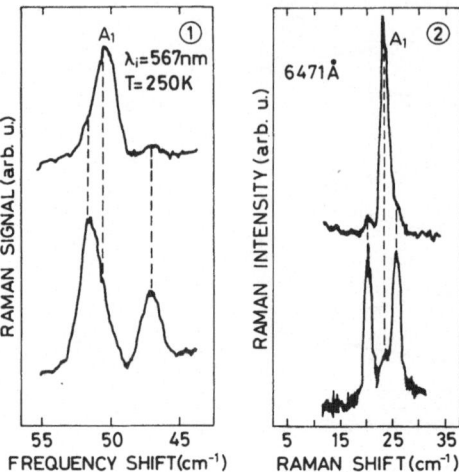

Fig. 3.32. Comparison between the Raman spectra obtained in backscattering (*bottom*) and in forward scattering (*top*) configurations around the lowest folded acoustic doublet for two different systems: (*1*) GaAs/AlAs [3.56]; (*2*) amorphous Si/SiN$_x$ [3.57]

[3.56]. In forward scattering, a single line lying between the two components of the backscattering doublet is observed. This line is attributed to the A_1 component of this doublet, the B_2 component being Raman forbidden by parity. Due to this selection rule, this method appears to be unable to give a pure experimental evidence of a zone center gap. Nevertheless, from the splitting between the forward scattering line and the center of the backscattering doublet, the authors obtained an estimate of the gap in both samples, again in correct agreement with the predictions of the elastic model. Considerable attention is paid in [3.56] to the dependence of the acoustic gaps on structural parameters. The strong oscillatory behavior of these quantities as a function of the ratio α is supported by the few available experimental results. Moreover, new insight is obtained on the selection rules, a topic to be treated in the next subsection. Very recently, similar experiments were reported [3.57] for amorphous structures. They are easier in this case because the substrate is transparent; the results illustrated in Fig. 3.32 are also in good agreement with the elastic model.

To summarize, a departure from the alternating elastic model has not been evidenced in most studies devoted to LA folded line frequencies. Only recently [3.123] some evidence for departure from the elastic continuum model (and better agreement with the alternating linear chain model) has been obtained for GaSb/AlSb superlattices. This effect should be easier to observe for TA modes. The bulk TA dispersion curves are known to be flat over a large region of the Brillouin zone (Fig. 3.1). Whereas these modes are forbidden in a backscattering experiment on a [001] surface, they have been observed as weak structures [3.95, 118] thanks either to departures from the ideal backscattering configuration or to defect-induced scattering. Similarly, some observation of zone edge disorder activated folded LA and TA modes has been reported and compared with the case of bulk disordered systems [3.143]. Recently, *He* et al. [3.100] extracted part of the folded TA dispersion curves from these structures, but without reaching sufficiently high energies to obtain evidence of sublinearities.

On the other hand, from the small amount of experiments on structures with other surface orientations [3.144], no reliable information on folded TA modes has been extracted to our knowledge.

It is interesting to note that, for vanishing in-plane wavevectors, folded TA and LA modes are decoupled since they have different symmetries. Thus, their respective dispersion curves cross. In contrast, new "internal" acoustic gaps appear for finite in-plane wavevectors which are out of the scope of light scattering experiments but have been observed by phonon spectroscopy using superconducting diodes [3.106, 107].

As concerns the opening of zone center and edges gaps, no real departure from the alternating elastic model has been observed till now. Due to the small amplitude of the acoustic modulation between the constituents of presently available systems, the folded frequencies are essentially sensitive to the periodicity, and unfortunately, not so much to the physical details of the problem. The few studies of this point have demonstrated the existence of acoustic gaps in rough agreement with the elastic model, but have been unable to provide accurate quantitative information on their magnitudes. Structures with larger acoustic modulation would be of interest to further investigate this problem.

b) The Intensities and Selection Rules of Folded Acoustic Modes

In contrast to their frequencies, whose main features have been understood from the beginning, the selection rules of lines which correspond to folded acoustic modes have generated some controversy. There is indeed a clear disagreement between the predicted selection rules for *zone center* Raman modes and the *backscattering* results. Both components of the folded doublets are systematically observed in parallel $z(x, x)\bar{z}$ configuration and not in the perpendicular $z(x, y)\bar{z}$ one, whereas the zone center selection rules predict the observation of one line (A_1) in parallel and the other (B_2) in perpendicular configuration. One thus has to explain why both lines are observed in the parallel (polarized) configuration but not in the perpendicular (depolarized) one. It has been suggested in [3.49] that the vanishing intensity in the latter case could be attributed to the macroscopic nature of the photoelastic process involved which introduces additional (cylindrical) symmetry: a strain along the z-axis produces a change in χ_{xx} and χ_{yy} but not in χ_{xy}.

The observation of the *alleged* B_2 modes in parallel configuration was attributed to resonance conditions in the early publications [3.49]. Its relation to the finite value of the scattering wavevector involved was recognized later. Conclusive proof of this fact was obtained in [3.56], the forward scattering spectra displaying a single component corresponding to the A_1 mode (see Fig. 3.32) and in [3.60], where the increasing relaxation of the zone center selection rules with increasing superlattice period was demonstrated.

Beyond the determination of the nature of the light scattering process, the quantitative description of the intensities has attracted a great deal of attention.

The modulation which underlies the observation of the folded acoustic modes can either appear in the elastic or in the photoelastic properties. In other words, it can be an intrinsic property of the object one looks at or a property of the method one uses to look at it. As we briefly analyzed in Sect. 3.2, this appears explicitly in the Fourier-transformed expression of the scattered intensities:

$$I_{i,q} \propto \left| \sum_n P_{-n} u_{i,n}^q (ng+q) \right|^2 \tag{3.69}$$

first introduced in [3.95].

The *u*-terms in (3.69) reflect the mechanical modulation and the *P*'s the photoelastic one. *Babiker* et al. [3.98] studied the predictions of such a model as a function of various sample parameters but neglected the photoelastic modulation. They considered, in particular, the case of the GaAs/GaAlAs structure but did not attempt to compare their calculations with experimental results. *Colvard* et al. [3.65] on the other hand, in view of the small amplitude of this acoustic modulation in the GaAs/AlAs systems as deduced from the folded line frequencies, made the complementary approximation of neglecting the mechanical modulation and investigated the effects issuing from the photoelastic modulation. They pointed out two features well suited to experimental proof: (i) the relative intensity of the folded lines and the true acoustic line and (ii) the vanishing intensity of the *even index* lines, for equal thickness of both constituting layers. Whereas feature (i) has not been considered till recently, feature (ii) was successfully checked by those authors. As illustrated in Fig. 3.33 the second doublet disappears in a sample consisting of the periodic stacking of 85 Å of GaAs and 88 Å of GaAlAs as long as the experiment is performed sufficiently far from electronic resonances. However, as can be seen in Fig. 3.24, such a vanishing intensity is predicted in the mechanical model even when neglecting the photoelastic modulation, as in [3.98]. The corresponding

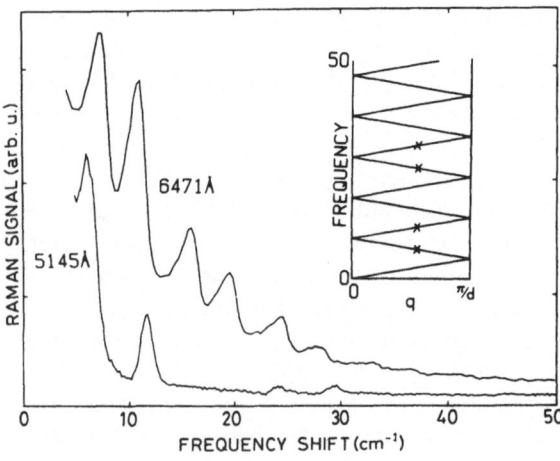

Fig. 3.33. Low frequency Raman spectra of a GaAs/GaAlAs super-lattice with nearly equal layer thicknesses (85/88 Å) obtained out of resonance (*lower trace*) and close to resonance (*upper trace*) [3.95]

thickness ratio α does not equal 0.5 in this case but the value at which the acoustic gaps vanish [see (3.14)]:

$$\alpha = \frac{v_B}{v_A + v_B} \qquad (3.70)$$

Since α amounts to 0.55 for GaAs/AlAs structures, the observation of a vanishing intensity around 0.5 is not a conclusive proof of the photoelastic mechanism. Nevertheless, we feel that the photoelastic modulation must dominate the physics of the problem: the small acoustic modulation could not induce the strong (compared to the optic modes) scattering observed for folded modes.

In order to obtain more conclusive evidence, a quantitative comparison of the predicted and measured line intensities is of interest. Besides the comparison between different folded orders, another feature was shown to be sensitive to the details of the process [3.60], namely the asymmetry of the lower doublet. As seen in Figs. 3.24–26, the asymmetry in the relative intensity of both components of the first doublet vanishes when neglecting the acoustic (mechanical) modulation. If the mechanical modulation is included, the asymmetry reverses upon switching on of a strong photoelastic modulation. The absence of asymmetry when neglecting the acoustic modulation is easy to understand as both components are related to the same Fourier coefficient of the photoelastic profile. Colvard et al. emphasized that their model was valid away from zone boundaries. They claimed that this assumption was correctly fulfilled in the usual backscattering conditions, where the mode frequencies are indeed insensitive to the opening of gaps. As demonstrated in detail in [3.60], asymmetries are actually often observed by backscattering on GaAs/AlAs structures. they depend strongly on the sample parameters and can be quantitatively explained by taking into account both modulations. The predictions one obtains in this framework were illustrated in Fig. 3.27 through the intensity of both lower folded branches ($v = -1$ and $+1$ modes) as a function of α and the wavevector q. At the SBZ center one recovers the parity selection rules, a single line (A_1) being active, while as q moves aways from the SBZ center, a continuous softening of the selection rules takes place. Figure 3.34 compares two series of experimental results and the predictions of the photoelastic model for different amplitudes of the photoelastic modulation. One finds the following qualitative results:

– the sign of the asymmetry supports the dominant role played by the photoelastic modulation
– the decrease in the amplitude of the variation of the asymmetry with α when increasing the sample period d reflects the softening of the zone center selection rules as the corresponding wavevector, normalized to the size of the Brillouin zone, moves away from zone center.

These asymmetries are not very sensitive to the exact value of the photoelastic modulation: a quantitative estimate is made in [3.60] (for $P_{AlAs}/P_{GaAs} \sim 0.1$). As concerns the relative intensity of the different doublets, some comparisons have

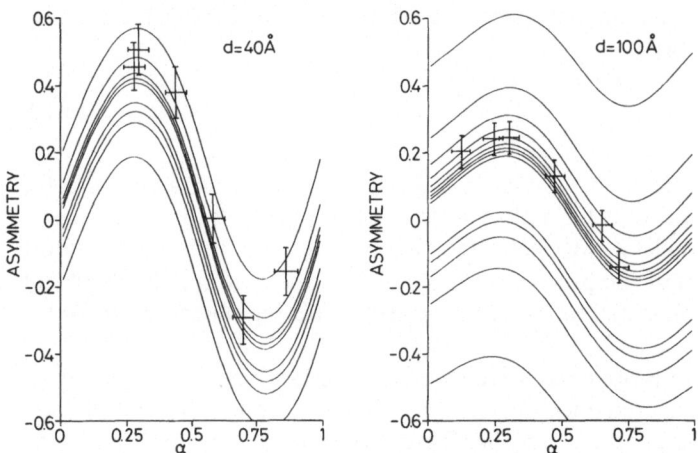

Fig. 3.34. Variation as a function of the thickness ratio α of the asymmetry of the lowest folded acoustic doublet measured on two series of GaAs/AlAs superlattices with two different fixed periods. The experimental results are compared with the predictions of the photoelastic model for different values of the photoelastic ratio (AlAs to GaAs) from -1 to 2 (see [3.60]). The experimental data correspond to ratio ≈ -0.2

been made for GaAs/AlAs or Si/Ge_xSi_{1-x} systems [3.131] which further support the dominance of the photoelastic mechanism for both types of structures. The relative intensities are not very sensitive to the exact value of the photoelastic ratio. They are, however, sensitive to the detail of the compositional profile [3.59].

A rather good method to obtain information about these properties is to compare the folded line intensities to the true acoustic one (Brillouin), as suggested in [3.65]. The very low frequency of the latter has, however, hindered such measurements. Even if on the Si/Ge system measurements with Raman spectrometers have detected these lines [3.128, 131], no attention has yet been focussed on this particular point. On the other hand, *He* et al. [3.100] have recently carefully analyzed the scattering intensities, including the Brillouin scattering, for GaAs/AlAs structures, with their Brillouin-Raman coupled set-up. They used a theoretical model which included the acoustic and photoelastic modulations and also the modulation of the optical properties of the samples. They thereby calculated the ratio P_{AlAs}/P_{GaAs} which depends on the incident wavelength. The value they found in the blue, 0.43, is in some disagreement with the one obtained in [3.60].

To summarize, a reasonable description has been obtained of the intensities of the acoustic lines on the basis of the coupled acoustic and photoelastic (and possible optical) modulation model. One must emphasize that the observation of the acoustic modes on superlattices is mainly due, like for the Brillouin effect of bulk crystals, to photoelastic modulation, and displays a close analogy to the observation of satellites in the x-ray diffraction patterns of superlattices. In the latter case, a competition takes place between a "sample modulation", namely the one of the lattice parameter along the growth axis, and a "coupling modulation", the one of the atomic scattering factors. On the nearly-lattice-matched superlattices (GaAs/AlAs) the latter is dominant and the intensity patterns and their variation with sample parameters are similar to the ones obtained in Raman scattering [3.145]. On the other hand, for strained-layers superlattices, the lattice parameter modulation becomes dominant and the diffraction patterns may look very different [3.146]. Such a situation has, unfortunately, never been encountered for phonons.

The limits of validity of the photoelastic model have been clearly indicated by *Klein* [3.99]. They lie in the dispersive character of the (complex) photoelastic tensor of the bulk constituents. Even if a good knowledge of these bulk properties exists, which is not often the case, is it justifiable to extrapolate them to structures displaying very different electronic properties and to large unfolded q's? Such a question emphasizes how careful one has to be when trying to interpret the intensities of folded acoustic modes as a function of incident wavelength.

3.4.3 Experimental Results on Optic Phonons

a) Introduction and Historical Aspects

In the analysis of the superlattice effects on the sample symmetry, we pointed out the two main consequences of the superperiodicity: (1) the modification of the zone center cubic modes and (2) the folding of cubic out-of-zone-center modes to the zone center. In our analysis of the experimental result in the acoustic range, we mainly focussed on the second aspect. When discussing the results in the frequency range of the bulk optic modes, one has to consider both features. The zone center optic modes are indeed easy to observe in cubic crystals, contrary to the acoustic ones, and they remain qualitatively unchanged in superlattices. Therefore, the observation of such modes in multilayer systems has often been possible, sometimes with a small shift in their frequencies, without giving any evidence of a superlattice effect. In the following we shall mainly discuss experimental work which carefully demonstrates, by means of the dependence on layer thicknesses, phonon *confinement* effects. Such results have been obtained mainly for GaAs/AlAs structures, where (i) the competing effects of strain are negligible and (ii) the bulk frequencies are reasonably well known. The oldest light scattering study to be found in the literature on superlattices [3.101] already discussed the dependence of the LO frequencies of GaAs and AlAs on the layer thicknesses in samples containing a few monolayers per period. Such studies have been repeated and extended by several groups [3.66b, 82, 112, 147–148].

Particularly convincing evidence of the confinement of optic phonons is obtained when new additional modes are observed in the optic phonon frequency range, due to the primitive cell multiplication. Such additional lines have also been observed mainly for the GaAs/AlAs system. While new lines were recorded in resonant conditions [3.71, 104] and attributed either to confinement or anisotropy effects, the first off-resonance observation [3.41] and quantitative demonstration [3.67] of this effect, was performed in 1983 on GaAs/Ga$_{1-x}$Al$_x$As samples. Somewhat similar structures have actually been previously observed at resonance [3.108] but without detailed assignment. Several subsequent studies have been devoted to these features mainly for pure GaAs/AlAs superlattices [3.42, 62–66a, 111, 149]. Recently, observations of confined modes have been reported for two different II–VI superlattices containing, respectively, CdTe and ZnTe [3.150] and CdTe and CdMnTe [3.127], and also for GaSb/AlSb systems [3.122, 123]. The observation of these *additional* optic modes is difficult, as compared to the folded acoustic ones, because of the low dispersion of the optic phonon branches. They must be resolved from the close, more intense main optic line. Since the properties of GaAs/Ga$_{1-x}$Al$_x$As and GaAs/AlAs structures present some important differences due to the overlapping or non-overlapping character of the optic phonon branches involved, we shall describe them separately in the two following subsections. In both cases, a good description of the frequencies is obtained by using the alternating linear chain model, and the zone center Raman selection rules are fulfilled.

A third subsection will be devoted to experimental results which stand out of the common framework of the work described so far, namely, scattering involving a non-vanishing in-plane wavevector. This condition was first intentionally obtained by *Zucker* et al. [3.54] using right angle scattering on a specially designed structure. Interesting selection rules were thereby observed.

A definitive understanding of the additional lines observed at resonances [3.71] and a detailed study of their line shape and dependence on sample parameters was obtained by *Sood* et al. [3.55]. These lines attributed to interface modes can be observed near resonance, according to these authors, due to some unspecified relaxation of the wavevector conservation. Similar results have been reported for GaAs/Ga$_{1-x}$Al$_x$As [3.151–153], CdTe/CdMnTe [3.127] and CdTe/ZnTe [3.150] structures. Related results for GaAs/Ga$_{1-x}$Al$_x$As structures have been obtained as well by *Lambin* et al. using high resolution electron energy loss spectroscopy [3.77]. An important feature in optic vibrations propagating off-axis is the role played by long range Coulomb forces.

b) The Optic Vibrations in GaAs/Ga$_{1-x}$Al$_x$As Superlattices

Figure 3.35 shows Raman spectra obtained in the frequency range of the GaAs LO mode for various GaAs/GaAlAs superlattices [3.67, 154]. The spectra are shown as two series. The first one corresponds to samples with all parameters fixed except the thickness of the GaAs layer. In the second one, the only varying parameter is the Al concentration in the Ga$_{1-x}$Al$_x$As layer. These spectra display four lines in the frequency range of the LO phonon of bulk GaAs, more precisely, at frequencies smaller or nearly equal to it. The upper line, whose shift from the bulk LO frequency is only noticeable for GaAs layer thicknesses smaller than 30 Å, always dominates the spectra. When the GaAs layer thickness is

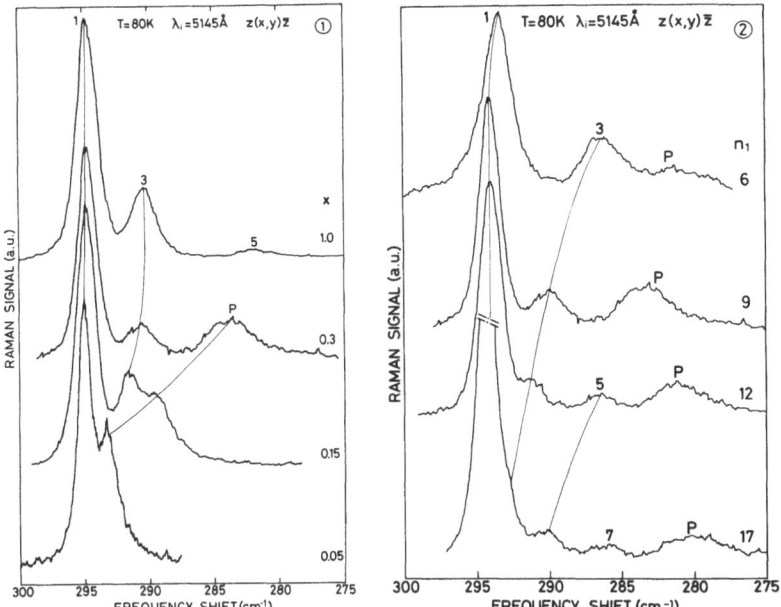

Fig. 3.35. Raman spectra in the GaAs optic modes frequency range obtained for two series of GaAs/Ga$_{1-x}$Al$_x$As superlattice with a single varying structural parameter: (1) the Al concentration x, [3.154]; (2) the number of GaAs monolayer in each period n_1 [3.67]

reduced, the lines labeled 1, 3, 5 shift towards lower frequency, whereas the line labeled P is almost unchanged. When the Al concentration is reduced, line P strongly shifts towards higher frequency whereas the other lines are pushed only slightly in the same direction. This behavior is rather similar to that of quantum levels in an electronic quantum well. The confined frequencies depend strongly on the well thickness and only slightly on its depth, at least as long as they are sufficiently below the top of the well. As pointed out in Sect. 3.2, such phonon quantum wells appear for partly non-overlapping optic phonon branches of the two bulk constituents, a feature approximately displayed by the pure GaAs band and the GaAs-type band of Ga$_{1-x}$Al$_x$As. However, contrary to the case of acoustic phonons, whose frequencies in the alloy can be well described by using the virtual crystal approximation, the optic phonons in these alloys display a qualitatively different behavior from that of pure compounds. Ga$_{1-x}$Al$_x$As is a prototype of two-mode-behavior [3.155]: over the whole concentration range, two different optic bands coexist. Their frequencies vary only slightly with x, each of them lying in the vicinity of the optic band of one of the constituting pure compounds. Moreover, their relative intensity qualitatively reflects the concentration of the corresponding compound. They have thus been identified as GaAs-type and AlAs-type optic bands of Ga$_{1-x}$Al$_x$As. This identification has been further supported by CPA calculations of the alloy lattice dynamics [3.156] which demonstrated that the local density of states at frequencies inside, say, the

GaAs-type band is large on the Ga — As bonds of the alloy. Moreover, these calculations provide "thick" (broadened) dispersion curves [3.157], an extension of the usual notion of sharp eigenstates in perfectly ordered crystals to moderately disordered systems. Since $Ga_{1-x}Al_xAs$ is indeed a weakly disordered system [3.158], these "thick" dispersion curves display only small broadenings, in good agreement with the results of light scattering. In [3.67] the following approximate description of the Raman spectra in the GaAs-type energy range of $GaAs/Ga_{1-x}Al_xAs$ superlattices (with $x \cong 0.3$) is thus introduced. In order to apply the simple alternating linear chain model described in Sect. 3.2.3, the authors fit both the pure GaAs and the GaAs-type LO dispersion curves in $Ga_{1-x}Al_xAs$ with linear chain models. In both cases, "effective masses" must be introduced and the free parameters are chosen so that the effective masses of As as well as the spring constants are the same in both compounds. This description greatly simplifies the dynamical properties of the alloy: the AlAs-type band is neglected as well as disorder in the GaAs-type band of the alloy. Better descriptions, which use, for instance, the CPA technique for superlattices, are beyond the present computational capabilities of most workers. On the other hand, replacing in the calculation the alloy layer by an ordered chain of the same average composition, as suggested by *Worlock* [3.159], is not very useful. Even if the zone center frequencies of, for example, the $Ga_{0.5}Al_{0.5}As$ alloy and $GaAlAs_2$ compound are nearly the same, their dispersion curves should be rather different. Using this rough model provides, however, a reasonable description of the experimental results. A comprehensive comparison is not easy because the frequencies depend on three different sample parameters: both layer thicknesses and the aluminium concentration x. As concerns the confined modes, their frequency strongly depends on the GaAs well thickness, slightly on its depth which is related to x, and very weakly on the barrier thickness. Among the bulk parameters introduced in the fit, three of them have a significant effect on the confined frequencies: both zone center LO modes, which are well known, and the curvature of the GaAs LO dispersion curve, which is extracted from the bulk neutron scattering data [3.160]. As we shall emphasize in the case of pure GaAs/AlAs superlattices, this feature is not known very well (recent and more accurate measurements for GaAs are shown in Fig. 3.1): it has been suggested that Raman scattering on superlattices may provide a rather good method to determine the bulk dispersion curves.

As concerns the selection rules and line intensities off-resonance, a good description is obtained using zone center Raman selection rules and the bond polarizability model. Contrary to the folded acoustic case, the finite value of the superlattice wavevector involved is insignificant. Confined optic modes are indeed not sensitive to the periodicity of the system but rather to the thickness of the individual layer in which they are confined. Since their penetration depth in the barriers is generally less than one monolayer, this statement is valid even for very small thicknesses. As for electronic *multiquantum wells*, the corresponding spectra are a superposition of the contribution of each quantum well and reflect the periodicity of the sample only through the superposition of these contributions. In terms of frequency, the dispersion curves for $k \parallel z$ are very flat and thus the results are independent of scattering wavevector.

The Raman selection rules predict the observation of the B_2 modes in $z(x, y)z$ and the A_1 modes in $z(x, x)z$ configuration. One should observe in $z(x, y)z$ configuration only odd parity modes, i.e., those belonging to the B_2 representation. The origin of the non-observation of the A_1 modes off-resonance has already been suggested in Sect. 3.2.3 within the framework of the bond polarizability model: the A_1 modes are forbidden for the exact cubic symmetry. Their activation, in a local description, comes only from the non-cubic sites in the structure, i.e., from the interface atoms. In usual superlattices, such atoms are not so frequent and their displacement is very small for confined modes. On the other hand, their activity is proportional to the polarizability modulation, contrary to that of the B_2 modes. The Raman activity of the latter can be described semi-quantitatively on the basis of the bond polarizability model. It is large when the contributions of the different bonds interfere constructively. This is the case in particular when a local wavevector is small, i.e., when the frequency of the modes is close to the zone center bulk frequency in a given layer. This explains why mode 1 is always intense due to the contribution of the well layers, and the other modes when they are close to the top of the barrier due to the contribution of the barrier layers (see Fig. 3.23), a fact which agrees with experimental observations. From an analysis of these intensities, one could estimate the relative magnitude of α_{xy} in both constituents. However, the description of an alloy as a pure effective compound is more questionable for eigendisplacements than for eigenenergies. These relative intensities have been experimentally shown to be strongly dependent on the incident wavelength, especially near resonance with electronic transitions, a fact that reveals the rough nature of the model.

c) The Optic Vibrations of GaAs/AlAs Structures

The optic phonons of these structures are similar in many ways to those of the GaAs/Ga$_{1-x}$Al$_x$As structures. From the lattice-dynamical point of view, they are simpler but also poorer in unusual features since the optic phonons are all confined, either to the GaAs or to the AlAs layers. Due to the large difference between the Ga and Al atomic masses there is no overlap between the optic phonon bands of the two compounds. On the other hand, this system is very interesting as it involves two pure compounds and thus, can be described without the drastic approximations introduced for the Ga$_{1-x}$Al$_x$As alloy.

Due to the large energy separation between the optical bands of GaAs and AlAs, the properties of the modes confined in, let us say, the GaAs layers, are almost independent of the details of the AlAs layers, including the dynamical properties of bulk AlAs. This has stimulated several groups to perform quantitative comparisons between calculated and measured frequencies [3.39, 42, 59, 62–66b, 101, 111–113, 147–149]. The common aim was to check the validity of the model for the GaAs modes, the bulk properties of this compound

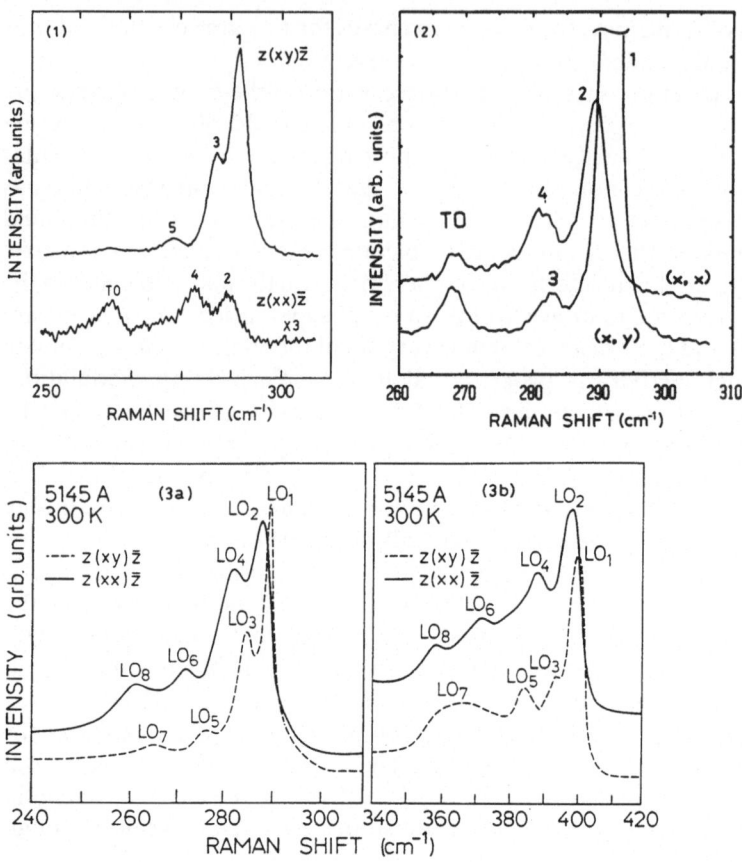

Fig. 3.36. Typical Raman spectra of confined optic modes in GaAs/AlAs superlattices obtained: (*1*) out of resonance, in the GaAs optical phonon range [3.111], (*2*) in resonant conditions in the same frequency range [3.108], (*3*) out of resonance for both the GaAs (*a*) and the AlAs (*b*) optical phonon ranges [3.149]

being assumed to be well known. In a second step, one could determine the bulk LO dispersion curve for AlAs which has not been determined so far. Figure 3.36 shows typical Raman spectra obtained on GaAs/AlAs structures by various groups, including both GaAs-type and AlAs-type modes of both A_1 and B_2 character. Contrary to the GaAs/Ga$_{1-x}$Al$_x$As structure, the A_1 modes have indeed been observed on GaAs/AlAs superlattices, first out of resonance as very weak structures [3.22, 111], then very clearly under resonance conditions [3.66a]. From these observations and the concomitant polarization selection rules, it was proved that the confined modes display alternating symmetries: the odd (even) indices corresponding to the B_2 (A_1) representation [3.66a]. Very recently, new Raman results have been obtained [3.149] giving evidence of well-resolved A_1 and B_2 modes, even out of resonance. Moreover, contrary to the previous works devoted to GaAs-type modes, these authors report the observation of all possible confined modes both in the GaAs and AlAs energy range. For instance, they report the observation in a $(GaAs)_8/(AlAs)_8$ structure, of eight GaAs and eight AlAs optic modes. These observations are rather impressive and they differ qualitatively from all previous reports. In the AlAs energy range, the spectra are usually weak, and higher order confined modes are thus difficult to resolve.

Let us now return to the analysis of the confined frequencies. Whereas some authors [3.39, 66b, 101, 111, 112, 147, 148] compared the measured frequencies with those calculated by using the alternating chain formalism as a function of the corresponding layer thickness, following the same approach as used for GaAs/GaAlAs structures, some others [3.42, 62–66a, 149, 150] took advantage of the strong confinement to introduce a direct comparison between the measured frequencies and the dispersion relation of the corresponding bulk compound. This method has the advantage of providing a comprehensive comparison between results depending on two parameters (the line index and the layer thickness) with a single curve (the bulk dispersion curve). The idea underlying this method was suggested in [3.65]: one can assign to each confined mode in a layer of given thickness an effective bulk wavevector depending on these two parameters. By analogy with a vibrating string, the effective wavevector was first taken to be that of (3.26) where $m = 1, 2, 3$, is the line index and d the layer thickness. *Sood* et al. [3.66a] plotted their experimental results obtained for three different samples and involving several confined modes together with the bulk neutron scattering results and emphasized some discrepancies. *Molinary* et al. [3.64], in a comment to [3.66a], demonstrated that calculated confined frequencies displayed a somewhat similar discrepancy with the bulk reference dispersion curve and attributed this discrepancy to superlattice effects. Finally, *Jusserand* et al. in another comment to the same paper [3.62], demonstrated that this discrepancy was more likely a boundary conditions effect than a superlattice effect. They derived from the alternating linear chain dispersion relation an approximate relation valid for strong confinement. It reduces to the definition of the effective bulk wavevector given in (3.25) which differs from (3.26). This difference originates from the microscopic nature of the problem and becomes important only for small layer thicknesses. It can be easily understood by considering that interfacial As atoms (separated by the period d) are not fixed but, instead, the first Al atoms in the barrier (distance between Al atoms both sides of the GaAs layer $d + a$). Using this more accurate expression reduces the discrepancy pointed out in [3.66], whereas more recent results [3.149] plotted vs q using (3.25) do not give any evidence of a discrepancy with bulk dispersion relations, as illustrated in Fig. 3.37.

The basic question which remains open is whether Raman scattering on thin layer superlattices can be used reliably to determine the dispersion relations of the phonons of the bulk constituents. Raman scattering does indeed provide a better frequency accuracy, but the validity of (3.25) for the determination of the equivalent wavevector is not always evident. The sensitivity of q to the interface quality may also be a serious problem. On the other hand, this method remains

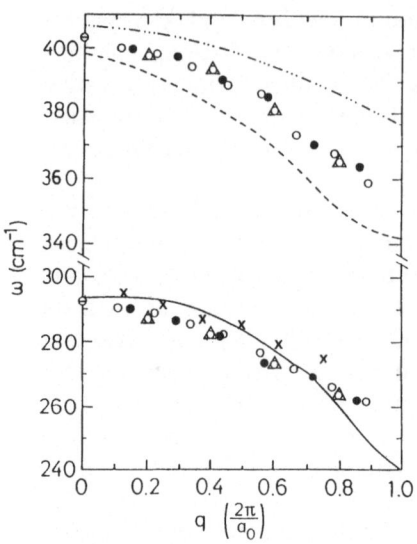

Fig. 3.37. Confined GaAs-type and AlAs-type LO frequencies measured for different GaAs/AlAs superlattices (*symbols*) plotted vs. q with the use of (3.25) compared to the GaAs bulk dispersion curve measured by neutron scattering (——) and two different calculations of the dispersion relations of AlAs (– – –, –··–) [3.149]

unique for compounds which are not available in large volume. For instance, the recent results on AlAs (see Fig. 3.37) should generate some new interest in the lattice dynamics of this frequently used and poorly known compound.

Very recent work [3.35a] shows that calculated and measured frequencies of GaAs/AlAs superlattices oriented along [100] and [012] can also be mapped on the corresponding bulk dispersion relations with (3.25). In these superlattices, however, longitudinal and transverse modes are mixed, a fact which leads to interesting effects (m odd-even mixing) in the eigenvectors.

d) Experimental Results Involving Out-of-Axis Vibrations

The single reported experiment which involves light propagation out of the superlattice axis [3.54] is illustrated in Fig. 3.38. The difficulty in obtaining these scattering conditions is due to the small thickness of the epitaxial layers (~ 1 μm), and the large refractive index which makes oblique incidence outside the sample nearly normal to that inside. The authors of [3.54] thus performed a right-angle scattering experiment, the incident light focussed as usual on the (001) surface of the sample and the scattered light being collected through the side (edge). This collection is possible with good efficiency thanks to two circumstances: (i) the incident wavelength is just at the band edge so that the structure is transparent to the *scattered* light and (ii) two cladding layers of $Al_x Ga_{1-x} As$ enclose the superlattice, which forms a waveguide for the scattered light. Near-resonant conditions are thus also involved in this experiment.

The spectra obtained in different polarization configurations are shown in Fig. 3.38 for a GaAs/GaAlAs sample with $d_1 \cong d_2 \cong 100$ Å. For such thicknesses, confinement effects on the optic phonon frequencies are not expected. They mainly emphasize the observation of puzzling selection rules given the local cubic symmetry: an LO (TO) line is observed in a configuration where a TO (LO)

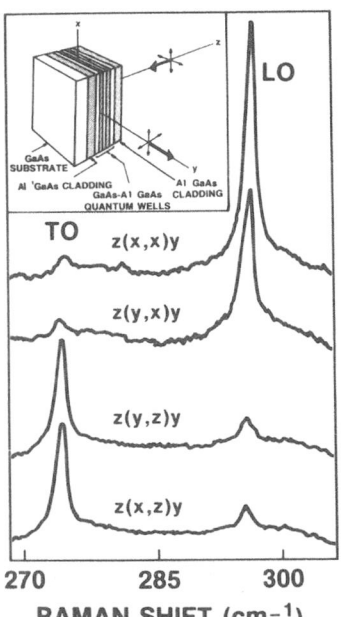

Fig. 3.38. Raman spectra of a GaAs/GaAlAs superlattice in the frequency range of optical phonons of GaAs obtained for different polarizations in the right angle scattering configuration illustrated in the inset together with the structure of the sample [3.54]. The superlattice is clad with GaAlAs so as to act as a waveguide for the scattered radiation

mode is Raman allowed. In these experiments, the wavevector of the phonon emitted in the scattering process is indeed oriented along y (the dispersion along z, and thus the corresponding wavevector component can be neglected). In the $z(y, x)z$ configuration, which allows the observation of a mode of eigendisplacement along u_z, it thus seems that a TO line should be observed, while in the $z(x, z)y$ configuration, which allows the observation of a mode of eigendisplacement along u_y, one might expect an LO line. The experimental results are just the opposite. This can be understood as due to the confinement of the vibrations. One must therefore take into account the *axial* local wavevector associated with the confined modes (3.25). As long as the in-plane wavevector component is smaller than the effective axial one (3.25) fixed by the layer thickness, the latter dominates and fixes the longitudinal or transverse character of the modes and an apparent, but understandable, breaking of the selection rules takes place.

The second series of experimental results concerning out-of-axis vibrations is of a different nature since the in-plane wavevector is not created purposely in a controlled manner, but appears due to some relaxation of the selection rules. Such relaxation is often observed near resonance and its origin is generally assigned to impurity assisted scattering processes (see Sect. 3.5). Such an effect has been first invoked by *Merlin* et al. [3.7] to explain the emergence near resonance of a new line between the TO and LO components of both the GaAs-type and the AlAs-type modes in a 14 Å AlAs superlattice (see Fig. 3.39). To explain this feature, they introduced the electrostatic anisotropic effective medium model (Sect. 3.2) and attributed these new modes to additional roots and poles in the effective dielectric constants (3.28). *Merlin* et al. [3.71] described,

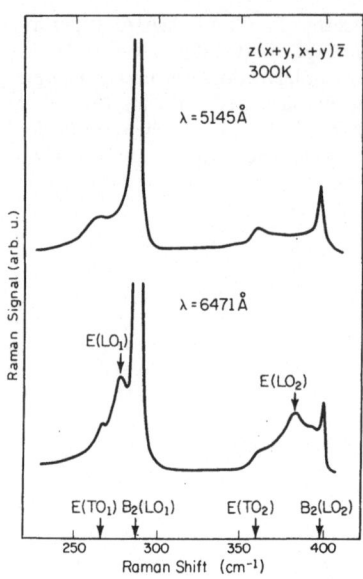

Fig. 3.39. Raman spectra of a GaAs/AlAs superlattice obtained out of resonance (*upper trace*) and at resonance (*lower trace*) [3.71]. The lines $E(LO_1)$ and $E(LO_2)$ which appear at resonance are believed to be due to interface modes. [3.55]

the corresponding vibrations as "bulk" confined modes propagating in the layer plane. Similar structures have indeed been later observed and re-interpreted in [3.55]. Their interpretation was based on the following observations:

- the AlAs and GaAs modes both display a resonant behavior near electronic transitions strongly confined to the GaAs layers,
- they appear as broad bands spread out between the TO and LO modes and display asymmetric lineshapes depending on the relative thickness of both GaAs and AlAs constituting layers.

A common feature of these observations and of the expression of the additional solutions introduced by *Merlin* et al. [3.71] is that the parameters of both layers are involved. These additional structures should thus be assigned to interface modes and not to bulk confined ones.

On the basis of the electrostatic model for layered structures we described in Sect. 3.2, *Sood* et al. [3.55] explained the lineshapes of the bands and their dependence on the layer thickness ratio. This is more clearly seen in the AlAs region, where the confined modes are less intense, as illustrated in Fig. 3.40. The asymmetric line shape is attributed to the different symmetry of the two interface modes existing in this frequency range (see Fig. 3.16). The antisymmetric component, which appears at lower or higher frequency depending on the relative thicknesses, is indeed not active via intra-band Fröhlich interaction (Sect. 3.5.1) which dominates the scattering process near resonance. The same analysis in the GaAs-type frequency range is not as clear, the spectra being dominated by the confined modes.

One must point out two main hypotheses of this description:

- the axial and in-plane selection rules must both be strongly relaxed for these modes, and not for bulk confined ones

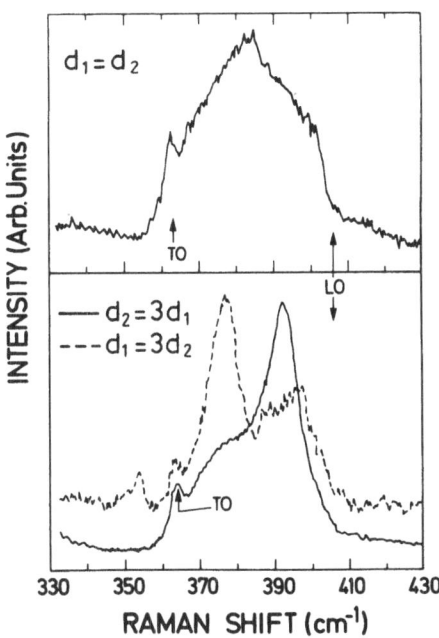

Fig. 3.40. Raman spectra in the frequency range of the optical phonons of bulk AlAs obtained at resonance in the $z(x,x)\bar{z}$ configuration for three GaAs/AlAs structures with different layer thickness ratios [3.55]

- there is no effect of coupling between the two families of modes although they are degenerate and some share the same symmetry (see Fig. 3.19).

Thus, some questions remain open in this novel field. Interface modes have been recently observed on other superlattice such as CdTe/ZnTe [3.160], diluted magnetic structures [3.127] and GaAs/Ga$_{1-x}$Al$_x$As [3.153]. Moreover, a magnetic field enhancement of the resonant Raman spectra on confined LO modes and an even stronger one on interface modes has been recently reported [3.151]. Also, an enhancement with decreasing laser power density has been observed [3.152]. These results seem to support the role played by the in-plane wavevector relaxation in these observations. Nevertheless, the need for a microscopic description of these features remains, even if a qualitative description is obtained using the electrostatic approximation [3.73a].

3.4.4 Light Scattering in Superlattices: Application to Sample Characterization

a) Introduction

In the former subsections we discussed the information that one can extract from light scattering experiments concerning the lattice dynamics of superlattices. As the main features are now reasonably established for the well characterized GaAs – AlAs system, it is interesting to look at the information that one can obtain from light scattering spectra on the structural parameters of superlattices:

- the dispersion curves of the constituting bulk compounds

- the thickness of the constituting layers or, more generally, the size and structure of the supercell
- the modifications of these properties due to strain when the two constituents are not lattice matched.

We have already considered the first point when extracting the AlAs sound velocity from the average frequency of folded acoustic doublets and when comparing the bulk optic dispersion curves with the frequencies of the modes confined to GaAs and AlAs. These methods are very useful for finding information about compounds which are difficult to produce as large samples. We already emphasized the sensitivity of these determinations to the accuracy in the structural parameters and to imperfections in the compositional profile. Here we shall consider the structural information one can extract from the Raman scattering spectra of superlattices and compare it with that obtained by the more usual x-ray diffraction method.

So far we have neglected the difference in lattice parameters between the two constituent cubic bulk compounds and also between them and the substrate. This approximation was justified for GaAs/AlAs structures grown on a GaAs substrate since the relative difference of the lattice parameters of GaAs and AlAs is less than $1\%_0$. We will analyze, in a second subsection, the effect of strain on the vibrations of the superlattice.

b) Structural Characterization Using the Vibrations of a Superlattice

The usual method to determine the structural parameters of superlattices is x-ray diffraction [3.145]. From the distance between the satellites which appear around the diffraction peaks of the average compound, an easy accurate and direct determination of the period of the sample is obtained. In perfectly lattice-matched structures, no more structural information can be deduced from these positions and the observation of these satellites originates only from the modulation of the atomic structure factor. Using an inverse Fourier transform, one can then deduce the structure profile from the satellite intensities with an accuracy depending on the number of observed satellites. Real superlattices are actually divided into nearly lattice-matched (mismatch $< 1\%$) ones and strained ones. In the former case, the previous analysis remains valid [3.161]. Moreover, the diffraction peak of the substrate is now slightly separated from the one for the average compound, a fact which enables us to determine its composition. Thanks to this additional feature, the thicknesses d_1 and d_2 of a simple GaAs/AlAs structure can be deduced by considering only the satellite positions. For strained layer superlattices, two modulations are involved and the x-ray diffraction patterns look very different [3.146]. A numerical fit of the positions and intensities is needed in this case.

The superlattice vibrations provide two probes of different nature, which potentially make light scattering an interesting tool to characterize the structural properties. The folded acoustic modes which propagate along the axis are mainly sensitive to the long range order. They thus provide information on the periodicity of the system and on the inner structure of the supercell, in a way rather similar to the x-ray satellites. Assuming perfect acoustic matching, the period of the sample can be easily deduced from the average doublet frequency. Taking into account the small acoustic mismatch displayed by presently existing superlattices makes this measurement less direct. In the experimentally derived quantity vd, the average sound velocity v (3.11) actually depends slightly on the details of the structure. Nevertheless, a good estimate of the period is obtained from this rapid and easy measurement. *Brugger* et al. used this method for Si/Ge structures [3.128] with an additional refinement. From the doublet splitting and the knowledge of the phonon wavevector, one can derive the superlattice velocity v and thus determine the period d with a better accuracy. Moreover, from the value of v, a rough estimate of the thickness ratio in a simple structure can be obtained. On the other hand, the determination of this ratio from the gap openings, which strongly depend on its value, is not accurate due to their small magnitude.

Thanks to the large modulation of the photoelastic coefficient, the line intensities are strongly sensitive to the thickness ratio, as we showed previously, and more generally to the inner structure of the supercell. This feature has been investigated in two different contexts: (i) the analysis of samples containing more than two layers by period and (ii) the characterization of interdiffusion profiles.

As concerns the first point, such complex samples have been first considered by *Nakayama* et al. [3.58]. They pointed out that for several GaAs/Ga$_{0.5}$Al$_{0.5}$As/AlAs structures the folded mode frequencies were strongly dependent on the period but not on the relative thicknesses of the three layers, whereas the relative intensities were strongly affected by the thickness ratio (see Fig. 3.41). This feature has been analyzed quantitatively in [3.60]. As illustrated in Fig. 3.42 for a $(GaAs)_{d_1}(AlAs)_{d_2}(GaAs)_{d_1}(AlAs)_{d_3}$ period, the calculated spectra only reproduce the line intensities when the details of the supercell are taken into account. A similar problem has recently attracted interest in connection with the Fibonacci superlattice [3.135–138], (see Chap. 5 of this volume). Raman intensities have indeed been demonstrated to be very sensitive to the long range quasiperiodic order in these systems.

Raman scattering studies of superlattices with smeared out interfaces were first reported in [3.65]. A decrease of the intensity of the lower acoustic doublet was observed when the superlattice was annealed at 850 K for an increasing length of time; the frequencies of the lines remained unchanged. This result was assigned to the diffusion of the interfaces which induces a decrease in the non-zero Fourier components of the profile. A similar effect was observed later [3.59] on analogous samples grown by MBE at different substrate temperatures. A simple quantitative analysis of the intensities was performed. By using an erf-function profile involving a free interface broadening parameter, one can describe the variation of the average intensity of both the first and second folded doublets by using the same broadening parameter. Some insight on the quality of the samples was thereby obtained for the first time using Raman scattering and

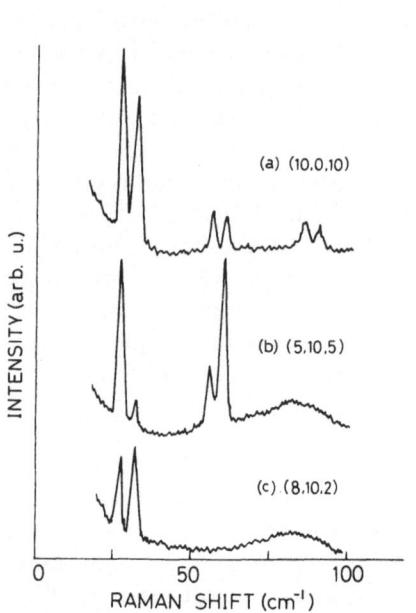

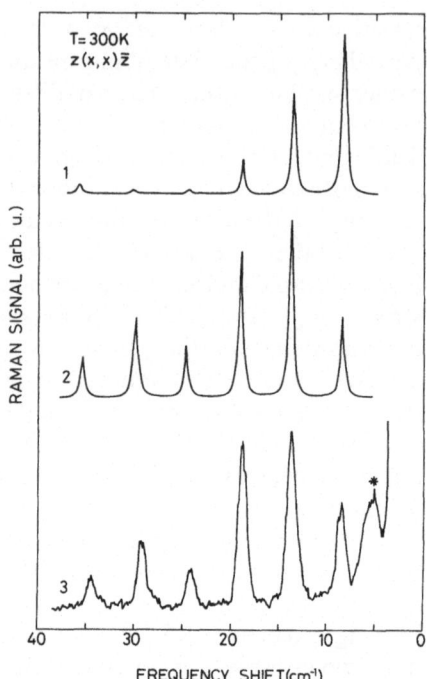

Fig. 3.41. Low frequency Raman spectra obtained for GaAs/GaAlAs/AlAs samples with equal periods and different individual layers thicknesses ($n1, n2, n3$ in monolayers) [3.58]

Fig. 3.42. Low frequency Raman spectra (*3*) obtained for a $(GaAs)_{d_1}/(AlAs)_{d_2}/(GaAs)_{d_1}/(AlAs)_{d_3}$ sample [3.60] compared with calculated spectra which either take into account the detail of the structure (*1*) or consider a $(GaAs)_{2d_1}/(AlAs)_{d_2+d_3}$ simplified cell (*2*)

the study was later extended to superlattices grown by organometallic vapor phase epitaxy (OMVPE) [3.114]. This analysis has been recently applied to annealed structures and correlated with the predictions of the diffusion theory and with x-ray measurements for the same samples [3.119]. However, as pointed out by *Klein* [3.99], the photoelastic model is easy to apply when the photoelastic coefficient varies linearly with x. In continuously varying GaAs/Ga$_{1-x}$Al$_x$As structures, this description is questionable as the usual incident energies lie between the frequency gaps of GaAs and AlAs and resonant features in the photoelastic profiles are expected.

To summarize, Raman scattering spectra of folded acoustic lines provide information similar to x-ray diffractometry. However, as the measured quantities are less directly connected to the structure parameters, lower accuracy can be expected. No real systematic comparison has been reported so far, and Raman scattering is certainly very useful when x-ray diffraction is not available or difficult to set up (for instance in the case of in situ characterization in high vacuum chambers).

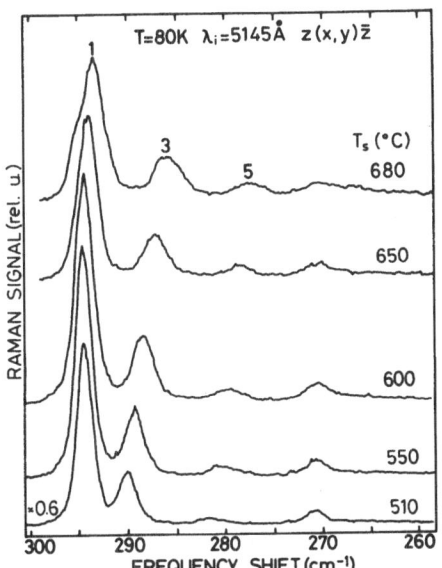

Fig. 3.43. Raman spectra in the GaAs optical phonon range for a series of GaAs/AlAs super-lattices with the same structural parameters but grown at different substrate temperature T_s [3.59]

Raman scattering also provides a second type of probe of the sample parameters through the confined optic vibrations. Contrary to the folded acoustic modes, they are sensitive to the local properties of the structure and give information on the potential well in which they are confined. This can be applied to determine individual layer thicknesses in superlattices and thus complete or check the data extracted from the folded acoustic lines. This has also been shown to be powerful for probing the shape of the interfaces [3.59]. Figure 3.43 shows the Raman spectra of phonons confined to the 25 Å thick GaAs layers of the same series of MBE sample used for folded acoustic intensity measurements. In all samples, three LO confined modes are observed whose intensity and width change only slightly from sample to sample. Their peaks shift towards lower frequencies, a fact which is assigned to the effective shrinkage of the layer thickness due to interface broadening. Because of the non-abrupt profile, the thickness seen by a confined mode depends on its frequency. As a consequence, the Raman spectra cannot be reproduced using the alternating linear chain model except for samples grown at low substrate temperature. A crude quantitative model has been introduced on the basis of an alternating linear chain with continuously varying effective parameters [3.59].

Using the erf-function profile allows the authors to fit the three confined mode positions with the same interface broadening parameter; the values obtained for this parameter are close to those deduced from the folded acoustic line intensities. While this method requires the knowledge of the corresponding bulk dispersion curve, the authors introduce a semi-quantitative test of the interface profile which only makes reference to the nearly parabolic shape of the dispersion curves involved. Under this assumption, the relative frequency separation between the confined modes directly reflects the shape of the well. For a rectangular (or parabolic) profile, the successive confined frequencies follow a quadratic (linear) progression. From the comparison of the line separation, an estimate of the profile abruptness is obtained. A similar analysis was not possible in the AlAs-type frequency range as only one confined mode was observed. From the recent Raman determination of the bulk dispersion curve [3.149], an analysis of the AlAs layer shape could be undertaken on the basis of the shift of this single confined frequency. As in the case of folded acoustic line intensities, a similar analysis has been applied to superlattices grown by OMVPE under different conditions [3.114] and recently, to annealed superlattices [3.119, 162].

c) Strained Layer Superlattices

Thus far, we have neglected the difference between the lattice parameters of the superlattice bulk constituents and that of the substrate. When "pseudomorphic growth" is achieved, the substrate, which is very thick compared to the epitaxial layer, imposes its lattice parameter in the directions perpendicular to the growth axis. The epitaxial layers then suffer a tetragonal elastic deformation. When the substrate is identical to one of the superlattice constituents (e.g. GaAs in GaAs/AlAs structures), the strain is localized in the other constituent (this can be changed by using a buffer layer between substrate and superlattice). These deformations accumulate elastic energy in the sample. Above a critical thickness, whose magnitude is not well understood, misfit dislocations appear which relax the strain in the layers. In superlattices two different critical thicknesses must be considered: the first one is related to the relaxation of the individual layers, one relative to the other, and the second one is related to the relaxation of the superlattice considered as a whole, relative to the substrate.

The strain due to lattice mismatch is invariant under the point group of the superlattice (D_{2d} for a [001]-grown superlattice). Hence, some shift of the LO and TO modes is expected without any additional splitting. Some change in the acoustic velocities is also predicted. This, together with the small lattice mismatch in the usual systems, explains why so little work has been devoted to this problem, only in the following structures:

– GaSb/AlSb [3.120–123], GaAs/InGaAs [3.154, 163]
– Si/Si$_{1-x}$Ge$_x$ [3.164, 165]
– some II–VI compounds: ZnTe/ZnSe [3.166], ZnTe/ZnS [3.167], CdTe/ZnTe [3.150], ZnSe/ZnS$_x$Se$_{1-x}$ [3.19]
– ZnSe on GaAs [3.168], ZnTe on GaAs [3.19].

In all these studies of superlattices except [3.150], the effect of strain is investigated on the LO line confined in layers whose thickness is too large or whose crystalline quality is too poor to induce a significant confinement effect. Raman scattering can then be used as a probe of the local strain exactly as for the global strain in a thick layer [3.169]. This interesting feature appears thanks to the local character of the Raman probe. Depending on the substrate, Raman

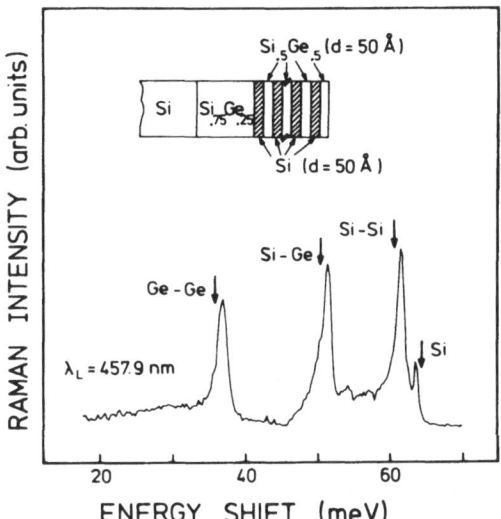

Fig. 3.44. Raman spectrum of the Si/SiGe structure described in the inset. *Arrows* indicate the corresponding frequencies in absence of strain [3.164]

shifts can be observed either for both compounds in opposite directions or restricted to the single strained one. The former case is illustrated in Fig. 3.44 in Si/Si$_{0.5}$Ge$_{0.5}$ structures grown on a Si$_{0.75}$Ge$_{0.25}$ buffer layer. The optic mode of the Si layers is shifted towards lower frequency whereas the three different modes of the alloy are shifted in the opposite direction. In the case of GaSb/AlSb structure grown on AlSb buffer layers, a Raman shift is only observed on the

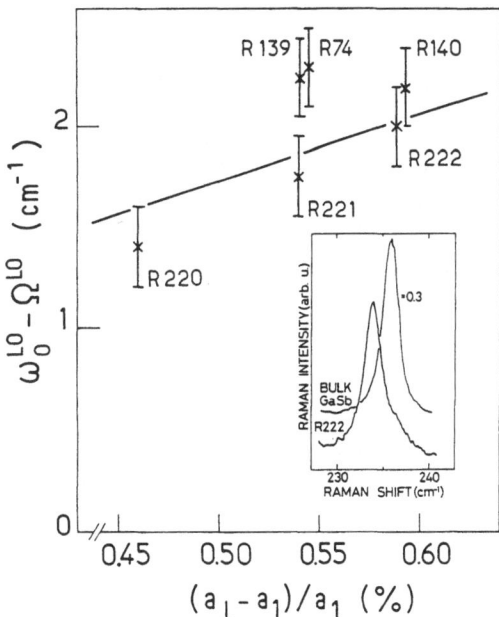

Fig. 3.45. Theoretical (———) and experimental values of the strain-induced shift of the GaSb-type LO frequency in GaSb/AlSb superlattices shown as a function the strain-induced change in lattice parameter of the GaSb layers along the superlattice axis. From [3.120]

GaSb LO line and can be quantitatively assigned to the elastic deformation (see Fig. 3.45). The strain has striking consequences in $GaAs/In_{1-x}Ga_xAs$ structures. As reported in [3.163], a single GaAs-like LO line is observed in this system, contrary to the prediction, based on unstrained data, of a GaAs-like band ordering very similar to that of $GaAs/Ga_{1-x}Al_xAs$. However, due to the strain, which almost exactly compensates the alloying effect, there is no modulation of the GaAs-like optic phonon frequency in this system [3.154].

Recently, *Menéndez* et al. [3.150] reported the observation at resonance of several confined modes in the highly strained CdTe – ZnTe system. They extract information on strain from a quantitative analysis of the confined frequencies with the alternating linear chain model and look for the corresponding shift of bulk optical branches. It would be interesting to investigate by these methods the strain-induced change in the shape of the optic dispersion curves or in the value of the sound velocity obtained from the zone folded frequencies in strained layer superlattices. We should also mention that *Olego* et al. [3.19] have been recently able to determine strain profiles. vs. depth in superlattices and in single epitaxial layers by measuring Raman spectra for different laser wavelengths (i.e., penetration depths). The strain has been shown to decrease from the interface to the substrate to the outermost free surface.

In order to quantify the effects of strain on the optical phonons of a zincblende-type bulk material, a dimensionless fourth rank tensor $\tilde{K}_{ij}$ (in contracted index notation) is defined [3.170]. The relative frequency shifts are obtained by contracting this tensor once with the strain and twice with the direction of the vibration under consideration. This tensor has, for the T_d point group, three independent components $\tilde{K}_{11}$, $\tilde{K}_{12}$, and $\tilde{K}_{44}$ which are sometimes replaced by three parameters p, q, r with the dimensions of a frequency squared [3.171]:

$$p = \tilde{K}_{11}\omega^2 , \qquad q = \tilde{K}_{12}\omega^2 , \qquad r = \tilde{K}_{44}\omega^2 . \tag{3.71}$$

These parameters are usually determined by means of Raman and IR measurements on samples subjected to uniaxial [3.170] or hydrostatic stress [3.169]. Results for zincblende-type materials are listed in Table 3.1. It is thus customary to split the strain into hydrostatic and uniaxial components. The effect of the hydrostatic component is described by the Grüneisen parameter γ related to $\tilde{K}_{ij}$ through

$$\gamma = -\frac{\tilde{K}_{11} + 2\tilde{K}_{12}}{6} . \tag{3.72}$$

The other two independent components of $\tilde{K}_{ij}$, $\tilde{K}_{11} - \tilde{K}_{12}$ and $\tilde{K}_{44}$, described the effects of a pure shear (traceless strain tensor) along [100] and [111], respectively. For lattice matched adjacent layers A, B perpendicular to [001] with different bulk lattice constants one find [3.165] for the "singlet" vibration along [001]:

$$\frac{\Delta\omega_s}{\omega} = \frac{\tilde{K}_{11}}{2}\varepsilon_{zz} + \frac{\tilde{K}_{12}}{2}(\varepsilon_{xx} + \varepsilon_{yy}) , \tag{3.73}$$

Table 3.1. Coefficients which determine the hydrostatic shift $(\tilde{K}_{11}+2\tilde{K}_{12})$ and the shear splittings $(\tilde{K}_{11}-\tilde{K}_{12}, \tilde{K}_{44})$ of diamond- and zincblende-type bulk semiconductors. In the cases where the coefficients for LO and TO phonons are given as equal, no sufficient experimental information is available to determine the separate values.

	$\tilde{K}_{11}+2\tilde{K}_{12}$		$\tilde{K}_{11}-\tilde{K}_{12}$		$\tilde{K}_{44}$	
	LO	TO	LO	TO	LO	TO
Si[a,b]	−5.5	−5.5	0.48	0.48	−0.61	−0.61
Ge[a,b]	−6.7	−6.7	0.46	0.46	−0.87	−0.87
AlSb[c]	−6.0	−5.2	0.97	0.55	−0.34	−0.71
GaP[a,d]	−6.5	−5.7	1.03	0.60	−0.50	−0.58
GaAs[a,e]	−7.4	−8.3	0.70	0.30	−0.53	−0.88
GaSb[f,g]	−7.3	−8.0	0.22	0.22	−1.08	−1.08
InP[a,h]	−7.4	−8.6	1.20	0.69	−0.18	−0.47
InAs[f,g]	−6.4	−7.3	0.57	0.57	−0.76	−0.76
InSb[f]	−7.0	−8.5	−	−	−	−
ZnS[a]	−6.0	−6.6	−	−	−	−
ZnSe[a,g]	−5.4	−8.4	1.24	1.24	−0.43	−0.43
ZnTe[a]	−7.2	−10.2	−	−	−	−

[a] [3.169]. [b] M. Chandrasekhar, J.B. Renucci, M. Cardona: Phys. Rev. B17, 1623 (1978).
[c] [3.173]. [d] I. Balslev: Phys. Stat. Sol. B61, 207 (1974).
[e] [3.170]. [f] K. Aoki, E. Anastassakis, M. Cardona: Phys. Rev. B30, 681 (1984).
[g] F. Cerdeira, C.J. Buchenauer, F.H. Pollak, M. Cardona: Phys. Rev. B5, 580 (1972).
[h] [3.172].

where ε_{zz} and $\varepsilon_{xx}=\varepsilon_{yy}$ are the components of the strain tensor related, for medium A, to the lattice mismatch $\varepsilon=(a_A-a_B)/\langle a \rangle$ through

$$\varepsilon_{xx}=\varepsilon_{yy}=-\varepsilon; \; \varepsilon_{zz}=-[2S_{12}/(S_{11}+S_{12})]\varepsilon ,\tag{3.74}$$

where S_{11} (>0) and S_{12} (<0) are the elastic compliance constants of A. We have assumed medium B to be unstrained, i.e., to be much thicker than A or to equal the substrate and match its lattice constant. Generalizations to other situations, e.g. A and B disconnected from the substrate, are straightforward (see Appendix 3.A).

The "doublet" vibration (perpendicular to [001]) shifts by an amount:

$$\frac{\Delta\omega_d}{\omega}=\frac{\tilde{K}_{11}}{2}\varepsilon_{xx}+\frac{\tilde{K}_{12}}{2}(\varepsilon_{yy}+\varepsilon_{zz}) .\tag{3.75}$$

We should note that, in principle, the parameters $\tilde{K}_{ij}$ are somewhat different for the LO and TO components of the optical phonons. Accurate determination of both sets of independent parameters, however, have only been recently performed for GaAs [3.170], InP [3.172], and AlSb [3.173]. This problem does not arise in the group IV materials (Ge, Si) since they have no LO – TO splittings.

We should report, in closing this section, that strained layer heterojunctions (InGaAs – GaAs – AlGaAs) have been successfully used to fabricate lasers which operate CW at room temperature (8250 Å) [3.174]. Effects of strain due to lattice mismatch on optical phonons have also been observed for single epitaxial layers, the most conspicuous case being that of silicon on sapphire (SOS) [3.175]. ZnSe on GaAs [3.168] and gray tin on InSb [3.176] have also been investigated.

3.5 Resonant Scattering

The electronic states of superlattices fall into two categories: Those with energies within the potential wells have wavefunctions confined to those wells with exponential decay into the barriers [3.177–180]. For sufficiently large barriers there is no interaction between wells and these states do not show any dispersion (band formation) along k_z. In this case, one speaks of multiple quantum wells (MQW). This concept is valid for states which correspond to those near the lowest band edges of the bulk material of the well (note that the well of the conduction band may be in material A while that of the valence band is in material B in the so-called type II superlattices). States of the well which lie above the barrier top are propagating states since they always find other states in the barrier to which to couple. Nevertheless, this coupling can be small, and resonant (nearly confined) states can result. For small period superlattices ($n_1 = n_2 \leqq 8$) all states disperse appreciably along k_z. For large well thicknesses the lowest (highest) conduction (valence) states do not appreciably disperse and one has a MQW. The energies and wavefunctions of these states are the same as for a single quantum well [3.181] except for the enhanced degeneracy.

The light scattering mechanism discussed in Sect. 3.3.3 ignores the dependence of the Raman tensor on laser frequency ω_i, although it could be easily assumed that the $\alpha_{\parallel, \perp}$ of (3.57) and its derivatives depend on ω_i. Even than, the resonance phenomena in which either $\hbar\omega_i$ or $\hbar\omega_s$ are equal to the energy $\hbar\omega_g$ of strong electronic inter-band transition (e. g., between valence and conduction states confined to the same material) would not be appropriately described: resonance phenomena appear both for $\omega_i = \omega_g$ and for $\omega_s = \omega_g$ (incoming and outgoing resonances, respectively). The theory of Sect. 3.3.3 implies that $\omega_i \simeq \omega_s$ and thus leads to only one resonance frequency for each ω_g. The assumptions of this "quasi-static" or "adiabatic" theory are justified whenever the phonon frequency ω_0 satisfies:

$$\omega_0 \ll |\omega_g - \omega_i + i\varGamma_g| \,, \tag{3.76}$$

where $\varGamma_g$ is the Lorentzian broadening of ω_g. Condition (3.76) usually holds at room temperature. At lower temperatures, however, it does not hold for optical phonons at the lowest ω_gs since $\varGamma_g$ becomes very small.

Resonance Raman scattering is rather rich in phenomenology even for bulk materials [3.10]. The theoretical interpretation is performed usually within the framework of the uncorrelated electron-hole approximation. Experimental results, however, sometimes reflect Coulomb interaction between these particles, i.e., the so-called excitonic effects [3.182].

3.5.1 Resonance Effects in the Bulk Constituents of Silicon and Zincblende-Type Superlattices

a) Resonant Electronic Transitions

The details of the electronic transitions which produce resonance in the Raman scattering by phonons in Si and GaAs-type materials is discussed in Sect. 2.2.4 ff. of [3.10]. Here we shall recall some general features and recent developments useful to the discussion of superlattices. The bulk resonances which have been mainly investigated occur at the so-called E_0, $E_0 + \Delta_0$, E_1, and $E_1 + \Delta_1$ gaps (see Fig. 2.37 of [3.10]). These gaps, transition energies, or critical points all have counterparts in superlattices. The E_0 gap, the lowest direct gap at the Γ ($k = 0$) point of the bulk BZ, takes place between p-like valence states (mainly anionic) and s-like conduction states (mainly cationic). The p-states are spin-orbit split by an amount Δ_0, a fact which gives rise to the $E_0 + \Delta_0$ gap. The strong E_1 and $E_1 + \Delta_1$ transitions are also spin-orbit partners. They take place along the four equivalent $\langle 111 \rangle$ directions between the split top valence bands and the lowest conduction band [Ref. 3.10; Fig. 2.37]. The standard Feynman diagram of resonant scattering by one phonon is shown in Fig. 3.46. One usually distinguishes between two-band and three-band terms in the scattering amplitude. In the former the phonon connects electronic states belonging to the same band ($c = c'$ or $v = v'$ in Fig. 3.46) while in the latter the phonon couples states belonging to different bands. Two-band terms are usually more strongly resonant than the three-band ones since the former have two energy denominators resonating at nearly the same frequency, while the latter only have one.

Expressions for the scattering efficiencies near the E_0, $E_0 + \Delta_0$, E_1 and $E_1 + \Delta_1$ gaps (also called critical points) are given in [3.10]. They contain two- and three-band terms and are based on the quasi-static approximation. They can be presented either in terms of analytic expressions for the combined densities of states at critical points [(2.194) of [3.10]] or, equivalently, of the dielectric function and its derivatives with respect to ω [(2.195, 201) of [3.10]]. These expressions can be easily transformed into others which do not require fulfillment of (3.76) [Ref. 3.1; Sect. 2.3.2]; see also [3.183, 184]). This is accomplished by replacing in the two-band terms the derivatives of $\chi(\omega_i)$ by the

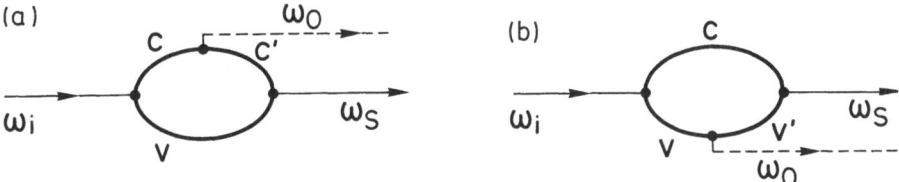

Fig. 3.46a, b. Feyman diagrams of resonant Raman scattering by one phonon (---) in semiconductors. The thin lines represent photons, the thick ones electrons and holes. The phonon is shown to couple to either the conduction (**a**) or the valence states (holes) (**b**)

finite difference ratios $[\chi(\omega_i) - \chi(\omega_s)]/\omega_0$ and the three-band terms $\chi(\omega_i)$ by the average of $\chi(\omega_i)$ and $\chi(\omega_s)$.

The expressions (2.194, 195) of [3.10] are based on the three-dimensional parabolic expansion of the E_0 and $E_0 + \Delta_0$ critical points. That for the E_1 and $E_1 + \Delta_1$ critical points [(2.201 of [3.10]), is based on a two-dimensional expansion of the energy differences between conduction and valence bands vs. **k**: this difference is constant *along* the [111] direction over a large region of the BZ. Hence, the longitudinal mass can be assumed to be infinite, and a cylindrically symmetric (i.e., two-dimensional) situation results. The equations so obtained for the E_1 and $E_1 + \Delta_1$ critical points can be easily transformed for application to the case of MQW [3.103].

We close this subsection by pointing out that the expressions mentioned apply to phonons at $\mathbf{k} \simeq 0$ i.e., they lead to scattering efficiencies independent of **k**. This is the so-called (dipole) allowed scattering. In the case of LO phonons, "forbidden" scattering of amplitude proportional to k, is induced by the electrostatic field which accompanies such phonons (Fröhlich mechanism, see Sect. 3.5.1c).

b) Deformation Potential Electron-Phonon Interaction

The matrix element of the electron-phonon Hamiltonian between electronic states is usually written in terms of coupling constants called deformation potentials (in eV, representing matrix elements per unit dimensionless phonon deformation; see for instance [Ref. 3.10; Eqs. (2.187, 199)]).

Resonant scattering at the E_0 and $E_0 + \Delta_0$ gaps is determined by a single deformation potential called d_0 ($\simeq +30$ eV for most materials treated here [3.185]). It is easy to see by using standard character tables for the T_d group of

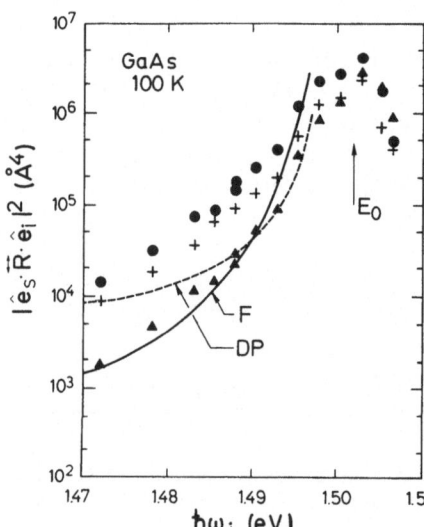

Fig. 3.47. Resonance measured for deformation potential (DP) scattering by phonons in GaAs in the allowed $e_i \parallel [100]$, $e_s \parallel [010]$ configuration, and forbidden scattering (F) obtained for $e_i \parallel e_s \parallel [100]$. From [3.183]. The solid and dashed curves are fits obtained with the theory of [3.186]

zincblende [3.80] that the Raman phonon of Γ_{15} (T_2) symmetry couples neither with the Γ_6 conduction band nor with the split-off valence band Γ_7. Thus, the $E_0 + \Delta_0$ gap only exhibits weakly resonant three-band terms related to the coupling of the Γ_7 valence band with its split-off mate Γ_8 (see structure at ~ 1.8 eV for GaAs in [Ref. 3.10; Fig, 2.8]). Two-band terms occur at the E_0 gap since the corresponding valence band (Γ_{15}) couples with itself via the Γ_{15} phonon. Thus, very strong resonances result, as shown in Fig. 3.47. The resonance at $E_0 + \Delta_0$, determined by three-band terms, has a strength proportional to $(d_0/\Delta_0)^2$. It should be much stronger for InP $(\Delta_0 \simeq 0.1$ eV) [3.187] than for GaSb $(\Delta_0 \simeq 0.8$ eV) [3.188].

The orbital wavefunctions of Γ_{15} states have p-like symmetry and thus vary under the operations of the T_d point group like the coordinates x, y, z. We represent them by X, Y, Z. The wavefunctions of the spin-split Γ_8 and Γ_7 states can be labeled in angular momentum notation:

$$\Gamma_8 \begin{cases} \left(\frac{3}{2}, \frac{3}{2}\right) = \frac{1}{\sqrt{2}} (X+iY)\uparrow \, ; & \left(\frac{3}{2}, -\frac{3}{2}\right) = \frac{1}{\sqrt{2}} (X-iY)\downarrow \\[2mm] \left(\frac{3}{2}, \frac{1}{2}\right) = \frac{1}{\sqrt{6}} (X+iY)\downarrow + \sqrt{\frac{2}{3}} Z\uparrow \, ; & \left(\frac{3}{2}, -\frac{1}{2}\right) = \frac{1}{\sqrt{6}} (X-iY)\uparrow - \sqrt{\frac{3}{2}} Z\downarrow \end{cases}$$

$$\Gamma_7 \quad \left(\frac{1}{2}, \frac{1}{2}\right) = \frac{1}{\sqrt{3}} (X+iY)\downarrow - \frac{1}{\sqrt{3}} Z\uparrow \, ; \quad \left(\frac{1}{2}, -\frac{1}{2}\right) = \frac{1}{\sqrt{3}} (X-iY)\uparrow + \frac{1}{\sqrt{3}} Z\downarrow .$$

$$(3.77)$$

The matrix elements of the electron-phonon Hamiltonian H_{ep} for Stokes scattering can be easily obtained by using the relationship (in atomic units [Ref. 3.10; Eq. (2.187)]):

$$\langle X|H_{ep}^Z|Y\rangle = \frac{d_0}{2a_0} \left(\frac{3}{2\omega_0 \mu N}\right)^{1/2} (n_B+1)^{1/2} \, , \qquad (3.78)$$

where H_{ep}^Z represents the Hamiltonian for a phonon polarized along z, μ the reduced mass of the primitive cell (PC), a_0 the lattice constant N the member of PCs, and n_B the Bose-Einstein factor. Other non-vanishing matrix elements of H_{ep} can be obtained from (3.78) through circular permutation.

The E_1 and $E_1 + \Delta_1$ resonances are determined by two deformation potentials, $d_{3,0}^5$ and $d_{1,0}^5$ [3.183, 184]. The former determines the three-band terms which couple the spin-orbit split valence states while the latter determine the two-band terms; the three-band terms usually dominate even at resonance [3.183, 184]. The E_1, $E_1 + \Delta_1$ resonances should be important in superlattices containing Ge, InSb, InAs, and GaSb since they occur in the region of standard ion lasers (2–3 eV). (For energies of these gaps see [3.96]). However, we shall not discuss these resonances any further since only one detailed report concerning them has appeared [3.189].

c) Fröhlich Interaction

The deformation potential is assumed to be independent of the k of the phonon and thus can be considered to be *local* in real space. In polar materials longitudinal phonons are accompanied by electric fields which contribute a long range (k-dependent) term H_{ep}^F, to the so-called Fröhlich Hamiltonian [Ref. 3.10; Eq. (2.212)], proportional to $|k|^{-1}$. The three-band terms arising from H_{ep}^F result in a small renormalization of the allowed Raman tensor [Ref. 3.10; Sect. 2.1.12] of [3.10]). The two-band terms, however, lead to a forbidden contribution [Ref. 3.10; Sect. 2.2.8] whose Raman tensor is usually diagonal and its scattering efficiency proportional to $|k|^2$. Expressions for this tensor as a function of band parameters are given in [3.10] for the adiabatic case. This restriction has been lifted in [3.190] and [3.191] for three- and two-dimensional critical points, respectively. The latter results can be adapted to an in-plane propagation in MQWs. The results are the same as those of [3.103].

The forbidden scattering discussed above is usually small because of its proportionality to $|k|^2$, which is basically zero for forward scattering and rather small for backscattering $|k| = 4\pi n/\lambda_i$. It can, nevertheless, become dominant at resonance. An enhancement mechanism, based on an additional interaction with charged impurities which increases the effective $|k|$, has been suggested [3.192]. This mechanism generates incoherent phonons with varying (but small) ks. The generation of incoherent phonons has been experimentally demonstrated [3.193]. A particularly reliable signature of the impurity-induced process is the fact that the *outgoing* resonance becomes dominant, while in the k-induced process structures of the same strength are seen both at the incoming and outgoing resonances [3.194].

d) Interference Between Fröhlich and Deformation Potential Scattering

The Raman tensors for Fröhlich, R_F, and deformation potential scattering, R_D, have the forms:

$$R_F = a_F \begin{pmatrix} 1 & & \\ & 1 & \\ & & 1 \end{pmatrix}, \quad R_D = a_D \begin{pmatrix} 0 & 1 & 0 \\ 1 & 0 & 0 \\ 0 & 0 & 0 \end{pmatrix} \tag{3.79}$$

for LO phonons propagating along z. It is easy to see that (3.79) leads to scattering for $e_i \| e_s \| [110]$ which is different to that for $e_i \| e_s \| [1\bar{1}0]$ as a result of the interference of the complex quantities a_F and a_D near resonance. The scattering efficiencies are proportional to $|a_D + a_F|^2$ and $|a_D - a_F|^2$ in the former and latter case, respectively. Such *interference* effects have been observed at $E_0 + \Delta_0$ [3.187, 188, 193] and at E_1, $E_1 + \Delta_1$ [3.194]. Similar effects have been reported at E_1, $E_1 + \Delta_1$ for MQWs [3.189].

3.5.2 Resonance Scattering by Phonons in Superlattices

The E_0 absorption of bulk zincblende materials transforms into a series of discrete lines when the material forms a quantum well (or MQW). The electronic states become localized to the wells and in the lowest conduction band at Γ a ladder with indices $l = 1, 2, 3 \ldots$, whose envelope wavefunctions are similar to the envelopes of the phonons of (3.34, 38), results. The top valence band develops into two such series: that of the heavy holes (HH) and that of the light holes (LH). In an "allowed" optical absorption process between such valence and conduction levels, l must be conserved: two such series develop, corresponding to the heavy and light hole ladders for $\Delta l = 0$. The absorption spectrum for uncorrelated electrons should thus be two such series of step functions. Excitonic effects actually sharpen up the leading edge of the step functions into series of peaks. Figure 3.48 shows the peaks of the heavy hole ladder of $GaAs/Al_{0.2}Ga_{0.8}As$ quantum wells. The light hole ladder is not seen, except possibly a shoulder on the high energy side of the $n = 1$ peak for $L_z = 140$ Å. Both ladders can be clearly observed in [3.196] for $Ga_{0.47}In_{0.53}As/Al_{0.48}In_{0.52}As$ MQWs. Each one of these absorption peaks is expected to lead to resonances in the Raman scattering cross section.

Such resonances were first observed in [3.103] for the heavy hole ladders associated with the E_0 edge of $GaAs/Ga_{1-x}Al_xAs$ superlattices and MQWs with $x = 0.1$ and 0.25. We show these observation in Fig. 3.49 for two MQWs (a, b) and two superlattices with subband widths δ_1. The solid curves in these figures represent fits using a theory which assumes uncorrelated electrons, the dashed curves the corresponding absorption spectra. No attempt was made in this work to determine the polarization selection rules. The fit (solid line) to the resonance profile $S(\omega_i)$ was obtained with the approximate expression:

$$S(\omega_i) \propto |\chi(\omega_i) - \chi(\omega_s)|^2 , \qquad (3.80)$$

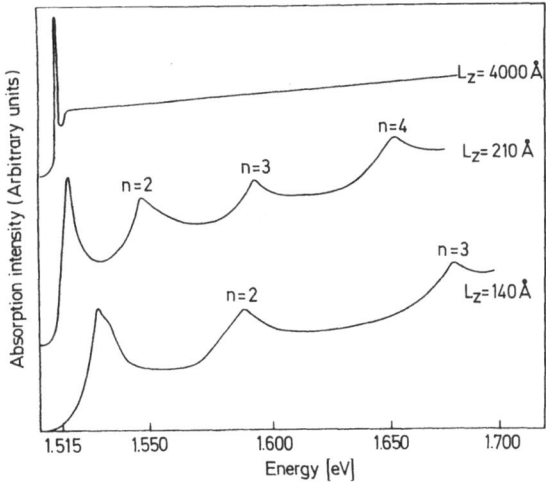

Fig. 3.48. Absorption spectra of $GaAs/Al_{0.2}Ga_{0.8}As$ quantum wells of three different widths d_1. From [3.195]

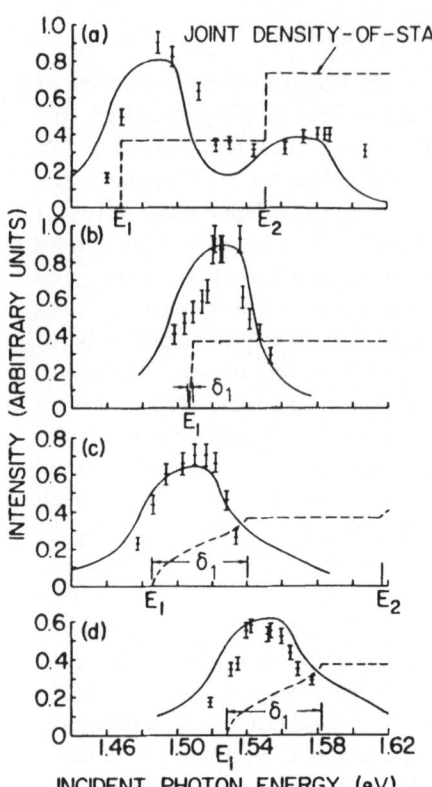

INCIDENT PHOTON ENERGY (eV)

Fig. 3.49. Resonant Raman scattering by LO phonons for two MQWs (**a, b**) and two superlattices (**c, d**) of GaAs/Al$_x$Ga$_{1-x}$As. $x = 0.25$ for (**a, b, d**), $x = 0.1$ for (**c**). The well widths d_1 are 100 (**a**), 52 (**b**), 39 (**c**), and 35 Å (**d**). The lines are theoretical fits (see text) [3.103]

which is justified, according to [3.103], very close to resonance, with $\chi(\omega)$ obtained from the two-dimensional expressions for uncorrelated electrons ((2.178) of [3.10]) in the case of the MQWs (a, b) and with a generalized three-dimensional expression for the true superlattices (c, d). The theory represents the data rather well. Similar measurements involving the $E_0 + \Delta_0$ localized levels associated with the $E_0 + \Delta_0$ gap are presented in [3.104] together with evidence for $l_v = 2 \to l_c = 2$ and $l_v = 3 \to l_c = 3$ transitions at the E_0 gap. These measurements were performed at room temperature.

Similar measurements, but performed at $T \simeq 2$ K, were reported in [3.197]. These temperatures allow the resolution of the incoming and outgoing components of the resonance (Fig. 3.50) associated with the $l = l_c = l_v = 1, 2, 3$ levels of the E_0 gap of GaAs in GaAs/Al$_{0.27}$Ga$_{0.73}$As MQWs. For $l = 1$ only the incoming resonance associated to the heavy hole (HH) ladder is seen: the outgoing one is swamped by the luminescence of the $l = 1$ light hole transition. For $n = 2$, both incoming and outgoing HH resonances are seen. In the region of ω_i between 1.75 and 1.85 eV, both $l = 2$ LH and $l = 3$ HH incoming and outgoing resonances are observed. In [3.198] a slightly different MQW allows the observation of the $l = 1$ incoming and outgoing HH resonances and the LH outgoing one. Interesting in all these measurements is the fact that the incoming

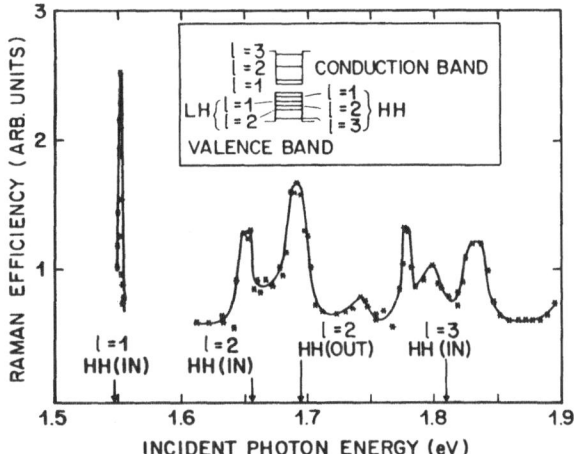

Fig. 3.50. Resonant profile for Raman scattering by GaAs-like LO phonons in an $(GaAs)_{36}(Al_{0.27}Ga_{0.73}As)_{73}$ MQW. The arrows indicate the exciton energies calculated for the corresponding single quantum wells [3.197]

resonances are somewhat weaker ($\sim$ one-half) than the outgoing ones. In [3.197] this is attributed to a three-band scattering mechanism, with the phonon coupling two different l-subbands ($l=2$ and 3 for the $l=2$ resonances of Fig. 3.50): the non-resonant denominator of the three-band process is smaller in this case for the outgoing than for the incoming resonance. This explanation is supported by the fact that sometimes a stronger incoming resonance is found [3.199] a fact that follows from the above model if the non-resonant state lies *below* the resonant one. In [3.199] resonances related to the lowest gap of the barrier (unconfined) are seen: not only for the LO phonons of the GaAs well but for the Ga-like and Al-like phonons of the alloy barrier.

An alternative explanation of the usually encountered dominance of outgoing resonances is given in [3.198]. It follows naturally and in a general way from the impurity enhanced Fröhlich mechanism [3.192]: for the outgoing resonance the other two energy denominators lie in the continuum of electronic excitations, a fact which introduces divergences in the k-space integrations. This does not happen for the incoming resonances. A fit with this model to the $l=2$ HH resonances is shown in Fig. 3.51. This mechanism has the advantage of its generality while that based on three-band models depends strongly on the details of the electronic structure. Three-band terms may become dominant under certain circumstances, especially when the separation between the two gaps involved equals the phonon frequency. In this case both the energy denominator may vanish (or nearly vanish) and double resonances occur [3.200, 200a] (Sect. 3.5.3b).

Resonances have also been observed for CdTe/CdMnTe MQWs [3.201]: incoming and outgoing (stronger) resonances appear for the $l=1$, HH transition while for $l=1$, LH transition the ratio of incoming/outgoing strengths may be reversed. In this case the MQWs strain enhances the HH−LH separation.

The scattering intensity at the peak of the $l=1$ and $l=2$ resonance has been measured vs. temperature in [3.202]. These measurements yield information on

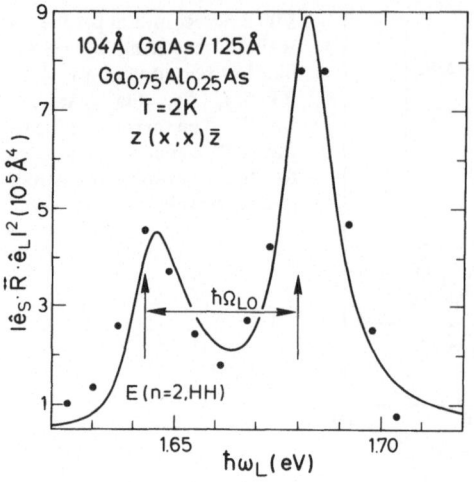

Fig. 3.51. Raman scattering resonance with $l = 2$ for a GaAs-Al$_{0.25}$Ga$_{0.75}$As MQW displaying the asymmetry between the incoming (smaller) and outgoing components. The points are experimental data, the line a theoretical fit based on the impurity enhanced Fröhlich mechanism [3.198]

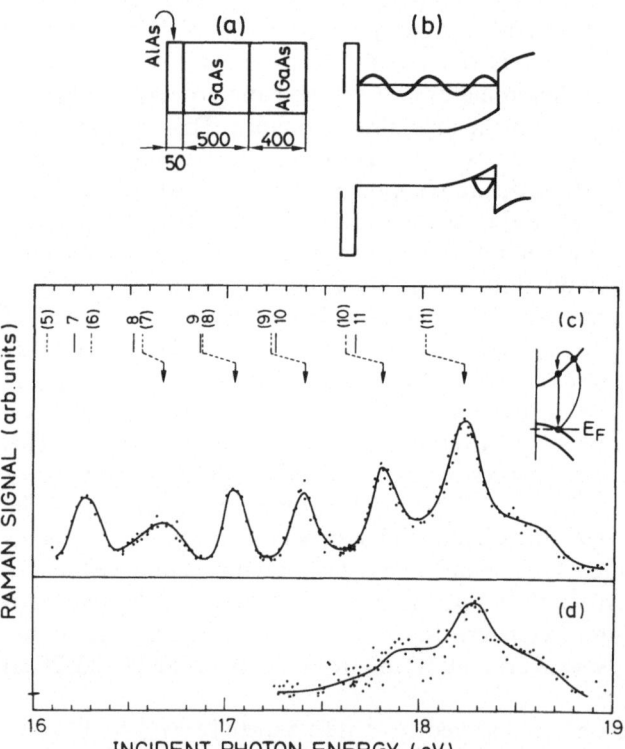

Fig. 3.52. (a) Period of the MQW of [3.203]. (b) Electronic structure of the well in real space. (c) Resonance scattering profile for the LO phonons of a GaAs well. (d) As in (c), for the GaAs-like LO phonons of the barrier to the right. The solid vertical bars indicate calculated positions of the $l_c \neq l_v = 1$ transitions and the accompanying number the corresponding l_c. The dashed lines show the position of the related outgoing resonances with l_c in brackets

the temperature dependence of the *homogeneous* linewidth of the corresponding excitons and its origin.

A particularly astute resonance experiment has been reported in [3.203]. The authors use a MQW consisting of asymmetric quantum wells composed of an AlAs barrier on the left, a 500 Å GaAs well, and a Be-doped $Al_{0.33}Ga_{0.67}As$ barrier (modulation doping) on the right (Fig. 3.52). The 500 Å GaAs localizes the heterojunction on the right, because the Be-doping produces a triangular well which localizes the electrons very close to the interface. This configuration lifts the $\Delta l = 0$ selection rule since the valence and conduction wavefunctions are not localized in the same region and the envelope functions of different l_c and l_v are not orthogonal. In this manner one sees *outgoing* resonances from $l_v = 1$ to l_c between 5 and 11 (see Fig. 3.52). These experiments can be used to map out details of the envelope functions of the confined states.

Most of the Raman work discussed so far was performed for MQWs with nearly k_z-independent electronic states (exception: Fig. 3.49c, d). A simple test of this condition can be made by measuring the ratio of the scattering intensities of the barrier modes to those of the well modes. Such a test has been applied to $(GaAs)_n/(AlAs)_n$ superlattices in [3.204]: this ratio increases from about 0.1 for $n \geq 10$ to 0.9 for $n = 2$. For $n = 1, 2$ both the GaAs- and AlAs-like phonons resonate at the E_0-like gap of the superlattices at ~ 2.15 eV [3.205], Fig. 3.53. A weaker resonance related either to a folded or an indirect gap appears at 1.92 eV for $n = 1$. Unfortunately, these resonances have only been measured at room temperature: strong luminescence makes low temperature observation hard. Similar results have been reported in [3.206] for a $(GaAs)_2/(AlAs)_1$ super-lattice.

The only report of resonances at the E_1 and $E_1 + \Delta_1$ gaps of superlattices has appeared in [3.189] for GaSb/AlSb samples. An increase in the E_1 gaps of bulk

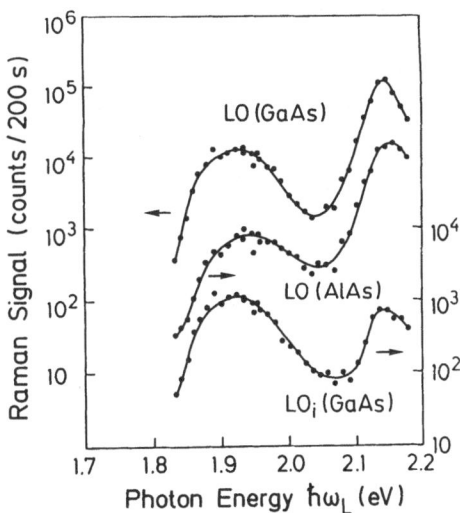

Fig. 3.53. Resonance of the Raman scattering of GaAs- and AlAs-like LO phonons in $(GaAs)_1(AlAs)_1$ superlattices. LO_i represents an additional peak above the main LO (GaAs) which becomes strong at the low frequency resonance (its strength has been divided by 10 in the plot). The origin of this peak is not clear. It may be a phonon away from $k = 0$

GaSb is observed, possibly related to partial confinement of the electronic states. Calculations of this confinement effect, however, indicate that it is much smaller than the observed one.

3.5.3 Scattering Mechanisms for the E_0-gap Related Transitions of Superlattices

a) LO-Phonons: Deformation Potential vs Fröhlich Interaction

In bulk zincblende-type material the Raman phonons have Γ_{15} symmetry. Consequently, the dipole-*allowed* Raman tensors have only off-diagonal components referred to the cubic axes. For backscattering at a [001] face the LO-phonons (Γ_{15}^z) have the Raman tensor of (3.60). These phonons generate in the D_{2d} group of the [001]-grown superlattice phonons of symmetries A_1 and B_2 (Sect. 3.3.2). Those of symmetry A_1 correspond to (3.34a, 38a) while those of B_2 symmetry are represented by (3.34b, 38b). The latter have a Raman tensor of the same form as the bulk (3.60), while the former have an *allowed* diagonal Raman tensor [diagonal components of (3.61)]. The diagonal tensor is forbidden in the bulk, thus it is expected to be small off-resonance. A dipole-*forbidden* diagonal tensor also appears in the bulk at resonance [3.10]: it is induced by the electrostatic Fröhlich interaction. The allowed tensor of the bulk is usually described by the deformation potential interaction. New deformation potentials, forbidden in the bulk, must thus be induced by the superlattice potentials: they are, however, expected to be small except for very small period superlattices. This fact has already been illustrated in Sect. 3.3.3 in connection with the bond polarizability model.

The confined B_2 modes are related to bulk modes through (3.38b) or, more accurately, (3.25) with m odd, the A_1-modes likewise through (3.25) for m even. The B_2 modes should scatter via a deformation potential but it is easy to see that they cannot lead to two-band terms: the matrix element for the coupling of an electronic state with itself must vanish since the product of the electronic state with itself has A_1 symmetry. A_1 phonons thus lead to two-band terms which should be strong (even dominant!) near resonances. They should appear in parallel scattering configurations while the B_2 phonons appear either in crossed polarizations parallel to the x and y axes or in parallel polarizations parallel to [110] (or [1$\bar{1}$0], see Sect. 3.5.1).

If n_1 and n_2 are not too small one can break up the matrix element of the *short-range* electron-phonon interaction (bulk deformation potential) into matrix elements of H_{ep} over the bulk PCs and a matrix element of the envelope functions (two electron states, one phonon). This is appropriate to B_2 phonons. At the E_0 gaps the phonons must then connect the $(\frac{3}{2}, \pm\frac{3}{2})$ valence bands with their $(\frac{3}{2}, \pm\frac{1}{2})$ counterparts (see (3.77), the quantization axis is z). In this manner three-band terms involving the HH or $(\frac{3}{2}, \pm\frac{3}{2})$, LH $(\frac{3}{2}, \pm\frac{1}{2})$, and conduction bands result. If the HH and LH splitting equals the phonon frequency, strong double resonance obtains [3.200].

A_1 symmetry forbidden scattering results in the bulk from the k-dependent Fröhlich interaction. In a superlattice of intermediate period (n_1, n_2 between 5 and 50) the periodicity fixes the value of k_z. Thus, for not too large values of $k_{x,y}$, the matrix element of the electron-phonon interaction becomes k-independent and can also be represented by a deformation potential. The resulting Raman resonances, for m even, are supposed to be very strong (two-band terms). In fact, although they should only appear for parallel incident and scattered polarizations, they can even mask the B_2 peaks (m odd) which should be observed in crossed (x, y) polarizations [3.65, 200a]. Because of the separation of B_2 and A_1 modes, the interference effects discussed in Sect. 3.51d do not take place. As n_1 increases, however, A_1 and B_2 modes begin to overlap within their linewidths and interferences should again reappear [3.206a].

We have, so far, discussed the k-conserving Fröhlich-like A_1 scattering. It implies, for backscattering, coupling to phonons with $k_{x,y} = 0$. As in the case of the bulk (Sect. 3.5.1c), impurity scattering may enhance these processes and allow coupling to phonons with $k_{x,y} \neq 0$ [3.200a]. This can be inferred from the fact that the outgoing resonance becomes stronger than the incoming one (see Fig. 3.54). Another Fröhlich-related mechanism, which appears even at $k_{x,y} = 0$, should exist in superlattices. In this case, in the bulk the contributions of electron and hole diagrams (Fig. 3.46a, b) cancel because of the opposite charges of electrons and holes. In MQWs the phonons may be strongly confined while the electrons (holes) may penetrate different amounts into the barrier, depending on

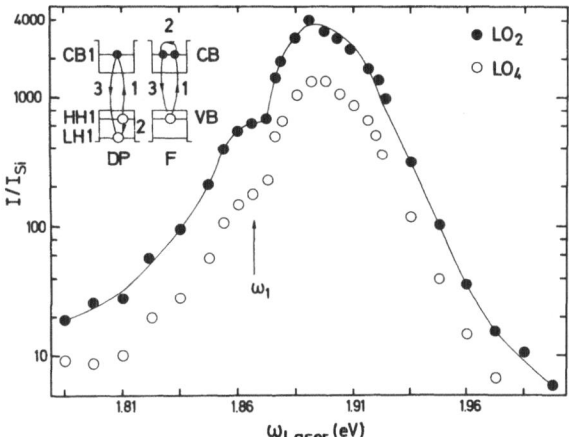

Fig. 3.54. Observed scattered intensity, measured with respect to that of the Raman phonon of Si, versus incident laser frequency for the $m = 2$ and $m = 4$ phonons of a $(GaAs)_7(AlAs)_7$ superlattice. The inset shows scattering diagrams for deformation potential (DP) and Fröhlich (F) electron-phonon interaction. The three steps indicated correspond to deformation potential (three-band) and Fröhlich (two-band) terms. For the F-case, a similar diagram with step 2 in the HH1 miniband, has to be subtracted. The resonant gap is ω_1: the outgoing resonance dominates [3.66a]

the masses and barrier heights. Thus, the matrix elements of H_{ep}^F do not cancel exactly, leading to "allowed" non-impurity-induced scattering for $k_{x,y} = 0$.

For superlattices of intermediate period, the confined modes resonate only with transitions between overlapping confined electronic states. The interface modes, however, extend to both sides of the interfaces and thus resonate with electronic transitions confined to either side [3.55].

b) Double Resonances

Several examples of double resonances in the scattering by LO phonons in quantum wells have been reported by the *Bell* group [3.200, 207, 208]. Similar effects have been observed in bulk GaAs samples under uniaxial stress [3.209, 210]: the stress-induced splitting of the Γ_8 valence bands equals, in these cases, the scattering phonon frequency.

The work of [3.208] was performed for a $(GaAs)_{10}$ *single* quantum well with $Ga_{0.3}Al_{0.7}As$ as a barrier, for which the $l = 1$ HH $- l = 1$ LH splitting $E_{1H} - E_{1L}$ equals the LO phonon energy $\hbar\omega_0$. As mentioned above, those two states are then coupled by the LO phonon via the deformation potential [see (3.77, 78)]:

$$\left\langle \frac{3}{2}, \frac{3}{2} \left| H_{ep}^z \right| \frac{3}{2}, -\frac{1}{2} \right\rangle = \frac{d_0}{a_0} \left(\frac{1}{2\omega_0 \mu N} \right)^{1/2} (n_B + 1)^{1/2} \tag{3.81}$$

and likewise for the coupling between $(\frac{3}{2}, -\frac{3}{2})$ and $(\frac{3}{2}, \frac{1}{2})$. The $(\frac{3}{2}, -\frac{1}{2})$ LH electron states can be excited to the conduction band by using a circularly polarized photon of $(+)$-polarization $(J_z = +1)$, while the excited conduction state can return to the $(\frac{3}{2}, \frac{3}{2})$ hole by emitting a $(-)$-polarized $(J_z = -1)$ photon (these polarizations are given with respect to fixed axes and not to the direction of propagation). Hence, the double resonant process should only occur for $(+)$-polarized incident and $(-)$-polarized scattered photons. A polarization ratio P:

$$P = \frac{I(+, -) - I(+, +)}{I(+, -) + I(+, +)} \tag{3.82}$$

equal to 0.79 is found, instead of the expected $P = 1$. The discrepancy is explained in [3.208] as due to loss of polarization in the virtual intermediate exciton state. In this work detailed scattered intensity measurements were performed in six different linear and circular polarization configurations. From the results, the authors derive information about exciton population and dephasing lifetimes.

In [3.200, 207] double resonances induced by the Fröhlich interaction are reported. They arise from coupling by the LO phonon between an $l = 1$ HH and an $l = 3$ HH state. The optical transitions involve the $l_c = 1$ state and thus the incoming one violates the $\Delta l = 0$ selection rule. This violation is interpreted by the authors as due to the mixing of $l_v = 3$ with $l_v = 1$ by the excitonic interaction. The polarization selection rules for this process are opposite to those for deformation potential coupling, as corresponds to the A_1 nature of the phonons involved: the

$(+, +)$ process is dominant. One finds for the corresponding polarization ratio:

$$P = \frac{I(+, +) - I(+, -)}{I(+, +) + I(+, +)} \tag{3.83}$$

$P = 0.885$, thus confirming the hypothesis of a Fröhlich interaction mechanism. The difference between this value of P and the expected $P = 1$ is attributed to lifetime processes in the intermediate excitonic excitation (dephasing time τ_2 and scattering time between the two excitonic states τ_1).

We should mention, in closing, an interesting recent observation [3.211] of doubly resonant Raman scattering in QWs in which one of the resonant states is the $l_v = 1$ continuum strongly modified by electron-phonon interaction. A very large number of peaks (at least 27) in the scattered intensity vs. ω_i are observed and assigned to various combinations of GaAs-like, AlAs-like and interface phonons. A requirement for these observations, which have not yet been completely theoretically analyzed, is the use of a very perfect quantum well: $n_1 = 9$ and also $n_1 = 8$ were used.

c) Photoelastic Mechanism

It has been well documented that the photoelastic mechanism is responsible for the Brillouin scattering in the bulk materials of interest here [Ref. 3.10; Sect. 2.3.4], [3.212]. In this case the phonon frequency ω_0 is very small and (3.76) is usually well satisfied even for $\omega_g = \omega_i$. For very sharp excitonic transitions at low T, however, polariton scattering may become important (Chap. 7 of *Light Scattering in Solids III*). By performing experiments for various scattering configurations it is actually possible to extract the three independent photoelastic constants p_{11}, p_{12}, and p_{44}, (Sect. 3.3.4) at least in the region of transparency [3.212, 213]. Most of the available experimental data [3.96] for p_{ij} apply to this region; some of them have been measured by Brillouin techniques and some by optical methods with the sample under static stress. Unfortunately, the latter measurements are usually confined to $p_{11} - p_{12}$ and p_{44} and do not give information on p_{11} and p_{12} separately: backscattering measurements for LA phonons (and their folded partners) in superlattices require the knowledge of p_{12} (Sect. 3.3.4).

We show in Fig. 3.55 the values of $p_{11} - p_{12}$ obtained with Brillouin and static piezo-birefringence techniques below the E_0 gap of ZnSe. The agreement between different methods is excellent. The theoretical fitting lines emphasize the strong resonant dispersion of these coefficients (and, correspondingly, the scattering efficiencies) at E_0. The separate values of p_{11}, p_{12}, and p_{44} obtained from Brillouin data are shown in Fig. 3.56. Note that p_{44} crosses zero before the resonance at E_0, while p_{11} and p_{12} do not (but $p_{11} - p_{12}$ does, Fig. 3.55), a fact common to most large gap materials of the family [3.215].

The data of Figs. 3.55 and 3.56 apply to the region below E_0. Photoelastic data above E_0 are rare. Not too precise results can be obtained by the piezo-

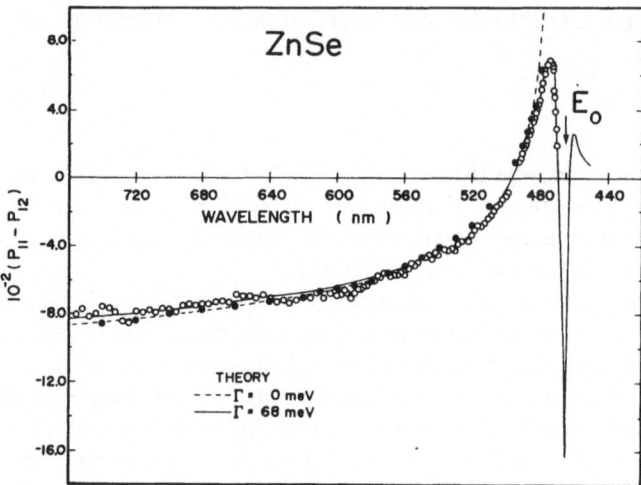

Fig. 3.55. Dispersion of the photoelastic coefficients $p_{11} - p_{12}$ of ZnSe at room temperature as measured by Brillouin scattering (○○○) and by piezo-birefringence techniques (●●●). The lines represent theoretical fits with expressions similar to those in [3.10]. Note the strong resonance at the E_0 edge which is preceeded by an anti-resonance (zero) [3.212]

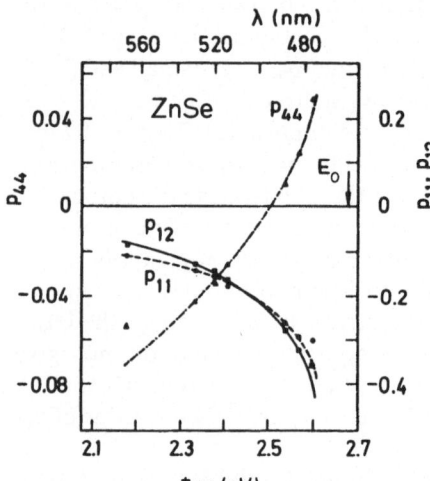

Fig. 3.56. Photoelastic coefficients p_{11}, p_{12}, and p_{44} of ZnSe at 295 K as measured by Brillouin scattering. The lines through the points are theoretical fits [3.214]

reflectance technique [3.216]. Kramers-Kronig analysis of piezo-reflectance data, or fits with theoretical expressions, can be used to obtain the related imaginary parts of p_{ij}. Data for GaAs, with details of the E_0, $E_0 + \Delta_0$, E_1, and $E_1 + \Delta_1$ resonances, are shown in Fig. 4 of [3.217] for the imaginary parts of p_{ij}: the real parts should be obtainable through Kramers-Kronig transformation. Accurate data for p_{12} below E_0 can be calculated from the $p_{11} - p_{12}$ results of [3.96] and those for $p_{11} + 2p_{12}$ given in Fig. 13 of [3.215]. These data, however, have not been used for the interpretation of light scattering in superlattices (the corresponding data for AlAs are not available)!

The data just discussed are usually fitted with expressions similar to those in Sect. 2.2.12 of [3.10]. Near E_0 and $E_0 + \Delta_0$ three deformation potentials (a, b, d) determine the resonant behavior [Ref. 3.10; Eqs. (2.255–257)] while four deformation potentials are needed to determine the behavior at E_1 and $E_1 + \Delta_1$. The photoelastic constants can be used near resonance for folded acoustic phonons provided their frequency fulfills (3.76), i.e., in the whole resonant range provided $\omega_0 \ll \Gamma_{\rm g}$. Otherwise (2.255–259) of [3.10] must be converted into those for finite ω_0 (with incoming and outgoing resonances) by using the prescription given in Sect. 3.5.1. These split resonances have yet to be observed for folded acoustic modes. On the whole, little quantitative work on resonant scattering by folded acoustic modes has been performed (see [3.123] for qualitative work on GaSb/AlSb superlattices).

d) Effects of Electric Fields on the Resonant Raman Scattering

The dielectric response and other optical properties of superlattices and MQWs are known to be affected by strong electric fields [3.218, 219]. These effects also appear in Raman scattering. Longitudinal fields distort the sinusoidal confined electronic wavefunctions as they lift the two-fold rotation axes in the super-lattice planes: violation of the $\Delta l = 0$ selection rule results. In particular, $\Delta l = \pm 1$ transitions are enhanced while those for $\Delta l = 0$ decrease in strength. Theoretical and experimental investigations of these effects in resonant Raman scattering can be found in [3.220, 221].

3.5.4 Resonant Scattering by Two Phonons

Resonant scattering by two and more phonons has been intensively studied for the bulk constituents of semiconductor superlattices [3.10]. One usually observes overtones of the LO phonons near Γ. They are induced by the Fröhlich mechanism, since the k of *each* phonon involved does not have to vanish (only the total sum does) these processes are dipole allowed. One also observes density of two phonon states and related effects, usually induced by deformation potential interaction. They are represented by Raman tensors of all possible symmetries: Γ_{15}, Γ_{12}, and Γ_1 (or, equivalently, T_2, E, A_1) in the zincblende structure.

Scattering by two phonons is also encountered in superlattices and QWs, especially near resonant conditions. Work for GaAs/Ga$_{1-x}$Al$_x$As systems has been published in [3.222–224]. Strong resonances appear in $e_{\rm i} \| e_{\rm s}$ scattering configurations and thus can be safely attributed to Fröhlich processes.

We show in Fig. 3.57 Raman spectra obtained for a (GaAs)$_7$/(AlAs)$_{21}$ MQW including two confined LO (GaAs-like) and mixed GaAs- and AlAs-like interface phonons. The resonant profile has a weak incoming (at the $l = 1$ gap, labeled ω_1) and a dominant outgoing peak: such behavior is standard in two-phonon scattering [3.187]. More complicated combinations of interface modes,

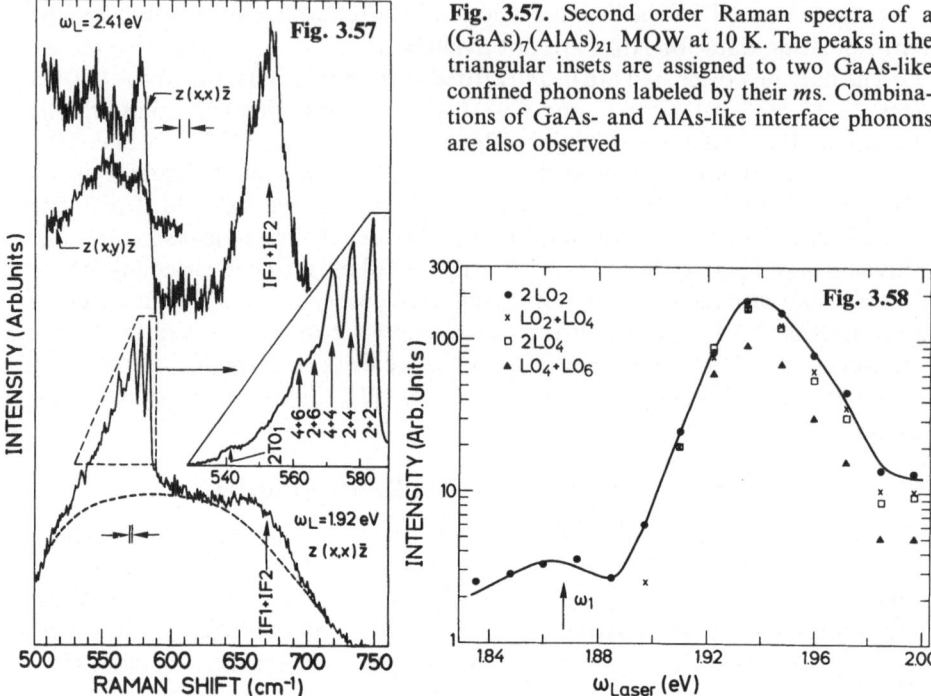

Fig. 3.57. Second order Raman spectra of a $(GaAs)_7(AlAs)_{21}$ MQW at 10 K. The peaks in the triangular insets are assigned to two GaAs-like confined phonons labeled by their ms. Combinations of GaAs- and AlAs-like interface phonons are also observed

Fig. 3.58. Profile of the resonances by two confined phonons of Fig. 3.57. The resonating electronic transition is indicated by ω_1. The main peak corresponds to the outgoing resonance, as is usual in the case of higher order Raman scattering

involving up to four phonons, were observed in [3.223] for a $(GaAs)_7/(AlAs)_7$ sample. Resonance profiles of the two-phonon peaks are shown in Fig. 3.58. Recently, triply resonant scattering by two phonons has been reported [3.200a].

3.5.5 Conclusions

We have discussed the resonant scattering mechanisms of the bulk constituents of semiconductor superlattices and shown how they can be carried over to the latter. Fröhlich interaction is dominant for LO-phonons at resonance, a fact which usually leads to a dominance in the outgoing peak of the resonance profile. Confined phonons of odd m resonate through a deformation potential while those of even m do it through a Fröhlich interaction. Interface phonons extend to both sites of the heterojunction and thus resonate at the gaps of both constituents. In some cases double resonances are observed.

While the photoelastic mechanism seems to be mainly responsible for scattering by acoustic phonons, it must be modified to take into account resonance phenomena: real and imaginary parts of the photoelastic constants (and also the finite phonon frequency) must be included. Scattering by two or more phonons, usually induced by the Fröhlich mechanism, can also be observed near resonance. It leads to dominant *outgoing* peaks in the resonance profiles and, under the proper conditions, to triple resonances [3.200a].

3.A Appendix

Strains in a pseudomorphic superlattice with a total thickness D_1 of material 1 and D_2 of material 2.

If the superlattice is attached pseudomorphically to a substrate of "infinite" thickness of material 1, we must take $D_1 = \infty$. We consider the general case of cubic materials with different elastic stiffness constants $c_{ij,1}$ and $c_{ij,2}$ and growth along either [100] or [111] for which the strains in the superlattice *plane* $\varepsilon_{\perp,1}$ and $\varepsilon_{\perp,2}$ are isotropic. Generalization to non-isotropic $\varepsilon_\perp$ (such as for [110] growth) is straightforward. By imposing the condition of pseudomorphism:

$$\Delta = \frac{a_2 - a_1}{\langle a \rangle} = \varepsilon_{\perp 1} - \varepsilon_{\perp 2} \tag{3.A.1}$$

and the condition of equal and opposite planar forces in materials 1 and 2 we obtain:

$$\varepsilon_{\perp,1} = \frac{Q_2 D_2}{Q_1 D_1 + Q_2 D_2} \Delta \tag{3.A.2}$$

$$\varepsilon_{\perp,2} = \frac{-Q_1 D_1}{Q_1 D_1 + Q_2 D_2} \Delta \tag{3.A.3}$$

where $Q_{1,2}$ are functions of the elastic compliance constants of each material. For the [100] superlattice case we have:

$$Q = c_{11} + c_{12} - \frac{2(c_{12})^2}{c_{11}} \tag{3.A.4}$$

The in-plane strain in each layer is given by

$$\varepsilon_\parallel = -\frac{2c_{12}}{c_{11}} \varepsilon_\perp \,. \tag{3.A.5}$$

For the [111] case we obtain

$$Q = \frac{2}{3}\left[c_{11} + 2c_{12} + c_{44} - \frac{(c_{11} + 2c_{12} - 2c_{44})^2}{c_{11} + 2c_{12} + 4c_{44}} \right]$$

$$\varepsilon_\parallel = -\frac{2(c_{11} + 2c_{12} - 2c_{44})}{c_{11} + 2c_{12} + 4c_{44}} \varepsilon_\perp \,. \tag{3.A.6}$$

Acknowledgements. It is a pleasure to acknowledge J. Menéndez, R. Merlin, D. Olego, D. Paquet, and A. Pinczuk for careful reading of the manuscript and numerous illuminating discussions.

References

3.1   M. Cardona (ed.): *Light Scattering in Solids I*, Topics Appl. Phys., Vol. 8 (Springer, Berlin, Heidelberg 1983) p. 31
3.2   M. Cardona, G. Güntherodt (eds.): *Light Scattering in Solids II*, Topics Appl. Phys., Vol. 50 (Springer, Berlin, Heidelberg 1983) p. 39
3.3   H. Vogt: In [3.2], p. 208
3.4   See, for instance, W. Marshall, S.W. Lovesey: *Theory of Thermal Neutron Scattering* (Clarendon, Oxford 1971)
3.5   H. Bilz, W. Kress: *Phonon Dispersion Relations in Insulators* (Springer, Berlin, Heidelberg 1979)
3.6   See, for instance, P.M. Fauchet: IEEE Circuits and Devices Mag. **2**, 37 (1986)
3.7   R.K. Chang, M.B. Long: In [3.2], p. 179
3.8   I. Tsang: Chapter 6 of this volume
3.9   D. von der Linde, J. Kuhl, H. Klingenberg: Phys. Rev. Lett. **44**, 1505 (1980)
3.10   M. Cardona: In [3.2], p. 19
3.11   M. Cardona: *Proc. SPIE Conf. #822* (San Diego, California, USA 1987) in press
3.12   M. Cardona: In *Proc. 1984 Seoul Int. Symp. Physics of Semiconductors and its Applications* (Korean Physical Society, Seoul 1985)
3.13   J.R. Sandercock: In *Light Scattering in Solids III*, ed. by M. Cardona, G. Güntherodt, Topics Appl. Phys., Vol. 51 (Springer, Berlin, Heidelberg 1982) p. 173;
       A.S. Pine: In [3.2], p. 253
3.14   M.H. Brodsky: In [3.1], p. 205
3.15   M. Cardona: *Phonon Physics* (World, Singapore 1982) p. 2, and references therein
3.16   N. Saint-Circq, R. Carles, J.B. Renucci, A. Zwick, M.A. Renucci: Solid State Commun. **39**, 1137 (1981)
3.17   D.W. Feldman, J. Parker, W. Choyke, L. Patrick: Phys. Rev. **173**, 787 (1978)
3.18   S. Mizushima, T. Shimanouchi: J. Am. Chem. Soc. **71**, 1320 (1949);
       W.L. Peticolas, G.W. Hibler, J.L. Lippert, A. Peterlin, H.G. Olf: Appl. Phys. Lett. **18**, 87 (1971)
3.19   D.J. Olego, K. Shahzad, J. Petruzello, D. Cammack: Phys. Rev. B**36**, 7674 (1987);
       D.J. Olego, K. Shahzad, D. Cammack, H. Cornelissen: Phys. Rev. B**38**, 5554 (1988)
3.20   M. Wolkenstein: Compt. Rend. Acad. Sci. URSS 32, 185 (1941)
3.21   M.V. Klein: IEEE J. QE-**22**, 1760 (1986)
3.22   B. Jusserand, D. Paquet: In *Semiconductor Heterojunctions and Superlattices*, ed. by G. Allan, G. Bastard, N. Boccara, M. Lannoo, M. Voos (Springer, Berlin, Heidelberg 1986) p. 108
3.23   M. Cardona: In *Lectures on Surface Science*, ed. by G.R. Castro, M. Cardona (Springer, Berlin, Heidelberg 1987) p. 2
3.24   F. Nizzoli, K.H. Rieder (eds.): *Dynamical Phenomena at Surfaces, Interfaces, and Superlattices*, Springer Ser. Surf. Sci. Vol. 3 (Springer, Berlin, Heidelberg 1984)
3.25   L. Esaki, R. Tsu: IBM J. Res. Dev. **14**, 61 (1970)
3.26   L.L. Chang, K. Ploog (eds.): *Molecular Beam Epitaxy and Heterojunctions* (M. Nijhoff, Dordrecht 1985)
3.27   K. Ploog, N.T. Linh (eds.): *Semiconductor Quantum Well Structures and Superlattices* (Editions de Physique, Les Ulis 1985)
3.28   E.H.C. Parker (ed.): *The Technology and Physics of Molecular Beam Epitaxy* (Plenum, New York 1985)

3.29   A.C. Gossard: *Treatise on Materials Science and Technology* (Academic, New York 1982) p. 13
3.30   K. Ploog: J. Cryst. Growth **79**, 887 (1986)
3.31a M.J. Ludowise: J. Appl. Phys. **58**, R31 (1985)
3.31b R.D. Dupuis: Science **226**, 623 (1984)
3.32   K. Kunc, R. Martin: Phys. Rev. Lett. **48**, 406 (1982);
       C. Falter: Physics Rep. **164**, 1 (1988)
3.33   G. Kanellis: Solid State Commun. **58**, 93 (1986); Phys. Rev. B**35**, 746 (1987)
3.34   K. Yip, Y.C. Chang: Phys. Rev. B**30**, 7037 (1984)
3.35   E. Richter, D. Strauch: Solid State Commun. **64**, 867 (1987)
       see also: E. Richter: Diplomarbeit Universität Regensburg, (1986) and unpublished work by these authors;
       see also: F. Ren, H. Chu, Y.C. Chang: Phys. Rev. Lett. **59**, 1841 (1987)
3.35a For calculations of (110) and (012) GaAs/AlAs superlattices and corresponding experiments see Z.V. Popović, M. Cardona, L. Tapfer, K. Ploog, E. Richter, D. Strauch: Appl. Phys. Lett. and Phys. Rev., to be published
3.36   M.T. Yin, M.L. Cohen: Phys. Rev. B**25**, 4317 (1982)
3.37   E. Molinari, A. Fasolino, K. Kunc: Superlattices and Microstructures **2**, 397 (1986)
3.38   R. Tsu, S.S. Jha: Appl. Phys. Lett. **20**, 16 (1972)
3.39   A.S. Barker, Jr., J.L. Merz, A.C. Gossard: Phys. Rev. B**17**, 3181 (1978)
3.40   A. Fasolino, E. Molinari, J.C. Maan: Phys. Rev. B**33**, 8889 (1986)
3.41   B. Jusserand, D. Paquet, J. Kervarec, A. Regreny: J. Phys. (Paris) **45–C5**, 145 (1984)
3.42   C. Colvard, T.A. Gant, M.V. Klein, R. Merlin, R. Fischer, H. Morkoc, A.C. Gossard: Phys. Rev. B**31**, 2080 (1985)
3.43   B. Jusserand: Unpublished
3.44   E.L. Albuquerque, M. Babiker, P. Fulco, S.R.P. Smith, D.R. Tilley: Unpublished
3.45   B. Djafari-Rouhani, J. Sapriel, F. Bonnouvrier: Superlattices and Microstructures **1**, 29 (1985)
3.46   L.M. Brekhovskikh: *Waves in Layered Media* (Academic, New York 1960)
3.47   R.L. Kronig, W.G. Penney: Proc. Roy. Soc. London A**130**, 499 (1930)
3.48   See for instance A.J. Dekker: *Solid State Physics* (Prentice Hall, Englewood Cliffs, N.J. 1957) p. 244
3.49   C. Colvard, R. Merlin, M.V. Klein, A.C. Gossard: Phys. Rev. Lett. **43**, 298 (1980)
3.50   S.M. Rytov, Akust. Zh. **2**, 71 (1956) [Sov. Phys. Acoust. **2**, 68 (1956)]
3.51   S.Y. Tong, A.A. Maradudin: Phys. Rev. **181**, 1318 (1969)
3.52   W.E. Jones, R. Fuchs: Phys. Rev. B**4**, 3581 (1971)
3.53   G. Kanellis, J.F. Morhange, M. Balkanski: Phys. Rev. B**28**, 3390, 3398, and 3406 (1983)
3.54   J.E. Zucker, A. Pinczuk, D.S. Chemla, A. Gossard, W. Wiegmann: Phys. Rev. Lett. **53**, 1280 (1984)
3.55   A.K. Sood, J. Menéndez, M. Cardona, K. Ploog: Phys. Rev. Lett. **54**, 2115 (1985)
3.56   B. Jusserand, F. Alexandre, J. Dubard, D. Paquet: Phys. Rev. B**33**, 2897 (1986)
3.57   P. Santos, L. Ley, J. Mebert, O. Koblinger: Phys. Rev. B**36**, 4858 (1987)
3.58   M. Nakayama, K. Kubota, S. Chika, H. Kato, N. Sano: Solid State Commun. **58**, 475 (1986)
3.59   B. Jusserand, F. Alexandre, D. Paquet, G. Le Roux: Appl. Phys. Lett. **47**, 301 (1986)
3.60   B. Jusserand, D. Paquet, F. Mollot, F. Alexandre, G. Le Roux: Phys. Rev. B**35**, 2808 (1987)
3.61   A. Fasolino, E. Molinari: J. Phys. (Paris), **48–C5**, 569 (1987)
3.62   B. Jusserand, D. Paquet: Phys. Rev. Lett. **56**, 1751 (1986)
3.63   A.K. Sood, J. Menéndez, M. Cardona, K. Ploog: Phys. Rev. Lett. **56**, 1753 (1986)
3.64   E. Molinari, A. Fasolino, K. Kunc: Phys. Lett. **56**, 1751 (1986)
3.65   C. Colvard, R. Fischer, T.A. Gant, M.V. Klein, R. Merlin, H. Morkoc, A.C. Gossard: Superlattices and Microstructures **1**, 81 (1985)
3.66a A.K. Sood, J. Menéndez, M. Cardona, K. Ploog: Phys. Rev. Lett. **54**, 2111 (1985)

3.66b A. Ishibashi, M. Itabashi, Y. Mori, K. Kaneko, S. Kawado, N. Watanabe: Phys. Rev. B33, 2887 (1986)
3.67 B. Jusserand, D. Paquet, A. Regreny: Phys. Rev. B30, 6245 (1984)
3.68 A. Nagaoui, B. Djafari-Rouhani: Surface Sci. 185, 125 (1987) and references therein
3.69 M. Born, K. Huang: Dynamical Theory of Crystal Lattices (Clarendon, Oxford 1954)
3.70 W.L. Mochan, M. del Castillo-Mussot, R.G. Barrera: Phys. Rev. B35, 1088 (1987); R.F. Wallis, R. Szenics, J.J. Quinn, G.F. Giuliani: Phys. Rev. B36, 1218 (1987)
3.70a For a recent observation of phonon polaritons in a GaAs/AlAs heterostructure see M. Nakayama, M. Ishida, N. Sano: Phys. Rev. B38, 6348 (1988)
3.71 R. Merlin, C. Colvard, M.V. Klein, H. Morkoc, A.Y. Cho, A.C. Gossard: Appl. Phys. Lett. 36, 43 (1980)
3.72 R. Fuchs, K.L. Kliever: Phys. Rev. 140A, 2076 (1965)
3.73 R.E. Camley, D.L. Mills: Phys. Rev. B29, 1695 (1984)
3.73a See Chap. 2, p. 25
3.74 M. Cardona: Am. J. Phys. 39, 1277 (1971)
3.75 G. Fasol, N. Mestres, H.P. Hughes, A. Fisher, K. Ploog: Phys. Rev. Lett. 56, 2517 (1986)
3.76 A very simple lattice dynamical model which includes both interface and confined modes and their interaction has been given by K. Huang, B.F. Zhu: Phys. Rev. B38, 2183 (1988); B38, 13377 (1988). Also, T. Tsuchiya, H. Akera, T. Ando: Phys. Rev. B, in press
3.77 Ph. Lambin, J.P. Vigneron, A.A. Lucas, P.A. Thiry, M. Liehr, J.J. Pireaux, R. Caudano, T.J. Kuech: Phys. Rev. Lett. 56, 1842 (1986)
3.78 See the appendix of J. Sapriel, J.C. Michel, J.C. Toledano, R. Vacher, J. Kervarec, A. Regreny: Phys. Rev. B28, 2007 (1983)
3.79 M. Tinkham: Group Theory and Quantum Mechanics (McGraw-Hill, New York 1964) p. 328
3.80 G. Koster: Solid State Physics, Vol. 5 (Academic, New York 1957); D.L. Rousseau, R.P. Bauman, S.P.S. Porto: J. Raman Spectr. 10, 253 (1981)
3.81 K. Kunc: Ann. Physiq. (Paris) 8, 319 (1973–1974)
3.82 T. Toriyama, N. Kobayashi, Y. Horikoshi: Jap. J. Appl. Phys. 25, 1895 (1986)
3.83 A.A. Maradudin: Festkörperprobleme/Advances in Solid State Physics XXI, ed. by J. Treusch (Vieweg, Dortmund 1981) p. 25
3.84 K. Kunc, H. Bilz: Solid State Commun. 19, 1027 (1976)
3.84a For recent work see B. Lou, R. Sudharsanan, S. Perkowitz: Phys. Rev. B38, 2212 (1988)
3.85 K.A. Maslin, T.J. Parker, N. Raj, D.R. Tilley, P.J. Dobson, D. Hilton, C.T.B. Foxon: Solid State Commun. 60, 461 (1986)
3.86 S. Go, H. Bilz, M. Cardona: Phys. Rev. Lett. 34, 580 (1975)
3.87 S. Go, H. Bilz, M. Cardona: Light Scattering in Solids, ed. by M. Balkanski, R.C.C. Leite, S.P.S. Porto (Flammarion, Paris 1976) p. 377
3.88 S. Nakashima, M. Balkanski: Phys. Rev. B34, 5801 (1986) and references therein
3.89 E. López-Cruz, M. Cardona, E. Martínez: Phys. Rev. B29, 5774 (1984)
3.90 E. López-Cruz, M. Cardona: Solid State Commun. 45, 787 (1983)
3.91 B. Jusserand, D. Paquet, A. Regreny: Superlattices and Microstructures 1, 61 (1985)
3.92 A.A. Maradudin, E. Burstein: Phys. Rev. 164, 1081 (1967)
3.93 D. Bermejo, S. Montero, M. Cardona, A. Muramatsu: Solid State Commun. 42, 153 (1982)
3.94 B. Zhu, K.A. Chao: Phys. Rev. B36, 4906 (1987)
3.95 M.V. Klein, C. Colvard, R. Fischer, H. Morkoc: J. Phys. (Paris) 45–C5, 131 (1984)
3.96 O. Madelung, M. Schulz, H. Weiss (eds.): Landoldt-Börnstein Tables, Vol. 17a, (Springer, Berlin, Heidelberg 1982) and references therein
3.97 For information on Si see: M.H. Grimsditch, E. Kisela, M. Cardona: Phys. Stat. Sol. B60, 135 (1980)
3.98 M. Babiker, D.R. Tilley, E.R. Albuquerque: J. Phys. C18, 1285 (1985)
3.99 M.V. Klein: In Proc. 10th Int. Conf. on Raman Spectroscopy, ed. by W.L. Peticolas, B. Hudson (University of Oregon, Eugene, Oregon 1986) p. 9–1

3.100 J. He, J. Sapriel, J. Chavignon, R. Azoulay, L. Dugrand, F. Mollot, B. Djafari-Rouhani, R. Vacher: J. Phys. (Paris), **48–C5**, 573 (1987)

3.101 J.L. Merz, A.S. Barker, Jr., A.C. Gossard: Appl. Phys. Lett. **31**, 117 (1977)

3.102 G. Abstreiter, A. Pinczuk, M. Cardona (eds.): *Light Scattering in Solids IV*, Topics Appl. Phys., Vol. 54 (Springer, Berlin, Heidelberg 1984) p. 5

3.103 P. Manuel, G.A. Sai-Halasz, L.L. Chang, Chin-an Chang, L. Esaki: Phys. Rev. Lett. **37**, 1701 (1976)

3.104 G.A. Sai-Halasz, A. Pinczuk, P.Y. Yu, E. Esaki: Solid State Commun. **25**, 381 (1978)

3.105 V. Narayanamurti, H.L. Stormer, M.A. Chin, A.C. Gossard, W. Wiegmann: Phys. Rev. Lett. **43**, 2012 (1979)

3.106 D.C. Hurley, S. Tamura, J.P. Wolfe, H. Morkoc: Phys. Rev. Lett. **58**, 2446 (1987); S. Tamura, J.P. Wolfe: Phys. Rev. **B35**, 2528 (1987)

3.107 P.V. Santos, J. Mebert, O. Koblinger, L. Ley: Phys. Rev. **B36**, 1306 (1987)

3.108 C. Colvard, R. Merlin, M.V. Klein, A.C. Gossard: J. Phys. (Paris) **42–C6**, 631 (1981)

3.109 J. Sapriel, J.C. Michel, J.C. Toledano, R. Vacher: J. Phys. (Paris) **45–C5**, 139 (1984)

3.110 B. Jusserand, D. Paquet, A. Regreny, J. Kervarec: Solid State Commun. **48**, 499 (1983)

3.111 M. Nakayama, K. Kubota, T. Kanata, H. Kato, S. Chika, N. Sano: Jap. J. Appl. Phys. **24**, 1331 (1985)

3.112 M. Nakayama, K. Kubota, H. Kato, S. Chika, N. Sano: Solid State Commun. **53**, 493 (1985)

3.113 B. Jusserand, D. Paquet, F. Alexandre, A. Regreny: Surface Sci. **174**, 94 (1986)

3.114 R. Azoulay, B. Jusserand, G. Le Roux, P. Ossard, L. Dugrand: J. Cryst. Growth **77**, 546 (1986)

3.115 B. Jusserand, D. Paquet, F. Mollot, F. Alexandre, G. Le Roux: Superlattices and Microstructures **2**, 465 (1986)

3.116 J. Sapriel, J. He, J. Chavignon, G. Le Roux, J. Burgeat, F. Alexandre, R. Azoulay, R. Vacher: In *Proc. 18th Int. Conf. on the Physics of Semiconductors*, ed. by O. Engström (World Scientific, Singapore 1987) p. 723

3.117 T.A. Gant, M.V. Klein, T. Henderson, H. Morkoc: In *Proc. 10th Int. Conf. Raman Spectroscopy*, ed. by W.L. Peticolas, B. Hudson (University of Oregon, Eugene, Oregon 1986) p. 9–2

3.118 J. Sapriel, J. Chavignon, F. Alexandre, R. Azoulay: Phys. Rev. **B34**, 7118 (1986)

3.119 D. Levi, S.L. Zhang, M.V. Klein, J. Kem, H. Morkoc: Phys. Rev. **B36**, 8032 (1987)

3.120 B. Jusserand, P. Voisin, M. Voos, L.L. Chang, E.E. Mendez, L. Esaki: Appl. Phys. Lett. **46**, 678 (1985)

3.121 G.P. Schwartz, G.J. Gualtieri, W.A. Sunder, L.A. Farrow, D.E. Aspnes, A.A. Studna: J. Vac. Sci. Technol. A**5**, 1500 (1986)

3.122 G.P. Schwartz, G.J. Gualtieri, W.A. Sunder, L.A. Farrow: Phys. Rev. **B36**, 4868 (1987)

3.123 P.V. Santos, A.K. Sood, M. Cardona, K. Ploog, Y. Ohmori, M. Okamoto: Phys. Rev. **B37**, 6381 (1988)

3.124 B. Jusserand: Unpublished

3.125 D.J. Lockwood, M.W.C. Dharma-wardana, W.T. Moore, R.L.S. Devine: Appl. Phys. Lett. **51**, 361 (1987)

3.126 S. Venugopalan, L.A. Kolodziejski, R.L. Gunshor, A.K. Ramdas: Appl. Phys. Lett. **45**, 974 (1984)

3.127 E.K. Suh, D.U. Bartholomew, A.K. Ramdas, S. Rodríguez, S. Venugopalan, L.A. Kolodziejski, R.L. Gunshor: Phys. Rev. **B36**, 4316 (1987)

3.128 H. Brugger, G. Abstreiter, H. Jorke, H.J. Herzog, E. Kasper: Phys. Rev. **B33**, 5928 (1986)

3.129 H. Brugger, H. Reiner, G. Abstreiter, H. Jorke, H.J. Herzog, E. Kasper: Superlattices and Microstructures **2**, 451 (1986)

3.130 M.W.C. Dharma-wardana, D.J. Lockwood, J.M. Baribeau, D.C. Houghton: Phys. Rev. **B34**, 3034 (1986)

3.131 D.J. Lockwood, M.W.C. Dharma-wardana, J.M. Baribeau, D.C. Houghton: Phys. Rev. **B35**, 2243 (1987)

3.132   P.V. Santos, M. Hundhausen, L. Ley: J. Non-Cryst. Solids **77–78**, 1069 (1985)
3.133   P.V. Santos, M. Hundhausen, L. Ley: Phys. Rev. B**33**, 1516 (1985)
3.134   P.V. Santos, L. Ley: Phys. Rev. B**36**, 3325 (1987)
3.135   B. Jusserand, F. Mollot, M.C. Joncour, B. Etienne: J. Phys. (Paris), **48–C5**, 577 (1987)
3.136   R. Merlin, K. Bajema, R. Clarke, F.Y. Juang, P.K. Bhattacharya: Phys. Rev. Lett. **55**, 1768 (1985)
3.137   M.W.C. Dharma-wardana, A.H. MacDonald, D.J. Lockwood, J.M. Baribeau, D.C. Houghton: Phys. Rev. Lett. **58**, 1761 (1987)
3.138   R. Merlin, K. Bajema, J. Nagle, K. Ploog: J. Phys. (Paris), **48–C5**, 503 (1987)
3.139   A. Magerl, H. Zabel: Phys. Rev. Lett. **46**, 444 (1981); see also: M.S. Dresselhaus, G. Dresselhaus: In *Light Scattering in Solids III*, ed. by M. Cardona, G. Güntherodt, Topics Appl. Phys., Vol. 51 (Springer, Berlin, Heidelberg 1982) p. 3
3.140   J. Sapriel, J.C. Michel, J. Chavignon, R. Vacher, A. Langford, F. Alexandre: In *Proc. Second Int. Conf. Phonon Physics*, ed. by J. Kollar, N. Kroo, N. Menyhard, T. Siklos (World Scientific, Singapore 1985) p. 526
3.140a For recent values of elastic constants and sound velocities of AlAb see Landolt-Börnstein Tables, Volume 22, ed. by O. Madelung and H. Schulz (Springer, Heidelberg 1987) p. 65
3.141   B. Jusserand: Ph. D. Thesis, Université Paris VI (1987)
3.142   A. Segmuller, A.E. Blakeslee: J. Appl. Cryst. **6**, 19 (1973)
3.143   See for instance: B. Jusserand, J. Sapriel: Phys. Rev. B**24**, 7194 (1981)
3.144   H. Brugger, G. Abstreiter: J. Phys. (Paris), **48–C5**, 661 (1988)
3.145   J. Kervarec, M. Baudet, J. Caulet, P. Auvray, J.Y. Emery, A. Regreny: J. Appl. Cryst. **17**, 196 (1984)
3.146   M. Quillec, L. Goldstein, G. Le Roux, J. Burgeat, J. Primot: Appl. Phys. Lett. **55**, 2904 (1984)
3.147   K. Kubota, M. Nakayama, H. Katoh, N. Sano: Solid State Commun. **49**, 157 (1984)
3.148   A. Ishibashi, Y. Mori, M. Itabashi, N. Watanabe: J. Appl. Phys. **58**, 2691 (1985)
3.149   Z.P. Wang, D.S. Jiang, K. Ploog: Solid State Commun. **65**, 661 (1988)
3.150   J. Menéndez, A. Pinczuk, J.P. Valladares, R.D. Feldman, R.F. Austin: Appl. Phys. Lett. **50**, 1101 (1987)
3.151   D. Gammon, R. Merlin, H. Morkoc: Phys. Rev. B**35**, 2552 (1987)
3.152   G. Ambrazevicius, M. Cardona, R. Merlin, K. Ploog: Solid State Commun. **65**, 1035 (1988)
3.153   A.K. Arora, A.K. Ramdas, M.R. Melloch, N. Otsuka: Phys. Rev. B**36**, 1021 (1987); A.K. Aora, E.-K. Suh, A.K. Ramdas, F.A. Chambers, A.L. Moretti: Phys. Rev. B**36**, 6142 (1987)
3.154   B. Jusserand, D. Paquet, F. Alexandre, M.W. Charane, A. Regreny: In *Proc. Second Int. Conf. Phonon Physics*, ed. by J. Kollar, N. Kroo, N. Menyhard, T. Siklos (World Scientific, Singapore 1985) p. 521
3.155   See for instance: I.F. Chang, S.S. Mitra: Adv. Phys. **20**, 359 (1971); L. Genzel, T.P. Martin, C.H. Perry: Phys. Stat. Solid. B**62**, 83 (1974)
3.156   R. Bonneville: Phys. Rev. B**24**, 1987 (1981) and references therein
3.157   P. Soven: Phys. Rev. **178**, 1136 (1969)
3.158   B. Jusserand, D. Paquet, K. Kunc: In *Proc. 17th Int. Conf. Physics of Semiconductors*, ed. by J.D. Chadi, W.A. Harrison (Springer, Berlin, Heidelberg, New York 1985) p. 1165
3.159   J.W. Worlock: In *Proc. Second Int. Conf. on Phonon Physics*, ed. by J. Kollar, N. Kroo, N. Menyhard, T. Siklos (World Scientific, Singapore 1985) p. 506
3.160   J.L.T. Waugh, G. Dolling: Phys. Rev. **132**, 2410 (1963)
3.161   R.M. Fleming, D.B. McWhan, A.C. Gossard, W. Wiegmann, R.A. Logan: J. Appl. Phys. **51**, 357 (1980)
3.162   Shu-Lin Zhang, D.H. Levi, T.A. Gant, M.V. Klein, J. Klem, H. Morkoc: In *Proc. 10th Int. Conf. Raman Spectroscopy*, ed. by W.L. Peticolas, B. Hudson (University of Oregon, Eugene, Oregon 1986) p. 9–4

3.163 M. Nakayama, K. Kubota, H. Kato, N. Sano: Solid State Commun. **51**, 343 (1984)
3.164 G. Abstreiter, H. Brugger, T. Wolf, H. Jorke, H.J. Herzog: Phys. Rev. Lett. **54**, 2441 (1985)
3.165 F. Cerdeira, A. Pinczuk, J.C. Bean, B. Batlogg, B.A. Wilson: Appl. Phys. Lett. **45**, 1138 (1984)
3.166 S. Nakashima, Y. Nakakura, H. Fujiyasu, K. Mochizuki: Appl. Phys. Lett. **48**, 236 (1986)
3.167 Le Hong Shon, K. Inoue, K. Murase, H. Fujiyasu, Y. Yamazaki: Solid State Commun. **62**, 621 (1987)
3.168 D.J. Olego: Appl. Phys. Lett. **51**, 1422 (1987); J. Vac. Sci. Technol. B**6**, 1193 (1988)
3.169 B.A. Weinstein, R. Zallen: In *Light Scattering in Solids IV*, ed. by M. Cardona, G. Güntherodt, Topics Appl. Phys., Vol. 54 (Springer, Berlin, Heidelberg 1984) p. 463
3.170 P. Wickboldt, E. Anastassakis, R. Sauer, M. Cardona: Phys. Rev. B**35**, 1362 (1987)
3.171 E. Anastassakis, A. Pinczuk, E. Burstein, M. Cardona, F.H. Pollak: Solid State Commun. **8**, 133 (1970)
3.172 E. Anastassakis, Y.S. Raptis, M. Hünermann, W. Richter, M. Cardona: Phys. Rev. B**38**, 7702 (1988)
3.173 E. Anastassakis, M. Cardona: Solid State Commun. **63**, 897 (1987)
3.174 Y.J. Yang, K.Y. Hsieh, R.M. Kolbas: Appl. Phys. Lett. **51**, 215 (1987)
3.175 S.P.J. Brueck, B.Y. Tsaur, J.C.C. Fan, D.V. Murphy, T.F. Deutsch, D.J. Silversmith: Appl. Phys. Lett. **40**, 895 (1982)
3.176 J. Menéndez, M. Cardona: Phys. Rev. B**29**, 2051 (1984)
3.177 H. Kamimura, T. Nakayama: *Comments on Condensed Matter Physics* (in press)
3.178 L. Brey, C. Tejedor: Phys. Rev. Lett. **59**, 1022 (1987)
3.179 C. Mailhiot, D.L. Smith: Phys. Rev. B**35**, 1242 (1987)
3.180 N. Hamada, S. Ohnishi: Superlattices and Microstructures **3**, 301 (1987)
3.181 T. Ando, A.B. Fowler, F. Stern: Rev. Mod. Phys. **54**, 437 (1982)
3.182 see, for instance: A.K. Sood, W. Kauschke, J. Menéndez, M. Cardona: Phys. Rev. B**35**, 2886 (1987);
       C. Trallero-Giner, A. Alexandrou, M. Cardona: Phys. Rev. B**38**, 10744 (1988)
3.183 A.K. Sood, G. Contreras, M. Cardona: Phys. Rev. B**31**, 3760 (1985)
3.184 M.I. Alonso, M. Cardona: Phys. Rev. B**37**, 10107 (1988)
3.185 L. Brey, N.E. Christensen, M. Cardona: Phys. Rev. B**36**, 2638 (1987)
3.186 R.M. Martin: Phys. Rev. B**3**, 676 (1971)
3.187 W. Kauschke, M. Cardona: Phys. Rev. B**33**, 5473 (1986)
3.188 W. Kauschke, M. Cardona: Phys. Rev. B**35**, 9619 (1987)
3.189 C. Tejedor, J.M. Calleja, F. Meseguer, E.E. Méndez, C.A. Chang, L. Esaki: Phys. Rev. B**32**, 5303 (1985)
3.190 R. Zeyher, C.S. Ting, J.L. Birman: Phys. Rev. B**10**, 1725 (1974)
3.191 W. Richter, R. Zeyher, M. Cardona: Phys. Rev. B**18**, 4312 (1978)
3.192 A.A. Gogolin, E.I. Rashba: Solid State Commun. **19**, 1177 (1976)
3.193 J. Menéndez, M. Cardona: Phys. Rev. Lett. **51**, 1297 (1983); Phys. Rev. B**31**, 3693 (1985)
3.194 W. Kauschke, N. Mestres, M. Cardona: Phys. Rev. B**36**, 7469 (1987)
3.195 R. Dingle, W. Wiegmann, C.H. Henry: Phys. Rev. Lett. **33**, 827 (1974)
3.196 W. Stolz, J.C. Mann, M. Altarelli, L. Tapfer, K. Ploog: Phys. Rev. B**36**, 4301 (1987)
3.197 J.E. Zucker, A. Pinczuk, D.S. Chemla, A. Gossard, W. Wiegmann: Phys. Rev. Lett. **51**, 1293 (1983);
       J.E. Zucker, A. Pinczuk, D.S. Chemla: Phys. Rev. B**38**, 4287 (1988)
3.198 W. Kauschke, A.K. Sood, M. Cardona, K. Ploog: Phys. Rev. B**36**, 1612 (1987)
3.199 J.E. Zucker, A. Pinczuk, D.S. Chemla, A. Gossard, W. Wiegmann: Phys. Rev. B**29**, 7065 (1984)
3.200 D.A. Kleinman, R.C. Miller, A.C. Gossard: Phys. Rev. B**35**, 664 (1987)
3.200a A. Alexandrou, M. Cardona, K. Ploog: Phys. Rev. B**38**, 2196 (1988)
3.201 S.K. Chang, H. Nakata, A.V. Nurmikko, R.L. Gunshar, L.A. Kolodziejski: Appl. Phys. Lett. **51**, 667 (1987)

3.202   J.E. Zucker, A. Pinczuk, D.S. Chemla, A.C. Gossard: Phys. Rev. B35, 2892 (1987)
3.203   T. Suemoto, G. Fasol, K. Ploog: Phys. Rev. B34, 6034 (1986)
3.204   A. Ishibashi, M. Itabashi, Y. Mori, N. Watanabe: Opto-Electron. Devices and Technologies 1, 51 (1986)
3.205   M. Cardona, T. Suemoto, N.E. Christensen, T. Isu, K. Ploog: Phys. Rev. B36, 5906 (1987)
3.206   N. Kobayashi, T. Toriyama, Y. Horikoshi: Appl. Phys. Lett. 50, 1811 (1987)
3.206a T.A. Gant, M. Delaney, H.V. Klein, R. Houdré, H. Morkoç: Phys. Rev., in press
3.207   R.C. Miller, D.A. Kleinman, A.C. Gossard: Solid State Commun. 60, 213 (1986)
3.208   R.C. Miller, D.A. Kleinman, S.K. Sputz: Phys. Rev. B34, 7444 (1986)
3.209   F. Cerdeira, E. Anastassakis, W. Kauschke, M. Cardona: Phys. Rev. Lett. 57, 3209 (1986)
3.210   A. Alexandrou, M. Cardona: Solid State Commun. 64, 1029 (1987)
3.211   A. Madhukar, P.D. Lao, W.C. Tang, M. Adam, F. Voillot: Phys. Rev. Lett. 59, 1313 (1987)
3.212   S. Adachi, C. Hamaguchi: Phys. Rev. B19, 938 (1979)
3.213   E. Käräjämäki, R. Laiko, T. Levola, B.H. Bairamov, A.V. Gol'tsev, T. Toporov: Phys. Rev. B29, 4508 (1984)
3.214   B.H. Bairamov, A.V. Golt'sev, V.V. Toporov, R. Laiho, T. Lavola: Phys. Rev. B33, 5875 (1986)
3.215   M. Cardona: In *Atomic Structure and Properties of Solids*, ed. by E. Burstein (Academic, New York 1972) p. 514
3.216   see, for instance, M. Cardona: *Modulation Spectroscopy* (Academic, New York 1969)
3.217   J.E. Wells, P. Handler: Phys. Rev. B3, 1319 (1971). The data of Fig. 4 in this reference are piezo-optical constants. They can be transformed into photoelastic ones by multiplying by elastic shiftness constants
3.218   R.T. Collins, K. v. Klitzing, K. Ploog: Phys. Rev. B33, 4378 (1986)
3.219   L. Viña, R.T. Collins, E.E. Méndez, W.I. Wang: Phys. Rev. B33, 5939 (1986)
3.220   C. Tejedor, A. Hernandez-Cabrera: Phys. Rev. B33, 7389 (1986)
3.221   C. Tejedor, J.M. Calleja, L. Brey, L. Viña, E.E. Méndez, W.I. Wang, M. Staines, M. Cardona: Phys. Rev. B36, 6054 (1987)
3.222   A.K. Sood, J. Menéndez, M. Cardona, K. Ploog: Phys. Rev. B32, 1412 (1985)
3.223   M.H. Meynadier, E. Finkman, M.D. Sturge, J.M. Worlock, M.C. Tamargo: Phys. Rev. B35, 2517 (1987)
3.224   V.V. Gridin, R. Besserman, K.P. Jain, M.V. Klein, H. Morkoc: Superlattices and Microstructures 3, 107 (1987)

4. Spectroscopy of Free Carrier Excitations in Semiconductor Quantum Wells

Aron Pinczuk and Gerhard Abstreiter

With 49 Figures

This chapter reviews inelastic light scattering studies of free carriers in semiconductor heterojunctions, multiple quantum well heterostructures, heterojunction superlattices, and in quantum wells of metal-insulator-semiconductor structures. These materials are fabricated by epitaxial growth techniques, such as molecular beam epitaxy, that allow deposition of ultra thin layers with control of composition and doping. These finely layered artificial structures, smooth on an atomic scale, have remarkable new properties. They arise from quantum confinement effects on the electron motion when the thickness of the layers or the range of quantum well potentials are smaller than the de Broglie wavelength. In most semiconductors this occurs for characteristic lengths smaller than ~ 1000 Å. Free carriers in these semiconductors are created by special doping techniques, by an externally applied voltage, or by photoexcitation. Under conditions of quantum confinement for electron motion along the direction normal to the layer, the free carriers have properties in common with those of an ideal 2D electron gas [4.1]. The great current interest in electron systems of reduced dimensionality in semiconductors is stimulated by discoveries of new physical phenomena and novel device applications. The papers in the Proceedings of the last two Conferences on the Electronic Properties of Two Dimensional Systems and a recent Special Issue of the IEEE Journal of Quantum Electronics provide an overview and references of recent research on electron systems of reduced dimensionality [4.2–4].

Inelastic light scattering is a powerful method to study the excitations of free carriers in semiconductors [4.5–7]. In recent years the method has been widely applied to studies of electrons confined in semiconductor quantum wells and superlattices. These applications are based on investigations of resonant inelastic light scattering by the electronic excitations in bulk n-GaAs [4.8–9]. The intensities measured in these studies indicated that the method is sufficiently sensitive to observe the excitations of free carriers with densities $n \lesssim 10^{11}$ cm^{-2}. This feature of resonant inelastic light scattering was highlighted in a proposal presented at the 14th International Conference on the Physics of Semiconductors [4.10]. The proposal was followed by the first observations of resonant light scattering by high mobility quasi-2D electron systems in modulation doped GaAs-(AlGa)As single heterojunctions [4.11] and multiple quantum wells [4.12]. These reports stimulated extensive light scattering research of quasi-2D electron systems that was reviewed in several publications [4.7, 13–15].

One of our aims is to convince the reader that inelastic light scattering is a versatile experimental tool to study 2 D electron systems and that the method has a broad impact in studies of electron systems of reduced dimensionality in semiconductors. We consider the basic conceptual framework required to interpret the experiments and focus the discussion on the more recent results. In Sect. 4.1 we present the theoretical background and a discussion of kinematics and selection rules. Section 4.2 considers the spectroscopy of 2 D electron gases. It reviews measurements of in-plane excitations with and without magnetic field. Section 4.2 also presents a discussion of research of inter-subband excitations in different quantum wells, including the cases of photoexcited systems and of space-charge induced potential wells. Section 4.3 deals with 2 D hole gases in Si and in GaAs. Hole gases require special consideration because of the degeneracy and non-parabolicity of the valence bands. Section 4.4 considers shallow impurities in quantum wells. The article is closed by concluding remarks in which we consider the impact of the light scattering method.

4.1 Mechanisms, Selection Rules, and Kinematics

The mechanisms and selection rules that apply to resonant inelastic light scattering by 2D electron systems in semiconductors have been considered in [4.10–16]. Within the framework of the effective mass approximation that gives an adequate description of electron states in semiconductors of reduced dimensionality [4.1, 4], the light scattering mechanisms are similar to those of 3 D systems [4.5–7]. The large resonant enhancements required for sufficient sensitivity are predicted for photon energies near the optical transitions of the free carriers. Examples of such optical transitions are discussed in Sects. 4.2 and 4.3 (see Figs. 4.5, 15, 17, 29, and 39). Much of the research reviewed in this chapter considers quantum wells in GaAs. We show in Fig. 4.1 the electron energy band structure of GaAs and several of the direct optical energy gaps. The free electrons occupy states at the Γ_6 conduction band minimum. For these carriers and shallow donor levels, the relevant optical resonances occur at the fundamental E_0 and spin-orbit split-off $E_0 + \Delta_0$ gaps. Their energies are $E_0 \simeq 1.5$ eV and $E_0 + \Delta_0 \simeq 1.9$ eV. For holes in Γ_8 valence states and shallow acceptor levels the most important optical resonance is at the E_0 gap. Holes in Si quantum wells occupy similar valence states. In this case large resonant enhancements are expected at $E_0 \simeq 3.4$ eV.

Figure 4.2 shows examples of transitions of a degenerate quasi-2D electron system. Each confined state of the quantum well is associated with a *subband* in the two dimensional wavevector space for electron motion in the plane. Figure 4.2a shows a *vertical inter-subband* transition between the two lowest subbands. If the two subbands are parallel the transition energy is $E_{01} = E_1 - E_0$ and independent of the wavevector K. Figure 4.2b represents an *intra-subband* transition with an in-plane wavevector transfer q. Such excitations have energies

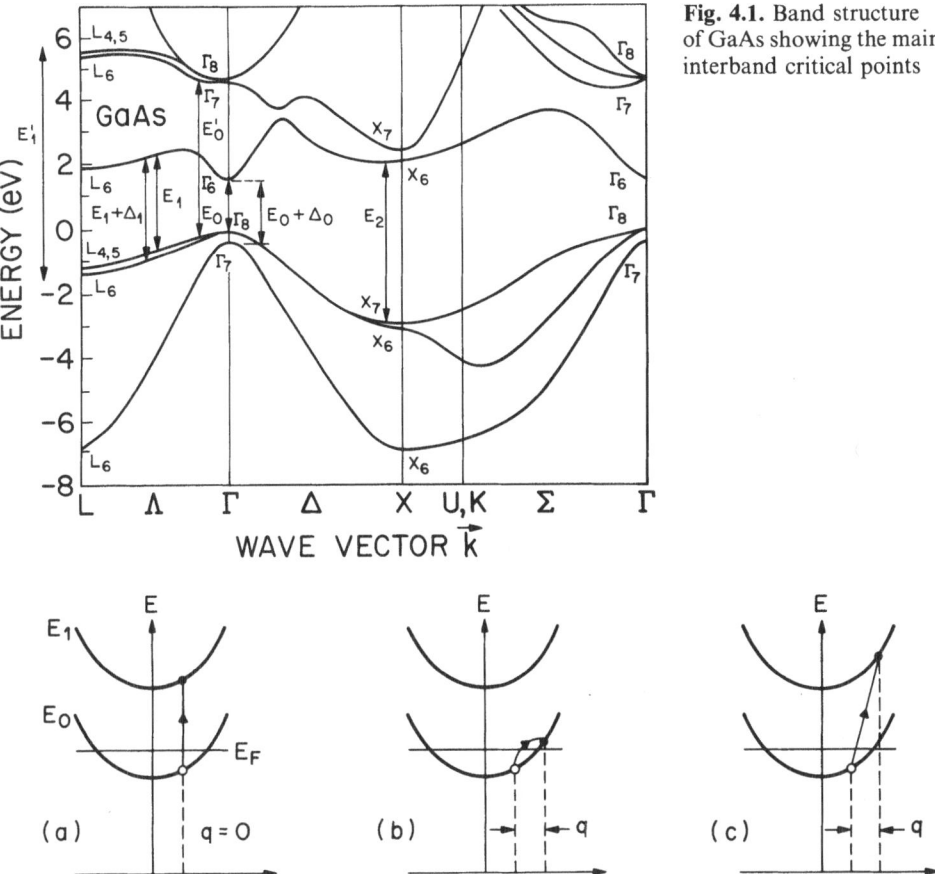

Fig. 4.1. Band structure of GaAs showing the main interband critical points

Fig. 4.2. Electronic excitations in quasi-2D systems (**a**) vertical inter-subband excitation; (**b**) intra-subband excitation; (**c**) non-vertical inter-subband excitation

$\hbar^2(K \cdot q + q^2/2)/m^*$, where m^* is the effective mass of the carries. At low temperatures there is a continuum of energies with a maximum at $K = k_F$, where k_F is the Fermi wavevector. Figure 4.2c shows a *non-vertical* inter-subband transition. These excitations have a continuum of energies centered at E_{01}. In a large magnetic field normal to the layer the kinetic energy of the in-plane motion is quantized into Landau levels. All the transitions are now discrete. They are inter-subband transitions, Landau level transitions, and combined transitions with simultaneous change of Landau levels and subbands.

Two types of light scattering spectra are measured. *Polarized* spectra are obtained with incident and scattered light polarizations parallel to each other. In *depolarized* spectra the two polarizations are orthogonal. Polarized spectra are interpreted as due to *charge-density excitations* and the depolarized spectra as arising from *spin-density excitations* [4.5–7, 10, 16–22]. These excitations appear as poles in charge-density and spin-density *response functions* [4.16–24]. The

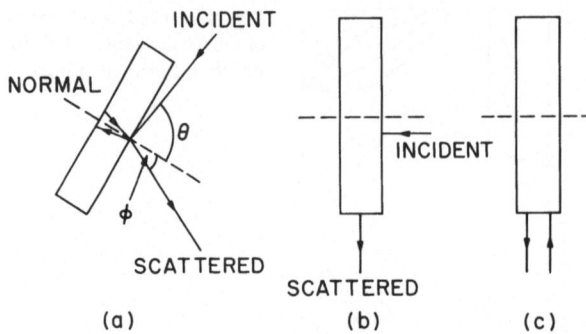

Fig. 4.3. Typical scattering geometries used in the investigations of electronic transitions

functions describe the response of the quasi-2D electron system to a spin-dependent external potential. The charge-density excitations are collective modes that are shifted to energies higher than those of the single-particle transitions shown in Fig. 4.2 by the effects of macroscopic electric fields. The spin-density excitations differ from the single-particle energies by many-body corrections (final state interactions) [4.18, 23, 24]. In the case of GaAs quantum wells in the absence of an external magnetic field the final state effects are relatively small [4.18, 25].

The capability to obtain spectra of single-particle (spin-density) and collective (charge-density) excitations has been a major factor in the impact of the inelastic light scattering method on spectroscopic studies of 2D electron systems. In GaAs quantum wells the selection rules for polarized and depolarized spectra have served as a basis for the interpretation of the results obtained with photon energies in resonance with the $E_0 + \Delta_0$ optical gaps. In spectra measured with photon energies near the fundamental E_0 gaps, the single-particle excitations may appear in both types of spectra [4.26–28].

The kinematics of inelastic light scattering depends on the geometry of the experiment. Figure 4.3 shows three geometries. For small energy shifts the moduli of k_L to k_S, the wavevectors of the incident and scattered light, are

$$|k_L| \simeq |k_S| = (2\pi/\lambda_L)\eta(\lambda_L) , \tag{4.1}$$

where λ_L is the laser wavelength and $\eta(\lambda_L)$ is the refractive index. The scattering wavevector is given by the vector difference of k_L and k_S. The nearly backscattering geometry of Fig. 4.3a is frequently used. Inside the sample light propagates along directions close to the normal to the plane of the 2D system. It is often convenient to set $\theta + \phi = 90°$. In this case the *in-plane* and *normal* components of the scattering wavevector are [4.29]

$$k = \frac{2\pi}{\lambda_L} (\cos\theta - \sin\theta) \quad \text{and} \tag{4.2a}$$

$$k_z = \frac{4\pi}{\lambda_L} \eta(\lambda_L)[1 - (1/2\eta)^2] , \tag{4.2b}$$

where z labels the direction along the normal to the plane. In the case of a more conventional backscattering geometry in which $\theta + \phi = 0°$ we have

$$k = (4\pi/\lambda_L) \sin \theta , \tag{4.3a}$$

$$k_z = (4\pi/\lambda_L) \eta(\lambda_L) [1 - \tfrac{1}{2}(k\lambda_L/4\pi\eta)^2] . \tag{4.3b}$$

In GaAs for the wavelengths of interest $\eta(\lambda_L) \simeq 3.6$. The wavevector component normal to the layers is within the range $5 \times 10^5 \text{ cm}^{-1} \lesssim k_z \lesssim 7 \times 10^5 \text{ cm}^{-1}$. The in-plane component k can be varied from a relatively small value, $\sim 10^4 \text{ cm}^{-1}$, up to a maximum of $\sim 2 \times 10^5 \text{ cm}^{-1}$. Larger values of k can be achieved with the geometries depicted in Figs. 4.3b and 4.3c. In these cases

$$k = \frac{2\pi}{\lambda_L} \eta(\lambda_L) \sim 2.5 \times 10^5 \text{ cm}^{-1} \quad \text{and} \tag{4.4}$$

$$k = \frac{4\pi}{\lambda_L} \eta(\lambda_L) \sim 5 \times 10^5 \text{ cm}^{-1} . \tag{4.5}$$

Translational symmetry in the plane of the 2D electron system requires that

$$k = q , \tag{4.6}$$

where q is the wavevector of the elementary excitation that appears in the light scattering spectrum. A rule of conservation of k_z could be applicable to *periodic multilayers*, like multiple quantum wells and superlattices. In this case there is a phase factor $\exp(iq_\beta ld)$ to the scattering amplitude from each layer (l is an integer that labels the layer and d is their spacing). The total scattering intensity is obtained by adding the contributions from all the layers [4.21].

$$I(q_\beta, k_z) \sim \left| \sum_{l=0}^{l=N-1} \exp[i(q_\beta - k_z)ld] \right|^2$$

$$= \frac{1 - \cos N(q_\beta - k_z)d}{1 - \cos(q_\beta - k_z)d} , \tag{4.7}$$

where N is the number of layers. For large N, $I(q_\beta, k_z)$ has a maximum at $q_\beta = k_z$.

Breakdown of the condition of wavevector conservation was found in electron systems with low carrier mobility [4.30]. The effect appears as a broadening of inter-subband transitions. Conversely, in multiple GaAs-(AlGa)As quantum wells of relatively high electron mobility ($\mu \gtrsim 5 \times 10^4$ cm^2/V s) conservation of wavevector has allowed the measurement of the dispersion of plasmons [4.29]. A surprising effect of breakdown of wavevector conservation has been reported in very high mobility ($\mu \gtrsim 2 \times 10^5$ cm^2/V s) quantum wells and heterojunctions under high magnetic field [4.31]. This effect is unexpected because in the absence of the magnetic field the spectra of collective modes indicate conservation of wavevector.

4.2 Two-Dimensional Electron Gases

4.2.1 In-Plane Excitations

In the absence of an external magnetic field, the in-plane excitations are derived from the transitions shown in Figs. 4.2b and 4.2c. The pure intra-subband excitations are the continuum of single-particle excitations and plasmons. At low temperatures the single-particle excitations of wavevector q have energies in the range

$$0 \leq E \leq \hbar^2 (qk_F + q^2/2)/m^* \ . \tag{4.8}$$

For a single 2D layer embedded in a medium with dielectric constant ε_0, the plasma oscillations have a frequency [4.1]

$$\omega_{2D}^2(q) = \frac{2\pi ne^2}{\varepsilon_0 m^*} q \ , \tag{4.9}$$

where n is the areal density. The plasma frequency of an *infinite* superlattice can be written as [4.32–34]

$$\omega_p^2(q, q_z) = \omega_{2D}^2(q) S(q, q_z) \ , \tag{4.10a}$$

$$S(q, q_z) = \frac{\sinh(qd)}{\cosh qd - \cos q_z d} \ , \tag{4.10b}$$

where q_z is the normal component of the mode wavevector and $S(q, q_z)$ is a structure factor of the superlattice with period d. For $qd < 1$ and $q_z d = 0$ the modes are quasi-3D: $\omega_p^2 = 4\pi ne^2/\varepsilon_0 m^* d$. For $qd < 1$ and $q_z d \neq 0$ there is an acoustic-like behavior given by

$$\omega_p(q, q_z) = v(q_z)q \quad \text{where} \tag{4.11}$$

$$v(q_z) = \left[\frac{2\pi ne^2 d}{\varepsilon_0 m^*} \frac{1}{1 - \cos q_z d} \right]^{1/2} \ . \tag{4.12}$$

An important property of these modes is the absence of Landau damping in GaAs, where $m^* = 0.07 m_0$ and $\varepsilon_0 = 13$.

In the presence of a high magnetic field normal to the layer, the pure in-plane excitations derive from electron transitions between Landau levels. The energies of the excitations are calculated from the poles in the charge-density and spin-density response functions [4.23, 24, 35]. These magnetic excitations of the 2D electron gas have been named *magnetic excitons* and *magnetoplasmons*. The magnetic excitons are the Landau level transitions shifted from the single-particle cyclotron energy by the final-state effects due to electron-electron interactions. The magnetoplasmons are the collective modes associated with Landau level transitions in which the electron-electron interactions produce final-state effects and a macroscopic electric field. As could be anticipated from Kohn's theorem [4.36], at $q=0$ the energies of the excitations are equal to the spacings between Landau levels. In the case of transitions between consecutive Landau levels the energy is $\hbar\omega_c$, where $\omega_c = eB/m^*c$ is the cyclotron frequency of the carriers.

For wavevectors $q \ll q_0 = 1/l_0$, where $l_0 = (\hbar c/eB)^{1/2}$ is the magnetic length, the magnetoplasmon frequencies are [4.23, 24, 33, 37]

$$\omega_p(q, B) = [\omega_p^2(q) + \omega_c^2]^{1/2} ,   \qquad (4.13)$$

where $\omega_p^2(q)$ is given either by (4.9) in the case of a single layer or by (4.10) in the case of a superlattice. At wavevectors $q \approx q_0$ the magnetoplasmon energies differ considerably from the prediction of (4.13) and the magnetic excitons can have energies lower than $\hbar\omega_c$ [4.23, 24, 35].

The most interesting range of wavevectors for studies of these excitations is in the vicinity of q_0. However, $q_0 \approx 10^6$ cm^{-1} for typical fields of $B \approx 6$ Tesla. Under conditions of wavevector conservation, this range is not accessible in light scattering and other optical experiments. Fortunately, breakdown of wavevector conservation in resonant inelastic light scattering experiments occurs even in the 2D electron gases of GaAs with the *highest* carrier mobilities ($\mu > 3 \times 10^5$ cm^2/V s) [4.31]. This surprising magnetic field dependence of the resonant light scattering cross sections of the 2D electron gas allows the direct observation of excitations of large in-plane wavevectors ($q > 10^6$ cm$^{-1} \approx q_0$).

Most of the inelastic light scattering studies of in-plane excitations have been carried out in modulation doped GaAs-(AlGa)As multiple quantum wells. Figure 4.4 shows the layer sequence in such structures. The undoped $(Al_xGa_{1-x})As$ and $(Al_{x'}Ga_{1-x'})As$ spacers with thicknesses d_3' separate the mobile electrons in the GaAs layers from the ionized donors and produce an enhancement of the electron mobility. The superlattice period is $d = d_1 + d_2$

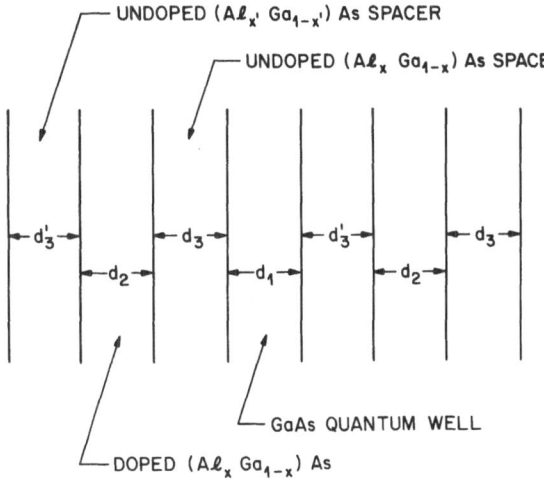

UNDOPED $(Al_{x'} Ga_{1-x'})$ As SPACER

UNDOPED $(Al_x Ga_{1-x})$ As SPACER

GaAs QUANTUM WELL

DOPED $(Al_x Ga_{1-x})$ As

Fig. 4.4. Sequence of layers in modulation-doped GaAs-(AlGa)As quantum wells

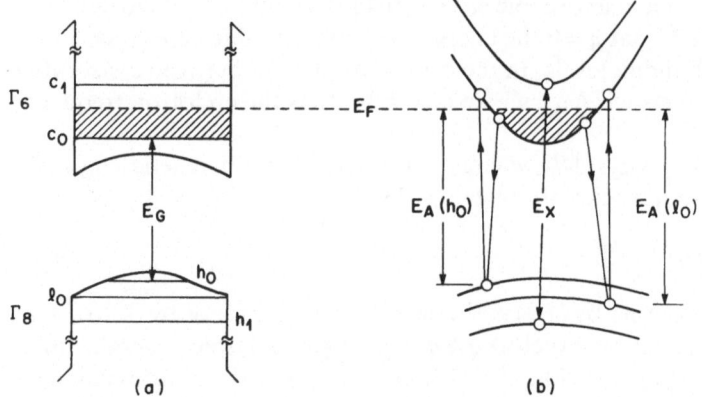

Fig. 4.5. Conduction and valence band energy diagram of a quantum well structure **(a)** real space; **(b)** k-space

$+d_3 + d_3'$. The quantum wells have mirror symmetry only when $x = x'$ and $d_3 = d_3'$.

The photon energies in these studies are in resonance with optical transitions near the fundamental optical gap. Figure 4.5a shows a symmetric quantum well and the states derived from the Γ_6 conduction band and the Γ_8 heavy and light valence bands of GaAs. The two lowest conduction band states are c_0 and c_1. The two higher heavy valence states are h_0 and h_1 and l_0 is the highest light valence state. Figure 4.5b describes these quantum well states as subbands in the 2D wavevector space. This figure also shows the optical transitions that enter in resonant inelastic light scattering processes by in-plane single-particle and collective excitations.

a) Results in the Absence of Magnetic Fields

i) Plasma Oscillations. The plasmons in n-type modulation doped multiple quantum well heterostructures were first measured in the light scattering experiments of *Olego* et al. [4.29]. Figure 4.6 shows the spectra obtained in a sample consisting of 20 periods in which the symmetric quantum well width is $d_1 = 262$ Å. The width of the central doped part of the $Al_{0.2}Ga_{0.8}As$ barrier is $d_2 = 317$ Å, and the undoped $Al_{0.2}Ga_{0.8}As$ spacers have widths $d_3 = d_3' = 163$ Å. The superlattice period is $d = d_1 + d_2 + 2d_3 = 905$ Å. The E_{01} and E_{01}' peaks shown in Fig. 4.6a are associated with single particle and collective excitations of the $c_0 \rightarrow c_1$ vertical inter-subband transitions. There is also a lower energy peak at 3.5 meV. Its remarkable behavior, displayed in Fig. 4.6b is the energy shift with the angle of incidence θ. This strong dependence on the in-plane scattering wavevector k [see (4.2)] led *Olego* et al. [4.29] to assign the low energy peak to the plasmon of the free carriers within the GaAs quantum wells. The quantitative analysis gave good agreement between measured plasmon energies and the predictions of (4.9–12).

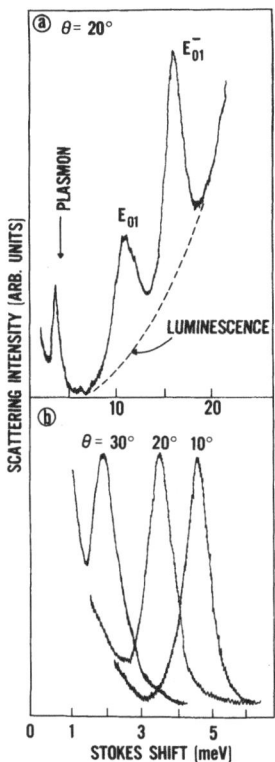

Fig. 4.6. (a) Nearly back scattering spectrum showing plasmons and inter-subband excitations in a modulation-doped multiple GaAs-(AlGa)As quantum well structure; (b) plasmon line for different angles of incidence [4.29]

The observation of the plasma oscillations and the determination of their dispersions demonstrated the novel applications of the inelastic light scattering method to studies of the in-plane motion of electrons confined in semiconductor quantum layers. The availability of this experimental tool stimulated further theoretical research of collective modes in semiconductor multilayers. *Giuliani* and *Quinn* [4.38] considered the surface plasmons of a semi-infinite superlattice. Unlike the 3 D surface plasmons, these modes are not subject to Landau damping in the long wavelength limit. This happens because quantization of the electron motion along the superlattice axis prevents the decay of these modes into electron-hole pairs. *Wu* et al. [4.39] have calculated the frequencies of surface plasmons on a lateral surface. *Jain* and *Allen* [4.40] have obtained the frequencies of plasmons in films with finite number of layers. They have shown the discrete character of the plasmon band (with respect to the possible choices of relative phases between the density fluctuations in different layers). In quasi-periodic semiconductor multilayers [4.41], plasma oscillations of unique properties have been predicted [4.42, 43]. Calculations in random systems have also been reported [4.44].

This theoretical research has stimulated further light scattering experiments. The discrete character of the plasmon band has been observed in samples with a relatively large number of 2D electron layers ($N = 14$) [4.45] and also in samples with few layers ($N=5$) [4.46]. Figure 4.7 shows spectra obtained from an *n*-type multiple symmetric quantum well sample with $N=14$ $d_3=d'_3=118$ Å, $x=x'=0.24$ and $d=778$ Å. The free electron density is $n=4\times10^{11}$ cm^{-2} in each layer and their mobility is $\mu=0.9\times10^5$ cm^2/V. The strong dependence of the peak positions on the in-plane components of the scattering wavevector identifies them as the plasmons from the multilayers. Unlike the spectra shown in Fig. 4.6, the modes appear as multiplets in which the strongest features (labeled S) are well-defined doublets. With other values of the normal component of the scattering wavevector k_z, obtained by changing the laser wavelength [see (4.2b)], it was possible to measure the dispersions of 7 of the 14 plasmon branches. The results or the measured mode dispersions are shown in Fig. 4.8.

In the interpretation of these results it is necessary to look into the behavior of plasmons in multilayers with large but finite N. We consider a *layered electron gas model*. In this model the multilayers are embedded in an infinite medium with dielectric constant ε_0 and the induced charge density fluctuation is written as [4.32–34]

$$\varrho(r,z,t)=\sum_{l=0}^{N-1}\varrho_l(r,t)\delta(z-z_l)\tag{4.14}$$

where r is the in-plane coordinate and

$$z_l=ld,\quad l=0....N-1\tag{4.15}$$

labels the positions of the N electron layers that represent the superlattice. Next we consider the Fourier transform

$$\varrho_l(r,t)=\frac{1}{(2\pi)^3}\int d^2q\,d\omega\,e^{i(q\cdot r-\omega t)}\varrho_l(q,\omega)\tag{4.16}$$

and use the equations of motion, Poisson's equation, and the equation of continuity to obtain [4.32]

$$\varrho_l(q,\omega)=\frac{\omega_{2D}^2}{\omega^2}\sum_{m=0}^{N-1}\varrho_m(q,\omega)\exp(-q|z_l-z_m|).\tag{4.17}$$

The solutions of (4.17) are evaluated with the Fourier expansion

$$\varrho_l(q,\omega)=\sum_\beta\exp(-iq_\beta z_l)\varrho_\beta(q,\omega),\quad\text{where}\tag{4.18}$$

$$q_\beta=\beta\frac{2\pi}{Nd},\quad\beta=0....,N-1,\tag{4.19}$$

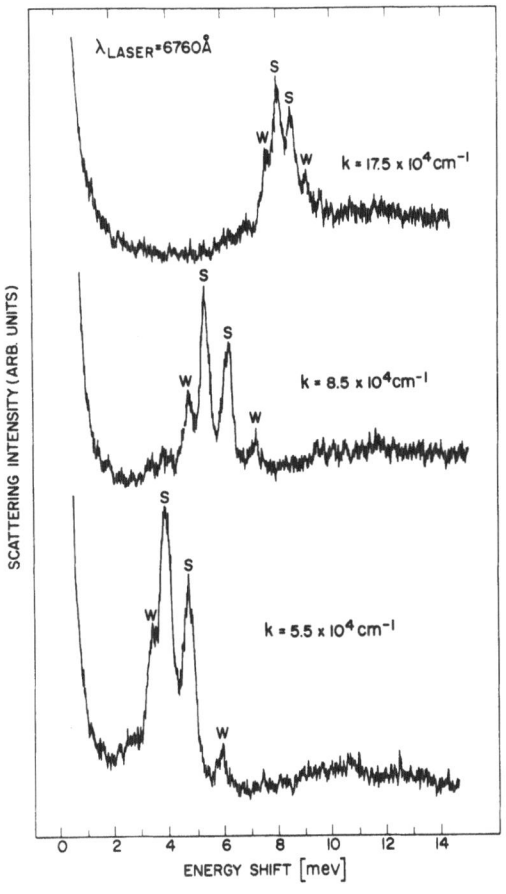

Fig. 4.7. Discrete plasmons at different values of the in-plane scattering wave vector. The spectra are excited with a laser wavelength $\lambda_L = 6760$ Å. The GaAs-(AlGa)As sample parameters are: $n = 4 \times 10^{11}$ cm^{-2}, $d = 778$ Å, $\mu = 90\,000$ cm^2/Vs. S and W identify strong and weak features [4.45]

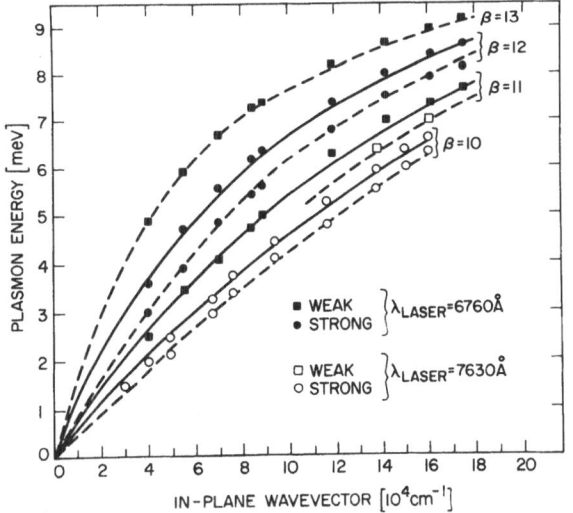

Fig. 4.8. Plasmon dispersion deduced from the spectra of Fig. 4.7 [4.45]

and the equation for ϱ_β is

$$\sum_\beta \exp(-iq_\beta z_l)\varrho_\beta = \frac{\omega_{2D}^2}{\omega^2}\sum_{\beta'}\exp(-iq_{\beta'}z_l)\varrho_{\beta'}F_{l\beta'}, \quad \text{where} \tag{4.20}$$

$$F_{l\beta'} = \sum_m \exp[-q|z_l-z_m|+iq_{\beta'}(z_l-z_m)]. \tag{4.21}$$

The solutions of (4.20) and (4.21) give the frequencies and the charge-density fluctuations of plasmons within the layered electron gas model. In the case $N\to\infty$, the sum in (4.21) is independent of l and given by [4.32]

$$F_{l\beta'} = F_{\beta'} = \frac{\sinh qd}{\cosh qd - \cos q_{\beta'}d} \tag{4.22}$$

leading to (4.10) for the plasmon frequencies of the infinite superlattice.

To find frequencies for finite N we consider the Fourier transformation [4.40]

$$\exp(-iq_{\beta'}z_l)F_{l\beta'} = \sum_\beta \exp(-iq_\beta z_l)F_{\beta\beta'} \quad \text{where} \tag{4.23}$$

$$F_{\beta'} = \frac{1}{N}\sum_l \exp(iq_\beta z_l)[F_{l\beta'}\exp(-iq_{\beta'}z_l)]. \tag{4.24}$$

Use of (4.23) in (4.20) leads to the following eigenvalue equation for the plasmon frequencies

$$\varrho_\beta = \frac{\omega_{2D}^2}{\omega^2}\sum_{\beta'}\varrho_{\beta'}F_{\beta\beta'}. \tag{4.25}$$

Jain and *Allen* [4.40] have derived an expression for $F_{\beta\beta'}$ that can be written as

$$F_{\beta\beta'} = S(q,q_\beta)\delta_{\beta\beta'} + f_{\beta\beta'} \quad \text{where} \tag{4.26}$$

$$f_{\beta\beta'} = \frac{1-\exp(-Nqd)}{2N}$$

$$\times\frac{1-\cosh(qd)[\exp(iq_\beta d)+\exp(-iq_{\beta'}d)]+\exp[i(q_\beta-q_{\beta'})d]}{[\cosh qd - \cos q_\beta d][\cosh qd - \cos q_{\beta'}d]}. \tag{4.27}$$

The eigenvalue equation can be written as

$$\sum_{\beta'}\left(\frac{\omega_{2D}^2(q)}{\omega^2-\omega_p^2(q,q_\beta)}f_{\beta\beta'}-\delta_{\beta\beta'}\right)\varrho_{\beta'}=0, \tag{4.28}$$

where $\omega_p^2(q,q_\beta)$ is given by (4.10) with $q_\beta=q_z$.

Equation (4.28) yields the frequencies of the N plasmon branches in terms of the $\omega_p(q,q_\beta)$, the frequencies of the infinite superlattice calculated at $q_z=q_\beta$. In a first-order approximation, the solutions of (4.28) are given by the splitting of the pair with q_β and $q_{N-\beta} = (2\pi/d) - q_\beta$ that is degenerate in the infinite superlattice [see (4.10) and (4.11)]. Within this approximation, the modes of the finite multilayers are *doublets* with oscillation amplitudes that have admixtures of wavevector components q_β and $(2\pi/d) - q_\beta$. The mode frequencies are given by

$$\omega_\pm^2(q, q_\beta) = \omega_p^2(q, q_\beta) + \omega_{2D}^2 [f_{\beta\beta} \pm (f_{\beta - \beta} f_{-\beta\beta})^{1/2}] , \qquad \text{where} \qquad (4.29)$$

$$f_{\beta\beta} = \frac{1 - \exp(-Nqd)}{N} \frac{1 - \cosh qd \cos q_\beta d}{[\cosh qd - \cos q_\beta d]^2} , \qquad (4.30)$$

$$f_{\beta - \beta} = \frac{1 - \exp(-Nqd)}{2N} \frac{1 - 2\cosh qd \exp(iq_\beta d) + \exp(izq_\beta d)}{[\cosh qd - \cos q_\beta d]^2} , \qquad \text{and} \quad (4.31)$$

$$f_{-\beta\beta} = \frac{1 - \exp(-Nqd)}{2N} \frac{1 - 2\cosh qd \exp(-iq_\beta d) + \exp(-izq_\beta d)}{[\cosh qd - \cos q_\beta d]^2} . \qquad (4.32)$$

The lifting of the degeneracy of the modes of the infinite superlattice is caused by the confinement of the charge-density fluctuations within the finite width $(N-1)d$. The oscillation amplitudes of the two modes can be written as

$$\varrho_- = A \cos [q_\beta(z_l - (N-1)d/2)] , \qquad (4.33a)$$

$$\varrho_+ = A \sin [q_\beta(z_l - (N-1)d/2] . \qquad (4.33b)$$

The lower energy mode is *symmetric* with respect to the mirror plane of the multilayers and the higher energy mode is *antisymmetric*.

The spectra shown in Fig. 4.7 are remarkable because the dominant structures consist of well-defined doublets. In [4.45] they have been interpreted quantitatively in terms of the plasmon pairs of (4.29–32). These results are displayed in Fig. 4.8. The seven plasmon branches shown in this figure are labeled by the β-values defined in (4.19). The dashed and full lines are the calculated plasmon dispersions. The fact that these doublets are the strongest features of the spectra indicates that for the relatively large $N = 14$ there is an approximate rule of conservation of the normal component of wavevector in accordance with (4.7). Figure 4.9 compares the approximate solutions of the amplitudes of the charge-density fluctuations given by (4.33) with the "exact" eigenvectors obtained by numerical evaluation of the solutions of (4.27–28) in the case of $N = 14$ and for a wavevector accessible in light scattering experiments [$\beta = 13$ in (4.19)]. These results show that the two oscillation amplitudes are very similar and represent further evidence that the "doublet" approximation defined by (4.29) is a good representation of the plasmons in the finite multilayers with relatively large value of N.

Fasol et al. [4.46a] measured light scattering spectra of discrete plasmons in the cases of smaller values of N. Figure 4.10 shows the spectra measured in a

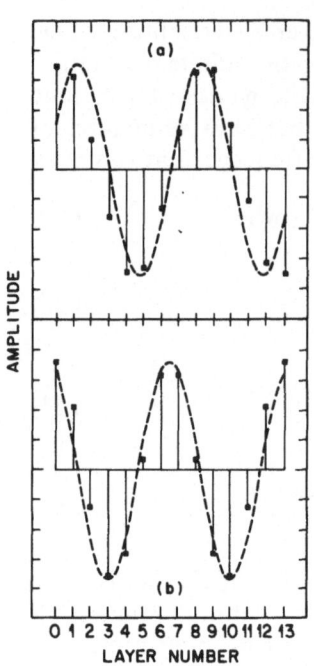

Fig. 4.9. Amplitude of charge-density fluctuations of plasmons with $N=14$ and $\beta=12$. The dots are the "exact" eigenvectors of (4.28) and the dashed lines are the approximate values given by (4.33)

AMPLITUDE

(a)

(b)

0 1 2 3 4 5 6 7 8 9 10 11 12 13
LAYER NUMBER

sample with $N=5$. These experiments were carried out in n-type modulation doped multiple GaAs-(AlGa)As heterostructures. The quantum wells are asymmetric, having $x' \neq x$ and $d_3 \neq d_3'$ (Fig. 4.4). As can be seen in Fig. 4.10a in the case of small N, almost all the five plasmon branches are simultaneously active in the light scattering spectra. The measured plasmon frequencies are in good agreement with the predicted value on the basis of the theory of *Jain* and *Allen* [4.40] [see also (4.27) and (4.28)]. Figure 4.10b shows that the depolarized spectra are dominated by a lower energy k-dependent scattering that has been assigned to spin-density single-particle excitations [4.46a].

Plasmon excitations with large q values are obtained with the scattering geometries shown in Fig. 4.3b and c. The right-angle scattering has been applied by *Sooryakumar* et al. [4.76] and is discussed in more detail in Sect. 4.2.2c. The maximum in-plane wave vector that can be achieved in this geometry is $q \lesssim 3 \times 10^5 \, \text{cm}^{-1}$. Even higher values are realized in back-scattering from cleavage planes with the wave vectors of incident and scattered light parallel to the layers (Fig. 4.3c). This geometry has been applied recently by *Egeler* et al. [4.46b] using a microscope Raman set up. The dispersion of multilayer plasmons has been measured up to $q \simeq 7 \times 10^5 \, \text{cm}^{-1}$. The observable mode has single-layer behavior in this limit. The realization of the scattering geometry shown in Fig. 4.3c also allows direct observation of in-plane and interface phonon excitations. First results have been obtained recently for the GaAs-(AlGa)As system [4.46b]. Such experiments are expected to provide deeper insight into in-plane excitations in multi quantum well structures.

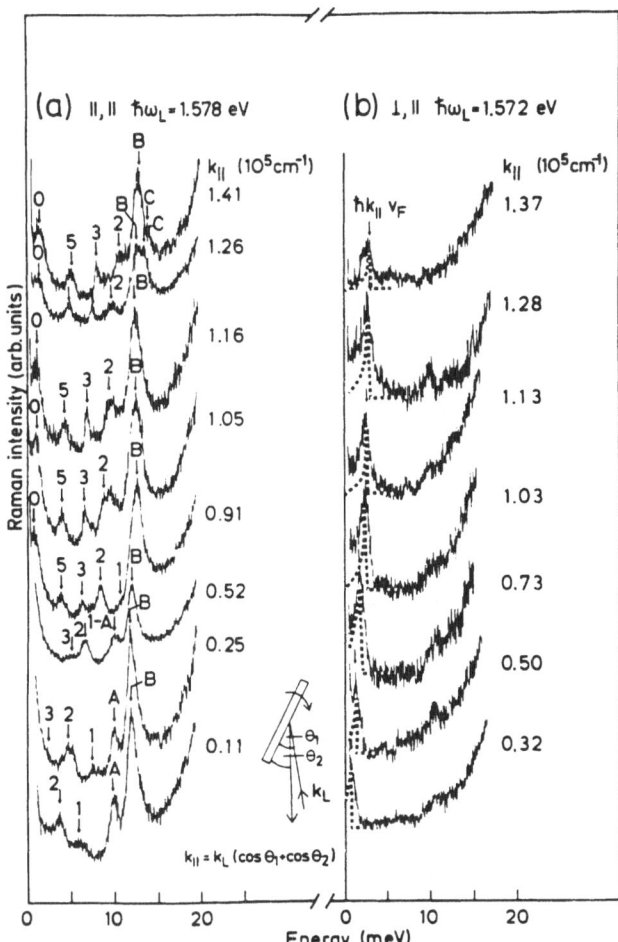

Fig. 4.10. Polarized and depolarized spectra of a five period GaAs-(AlGa)As multiple hetero-structure for different in-plane wave vectors [4.46a]

ii) Single Particle Excitations. In the absence of significant effects of electron-electron interactions the lineshapes of light scattering spectra of single particle excitations are expected to follow the density of states for particle-hole pair excitations at wavevector q and frequency ω. In the case of infinite electron relaxation time the lineshape function for the light scattering spectra can be written as [4.5–7]

$$I(q, \omega) \sim \text{Im} \{\chi(q, \omega)\} \tag{4.34}$$

where $\chi(q, \omega)$ is the electric susceptibility function of the electron gas. The susceptibility of an ideal 2D electron gas embedded in a uniform dielectric medium has been calculated by *Stern* [4.1, 47]. $\text{Im} \{\chi(q, \omega)\}$ has a peak at $\omega = q v_F$

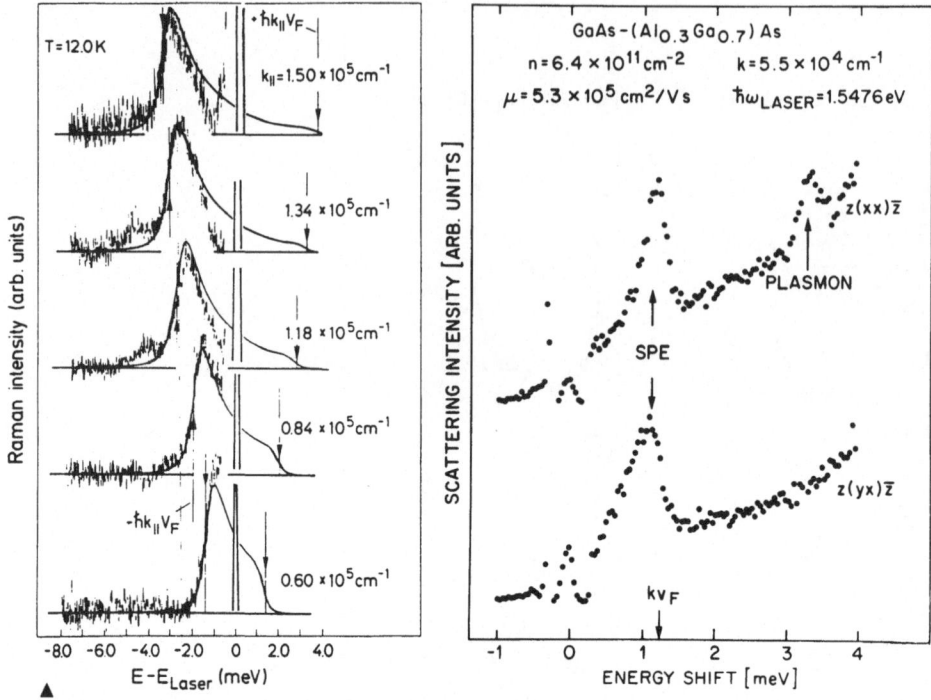

Fig. 4.11. Comparison of depolarized spectra with single-particle excitation theory [4.48]

Fig. 4.12. Polarized and depolarized spectra of symmetric modulation-doped multiple quantum wells in resonance with $l_0 \Rightarrow c_0$ optical transitions. The peaks of single-particle and plasmon excitations are shown [4.28]

(v_F is the Fermi velocity) and a cutoff at $\omega = qv_F + \hbar q^2/2m^*$. The calculated peak is considerably sharper than that of the Lindhard function of the 3D electron gas.

Inelastic light scattering by the in-plane single particle excitations were first reported by *Fasol* et al. [4.46a, 48]. These experiments were carried out in the same modulation doped asymmetric quantum well samples that were used for the measurements of discrete plasmon branches. Figure 4.11 shows measured spectra. These features are observed in depolarized spectra assigned to spin-density fluctuations [4.7]. Also shown in the figure are the lineshape fits from (4.34). In this analysis the finite electron relaxation time is included in an approximation due to *Mermin* [4.7, 49]. The value of the relaxation time is adjusted to yield a good fit to the measured spectra. The adjusted values of the relaxation times, however, are about three times larger than those obtained from mobility measurements.

Figure 4.12 shows typical inelastic light scattering spectra of in-plane excitations obtained with photon energies in resonance with the states of the fundamental gap [4.28]. These spectra are obtained in modulation doped symmetric GaAs quantum wells with incident photon energies in resonance with the transitions near $E_A(l_0)$ (Fig. 4.5). The polarized and depolarized spectra are labeled $z(xx)\bar{z}$ respectively, where $z = (001)$, $y = (010)$ and $x = (001)$ directions. In Fig. 4.12 we see that the bands of single-particle excitations (labeled SPE) appear in depolarized as well as polarized spectra. This is in contrast with the results of previous work described above. It can also be seen in Fig. 4.12 that the bands of the SPE are somewhat different in the two types of spectra. This result is surprising and introduces an element of uncertainty in any lineshape analysis of SPE spectra with electron relaxation times.

The new selection rules for the observation of spectra of single-particle excitations appear to be related to the strong resonant enhancements for photon energies very close to the fundamental optical gaps. The large resonant enhancements allow light scattering measurements with low incident power levels (less than 1 W/cm^2), which minimizes heating of the free carriers by the laser beam. The electron temperatures determined from Stokes-anti-Stokes intensity ratios are less than 3 K (see Fig. 4.12).

b) Results in High Magnetic Fields

The first observations of resonant inelastic light scattering by the 2 D electron gas in high magnetic fields were reported in 1981 [4.50, 51]. In these experiments the magnetic fields were along the normal to the plane and the incident photon energies were in resonance with the states of the $E_0 + \Delta_0$ optical gaps. The depolarized spectra show Landau level transitions at energy $\hbar\omega_c$ with an electron effective mass $m^* = 0.068 m_0$. Bands that appear at a somewhat higher energy have been tentatively assigned to magnetoplasmons. However, their energies could not be interpreted with (4.13)[4.52].

Pinczuk et al. reported new inelastic light scattering measurements in high magnetic fields [4.37]. These experiments were carried out in somewhat higher mobility modulation doped GaAs-(AlGa)As multiple quantum wells and with photon energies in resonance with the E_0 optical gaps. Figure 4.13 shows spectra measured in a sample consisting of 15 periods in which $d_1 = 270$ Å, $x' = x$, $d_3 = d_3'$ and $d = 980$ Å. The electron density is $n = 8.2 \times 10^{11}$ cm^{-2} and their mobility is $\mu = 1.3 \times 10^5$ cm^2/V s. In these measurements the magnetic fields are normal to the layers and the sample is immersed in superfluid ^{4}He. The spectra were excited with photon energies in the range 1.54–1.58 eV. They are close to the fundamental gap at ~ 1.51 eV. For $B = 0$ the depolarized spectrum labeled $z(y'x')\bar{z}$ shows the inter-subband transition at E_{01} that gives the spacing between the two lowest subbands. The polarized $z(x'x')\bar{z}$ spectrum shows two additional peaks labeled I_- and P that are assigned to a collective inter-subband excitation and to the plasmon of the multilayers. The spectra shown for $B \neq 0$ have two new bands. For $B \leq 4$ Tesla they are at $E_{01} - \hbar\omega_c$ and $\hbar\omega_c$ and for $B \geq 4$ Tesla, at $\hbar\omega_c - E_{01}$ and $\hbar\omega_c$.

The intensities in these spectra are strongly dpendent on magntic field and incident photon energy. With changes in field and photon energy it was also

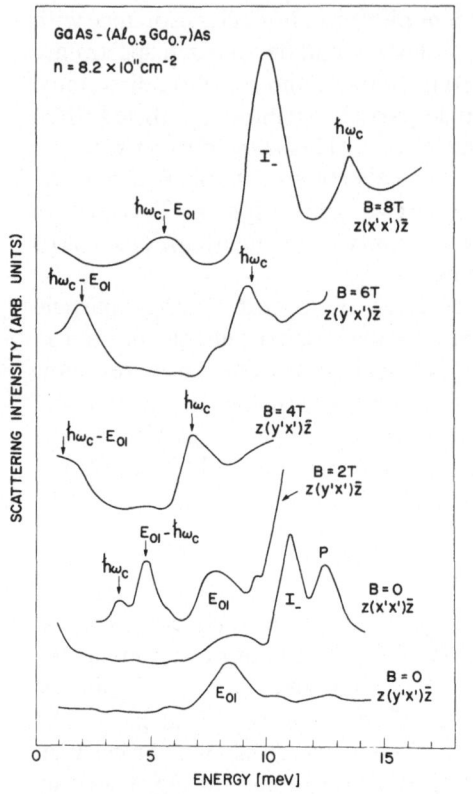

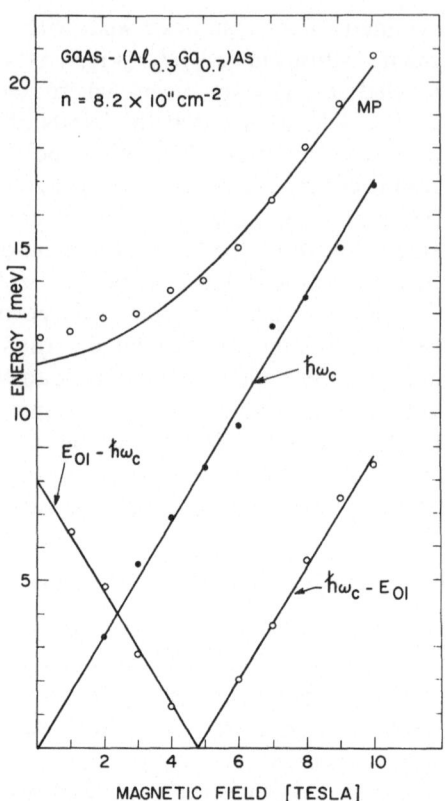

Fig. 4.13. Spectra of symmetric modulation-doped multiple quantum wells for different values of magnetic field B applied along the direction perpendicular to the wells [4.37]

Fig. 4.14. Energies of combined Landau level inter-subband transitions of electrons in the multiple quantum wells as function of magnetic field. MP is the magnetoplasma mode of the multilayers [4.37]

possible to identify another mode that was assigned to the magnetoplasmon of the multilayers [4.37]. Figure 4.14 shows the positions of the transitions at $E_{01} - \hbar\omega_c, \hbar\omega_c - E_{01}, \hbar\omega_c$ and the magnetoplasmon (labeled MP) as a function of the magnetic field. The lines that fit the $\hbar\omega_c$, $E_{01} - \hbar\omega_c$, and $\hbar\omega_c - E_{01}$ data are calculated with $m^* = 0.068 m_0$ and the measured value of $E_{01} = 8$ meV. The magnetoplasmon data were fitted with (4.13) and the 3D plasma frequency of the multilayers

$$\omega_p^2(0) = \frac{4\pi n e^2}{\varepsilon_0 m^* d} , \qquad (4.35)$$

where $\varepsilon_0 = 13$ for GaAs. This choice of plasma frequency was justified because $k_z d \simeq 6 \simeq 2\pi$. In this case, the condition of conservation of the normal component of wavevector (4.7) allows coupling to the quasi-3D mode with

$q_\beta = (2\pi/d) \equiv 0$, in which all the planes oscillate in phase. For $B > 4$ Tesla there is good agreement between measured and calculated energies. The discrepancies at lower field could arise from resonant coupling between the magnetoplasmon and the collective inter-subband excitation I_- [4.19, 53, 54].

Previously to the results shown in Figs. 4.13 and 4.14, combined transitions, in which the electrons simultaneously change Landau level and subband states, were measured in Si inversion layers by means of infrared optical absorption experiments [4.55]. In these experiments *tilted* magnetic fields cause coupling between the in-plane and normal to the plane electron motions that allow a simultaneous change in subband index and Landau level quantum number. However, in the light scattering experiments described above the combined transitions are observed with magnetic fields normal to the plane. The new selection rules that apply in the resonant inelastic light scattering experiments were attributed to the mixing between subband and in-plane motions in the valence states that participate in the optical transitions of the resonant inelastic light scattering processes.

Figure 4.15 shows the two virtual optical transitions in resonant inelastic light scattering at energies $\hbar\omega_c$, $E_{01} - \hbar\omega_c$, and $\hbar\omega_c - E_{01}$. The conduction state of subband index c_0 and Landau level quantum number l is represented by $|c_0, l\rangle$ and $|v, m\rangle$ represents the intermediate states derived from the Γ_8 light and heavy valence subbands, where v is the subband index (e.g., h_0, l_0, or h_1) and m is the Landau level quantum number. For $B = 0$ Figs. 4.15b and 4.15c describe Stokes and anti-Stokes processes of inelastic light scattering by inter-subband excitations. They exist because away from the Brillouin zone center there is extensive coupling and mixing between the valence subbands [4.56–60]. The mixing allows, *within the dipole approximation*, a change in conduction subband index in the processes shown in Figs. 4.15b and 4.15c. The observation of combined transitions for $B \neq 0$ can be interpreted by the additional valence band mixing introduced by the magnetic field. In effective-mass theories of the valence subbands in a normal magnetic field [4.56, 57, 59] the states $|v, m\rangle$ are admixtures

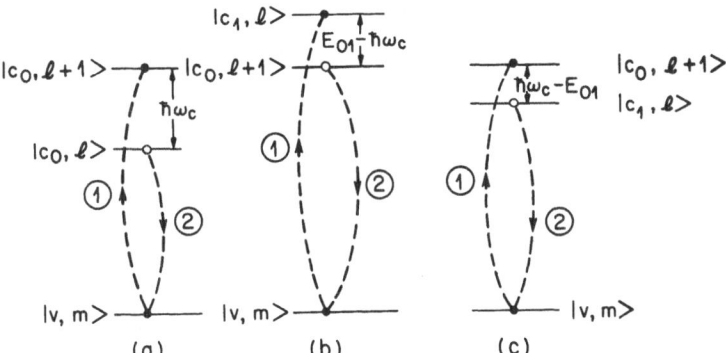

Fig. 4.15. Optical transitions in resonant inelastic light scattering by electrons in high magnetic fields. The numbers indicate the order of the transitions

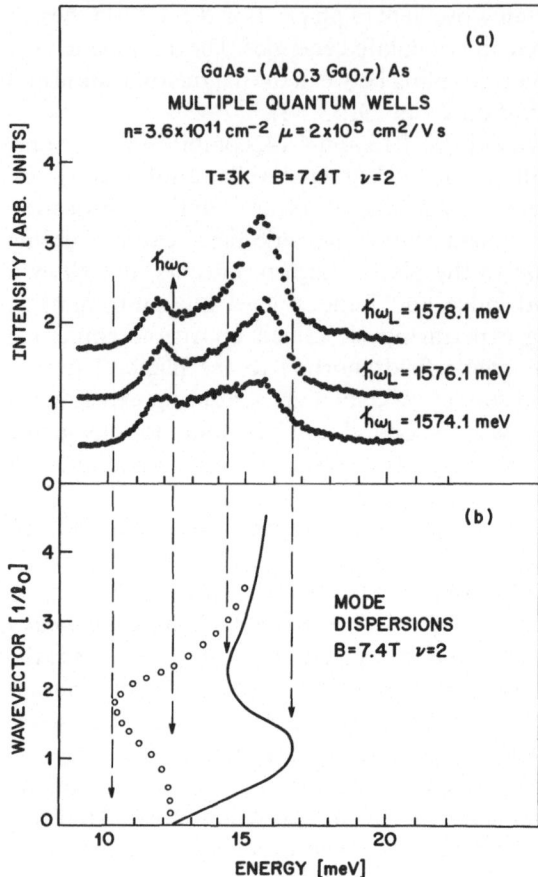

Fig. 4.16. (a) Resonant inelastic light scattering spectra of large wave vector magnetoplasmons of the electron gas in GaAs quantum wells. $\hbar\omega_L$ is the incident photon energy. (b) Calculated mode dispersions for $V = 2\pi l_0^2 n = 2$. The full line is the magnetoplasmon. The circles correspond to spin-density excitations [4.61]

of light and heavy subbands with harmonic oscillator wavefunctions having quantum numbers l' in the range $m - 1 \leq l' \leq m + 2$. These mixed character eigenfunctions allow resonant inelastic light scattering by new transitions including the excitations at energies $\hbar\omega_c$, $E_{01} - \hbar\omega_c$, and $\hbar\omega_c - E_{01}$.

Light scattering spectra by magnetoplasmons have been recently measured in higher mobility $(2 \times 10^5 \leq \mu \leq 5 \times 10^5 \text{ cm}^2/\text{V s})$ n-type GaAs-(AlGa)As multiple quantum wells [4.31]. Figure 4.16a shows spectra measured in a sample with symmetric quantum wells at a field of 7.4 Tesla and Landau level filling factor $v = 2$ (the Landau level filling factor is related to the density by the expression $v = n/2\pi l_0^2$). At this filling factor only the lowest Landau levels with the two values of spin are occupied by the electrons. The measurements were carried out with incident photon energies in resonance with higher quantum well states derived from the fundamental gap of GaAs. The spectral lineshape is a scattering continuum with an onset close to $\hbar\omega_c$ and a cutoff about 5 meV higher. The slight dependence of the lineshape on incident photon energy is due to the strong resonant enhancement of the cross section which is also dependent on incident and scattered photon energies.

The spectra of Fig. 4.16a have been measured in the conventional backscattering geometry of Fig. 4.3a with $\theta = 0$ and $\phi \lesssim 15°$. In this geometry, the in-plane component of the scattering wavevector is $k \lesssim 2 \times 10^4$ cm^{-1}. The superlattice consists of a large number of layers ($N = 50$) with $d = 750$ Å. This gives $kd \lesssim 0.15$ and $k_z d = 4.3$. Under conditions of wavevector conservation we expect coupling to magnetoplasmon modes at energies given by (4.10) and (4.13). Such modes have energies that are shifted from $\hbar\omega_c$ by less than 0.8 meV. The much larger shifts seen in Fig. 4.16c must be interpreted with breakdown of wavevector conservation as magnetoplasmons with $q \gtrsim 10^6$ cm^{-1}. At these large wavevectors, $qd > 1$, coupling between layers is negligible, as can be seen in (4.10b), and the magnetoplasmons of the superlattice are identical to those of a single layer. Spectra measured recently in single GaAs-(AlGa)As heterojunctions, similar to those in Fig. 4.16a, support this interpretation [4.61].

Comparison between the measured spectra with calculated magnetoplasmon energies gives further support to the interpretation of the spectra in Fig. 4.16 as large wavevector magnetoplasmons. In the absence of wavevector conservation, we expect structure at the energies of the peaks in the density of states. Figure 4.16b shows the calculated dispersions for the magnetoplasmon and the magnetic exciton. The critical points in the magnetoplasmon dispersion occur at wavevectors $q \sim q_0 = (1/l_0) \simeq 10^6$ cm^{-1} and $q \sim 2q_0$. There is good agreement between the measured energies and the calculated positions of the critical points in the magnetoplasmon dispersion.

Resonant light scattering spectra of large wavevector magnetoplasmons have also been measured in single GaAs-(AlGa)As heterojunctions with much larger electron mobility ($\mu \gtrsim 10^6$ cm^2/V s) [4.61]. The results are very similar to those shown in Fig. 4.16. Breakdown of wavevector conservation in resonant inelastic light scattering by the 2D electron gas in a high magnetic field allows the observation of modes that are not easily accessible to experimental examination. It is surprising that the strengths of spectra like those shown in Fig. 4.16a do not depend in any remarkable way on the value of the zero field electron mobility.

4.2.2 Inter-subband Excitations

This section considers the spectroscopy of the inter-subband excitations that derive from the transitions shown in Figs. 4.2a and 4.2c. The first observations of inter-subband excitations were reported in modulation doped GaAs-(AlGa)As single heterojunctions [4.11] and multiple quantum wells [4.12]. In these results the spectral features were relatively broad as a consequence of the low electron mobility available at that time. Advances in molecular beam epitaxy and in the modulation doping procedure led to very high carrier mobilities. In the higher mobility samples the light scattering spectra also demonstrated the first clear separation between spin-density excitations (in depolarized spectra) and charge-density excitations (in polarized spectra) [4.62]. The photon energies in these studies were in resonance with the $E_0 + \Delta_0$ optical gap of GaAs.

Light scattering spectra of inter-subband excitations are also obtained with photon energies in resonance with the transitions between states of the fundamental optical gap E_0 [4.28]. Figure 4.17a shows a symmetric quantum well and states derived from the Γ_6 conduction band, the Γ_8 heavy and light valence bands, and the Γ_7 spin-orbit split-off valence band. Figure 4.17b shows the optical transitions in inelastic light scattering by the $c_0 \rightarrow c_1$ inter-subband excitations. The spectra shown in Fig. 4.18 were obtained with a laser photon energy of $\hbar\omega_L = 1.580$ eV, under the resonance conditions described in Fig. 4.17b. These results exemplify the assignment of the spectra to spin-density and charge-density excitations by means of polarization selection rules. The depolarized $z(y'x')\bar{z}$ spectrum shows a single band assigned to the lowest spin-density inter-subband excitation. Many-body corrections to spin-density energies are small in GaAs quantum wells [4.25]. Therefore E_{01} represents an energy close to the spacing between the two lowest conduction subbands. In the polarized $z(x'x')\bar{z}$ spectrum the peaks labeled I_+ and I_- at energies E_+ and E_- are assigned to the coupled-collective inter-subband – LO phonon excitations proposed by *Burstein* et al. [4.16].

The sample used to obtain the spectra of Fig. 4.18 has symmetric wells in which $d_1 = 150$ Å. In addition $d_2 = 200$ Å, $d_3 = d_3' = 200$ Å and $x_3 = x_3' = 0.3$. In symmetric quantum wells the $c_0 \rightarrow c_1$ excitations are of odd parity with respect of the mirror planes and the LO phonon taking part in the $I_\pm$ coupled modes has a charge density fluctuation of the same parity (the symmetries of phonons have been considered in [4.63] and in Chap. 2 of this volume). The LO peak in Fig. 4.18 occurs at 36.4 meV, slightly lower than the LO phonon of bulk GaAs (36.7 meV). The peak may arise from the top wells depleted of free electrons. It could also be due to LO phonons of opposite parity that couple to the higher energy $c_0 \rightarrow c_2$ inter-subband transitions of even parity.

The differences in the energies of single-particle and collective excitations are a direct manifestation of the macroscopic electric fields that occur in the charge-density fluctuations of inter-subband transitions [4.64]. Such effects, known as depolarization shifts, have been considered theoretically in several papers [4.20, 65, 66]. In the simplest case the energy of the charge density excitation can be written as

$$E_{01}^{*2} = E_{01}^2 + E_p^2 , \tag{4.36}$$

where E_p is an effective plasma energy given by

$$E_p = [8\pi n e^2 E_{01} L_{01} / \varepsilon_\infty]^{1/2} . \tag{4.37}$$

In (4.37) ε_∞ is the high frequency dielectric constant of the medium and L_{01} is the Coulomb matrix element [4.65, 66]. In polar semiconductors like GaAs

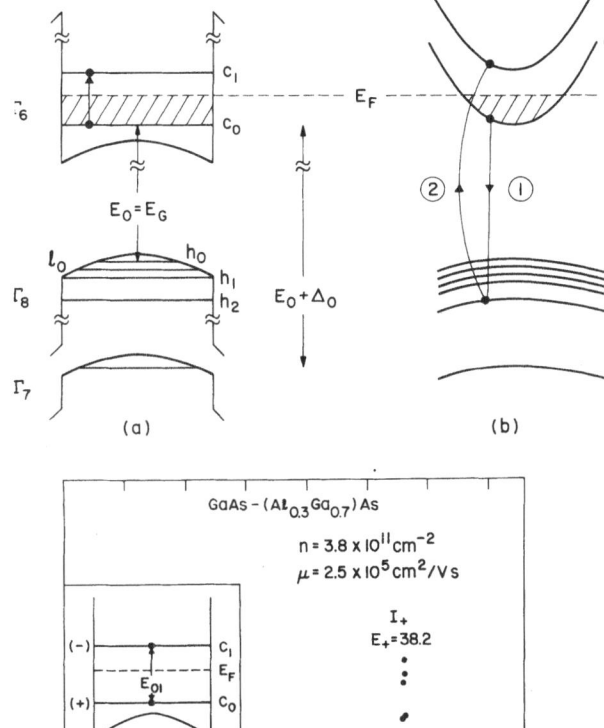

Fig. 4.17a, b. Schematic representation of quantum well energy levels and subbands in n-type GaAs-(AlGa)As symmetric quantum wells. (**a**) Fundamental and spin-orbit split-off energy gaps. The lowest energy inter-subband transition is also shown. (**b**) The two optical transitions in inelastic light scattering when the resonance is with the $h_2 \Rightarrow c_1$ optical transition

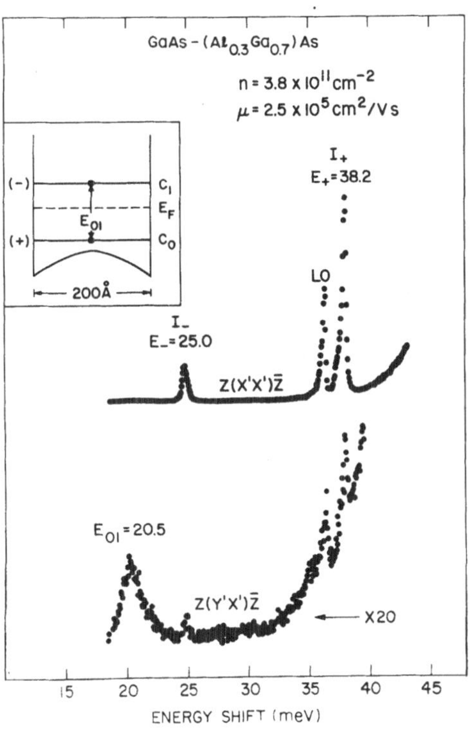

Fig. 4.18. Polarized and depolarized spectra of inter-subband excitations in multiple quantum well structures [4.28]

collective inter-subband excitations are coupled to optical phonons and the coupled mode frequencies are given by [4.13–16, 20, 62]

$$1 - \frac{E_{TO}^2 - E_{\pm}^2}{E_{LO}^2 - E_{\pm}^2} \times \frac{E_p^2}{E_{\pm}^2 - E_{01}^2} = 0 \; , \tag{4.38}$$

where E_{LO} and E_{TO} are the energies of LO and TO phonons. The solutions of (4.38) are

$$
\begin{aligned}
E_\pm^2 = {}& \tfrac{1}{2}(E_{01}^2 + E_{LO}^2 + E_p^2) \pm \tfrac{1}{2}[(E_{01}^2 + E_{LO}^2 + E_p^2)^2 \\
& - 4(E_{01}^2 E_{LO}^2 + E_{TO}^2 E_p^2)]^{1/2} .
\end{aligned}
\tag{4.39}
$$

For $E_{01} \ll E_{LO}$ one finds $E_- \simeq E_{01}^*$ and for $E_{01} \gg E_{LO}$ the high energy mode is $E_+ \simeq E_{01}^*$.

The Coulomb matrix element L_{01} that describes the depolarization shift is given by

$$
L_{01} = \int_{-\infty}^{\infty} dz \left[\int_{-\infty}^{z} dz' \psi_1(z') \psi_0(z) \right]^2 ,
\tag{4.40}
$$

where $\psi_0(z)$ and $\psi_1(z)$ are the envelope functions of the two subbands. The value of L_{01} is the only adjustable parameter in the analysis of the coupled inter-subband-LO phonon modes. The values of this Coulomb overlap integral have been determined experimentally from spectra like those in Fig. 4.18 and in many GaAs quantum wells with different parameters. They are in good agreement with the values obtained by model calculations of the subband structure and envelope functions.

More general equations for collective inter-subband excitations were given in [4.7, 13, 19, 67–70]. They consider the cases when more than one subband is occupied by carriers and where coupling exists between inter-subband transitions $i \rightarrow m$ and $j \rightarrow n$. *Tsellis* and *Quinn* [4.19, 67] also included finite values of q (the non-vertical inter-subband transitions of Fig. 4.2c) and, in the case of an infinite superlattice, finite values of q_z. The energies of these collective modes are the solutions of [4.19]

$$
\det |\delta_{im}\delta_{jn} - [V_{ijmn}(q) + T(q, q_z)\tilde{V}_{ijmn}(q)]\pi_{mn}(q)| ,
\tag{4.41}
$$

where

$$
\pi_{mn}(q) = 2\,\frac{f(E_n) - f(E_m)}{E_n - E_m - \hbar\omega} ,
\tag{4.42}
$$

$$
V_{ijmn}(q) = \frac{2\pi e^2}{\varepsilon_\infty q} \int_0^d dz \int_0^d dz' \psi_i(z)\psi_j(z) \exp(-q|z - z'|)\psi_m(z')\psi_n(z') ,
\tag{4.43}
$$

$$
T(q, q_z) = \frac{\sin(qd) - \cosh(qd) + \cos(q_z d)}{\cosh(qd) - \cos(q_z d)} ,
\tag{4.44}
$$

and $\tilde{V}_{ijmn}(q)$ is similar to $V_{ijmn}(q)$ with $|z - z'|$ replaced by $(z - z')$. The q-dependence of the single particle energies, E_n and E_m, has not been explicitly written in (4.42) for the irreducible polarization π_{mn}. Coupling to optical phonons is included as in the two subbands case [4.13, 64]. This is done in the Coulomb interactions $V_{ijmn}(q)$ and $\bar{V}_{ijmn}(q)$, where the background dielectric constant ε_∞ is replaced by the dielectric function of the polar lattice

$$\varepsilon_L(\omega) = \varepsilon_\infty (E_{LO}^2 - \hbar^2\omega^2)/(E_{TO}^2 - \hbar^2\omega^2) . \tag{4.45}$$

a) Results for Different Quantum Well Shapes

Light scattering by single particle and collective inter-subband excitations has been studied in GaAs structures with different quantum well shapes. Examples are shown in Fig. 4.19. The spectra of spin-density excitations are the solid lines and the dashed lines are due to the charge-density excitations. The figure also shows the profiles of the conduction band quantum wells and the possible inter-subband transitions. The different quantum well shapes are created by molecular beam epitaxy using specific doping and composition sequences. A symmetric modulation doped double heterostructure (DHS) leads to the potential well shown in the upper part of the figure. There is additional band bending caused by the charge transfer associated with the selective doping. The two lowest

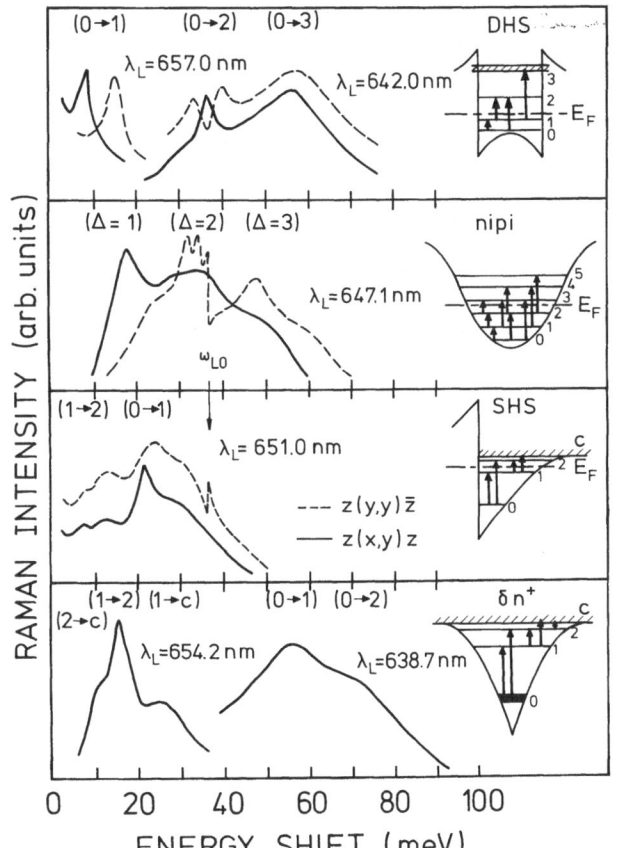

Fig. 4.19. Polarized and depolarized spectra of inter-subband excitations in various potential wells achieved in different GaAs quantum well and superlattice structures

subbands are occupied. Their spacing, E_{01}, is much smaller than the LO phonon energy. In the charge-density spectrum E_{01}^* is given by (4.35) and (4.36). Because E_{02} is very close to E_{LO}, a coupled modes doublet is seen in the charge-density spectrum. The highest subband is already close to the top of the well and gives rise to the broad transition at E_{03}. In the three other cases shown in Fig. 4.19 the potential wells are nearly parabolic (in the doping n-i-p-i superlattice) or nearly triangular as in the single heterostructure (SHS) and in the doping spike. In each of these cases the energies in the spectra of spin-density excitations are close to the subband spacings.

Some of the spectra shown in Fig. 4.19 are split into two parts. Each section is excited with different laser photon energies. This is necessary because of the sharp resonant enhancement close to the $E_0 + \Delta_0$ optical gap of GaAs. The basic matrix element for resonant light scattering by inter-subband excitations was considered in [4.16], and some of its most remarkable properties were discussed in [4.7]. It involves optical transition matrix elements from the spin-orbit split-off Γ_7 valence band state to the empty excited conduction subband and from the occupied conduction subband to the valence state (Figs. 4.1 and 4.17): The resonant denominator depends on the energy of the excited conduction subband state [4.7, 16]. Therefore, the relative intensities of different transitions depend on incident photon energy. To illustrate this feature we show in Fig. 4.20 the resonant enhancement profiles of $0 \to 1$ and $0 \to 2$ spin-density excitations in a symmetric GaAs-(AlGa)As multiple quantum well structure. The intensities of the even parity transition E_{02} are larger than those of the odd parity transition E_{01}. In symmetric quantum wells odd parity transitions are forbidden in light scattering. The observation of the E_{01} transitions has been explained by selection

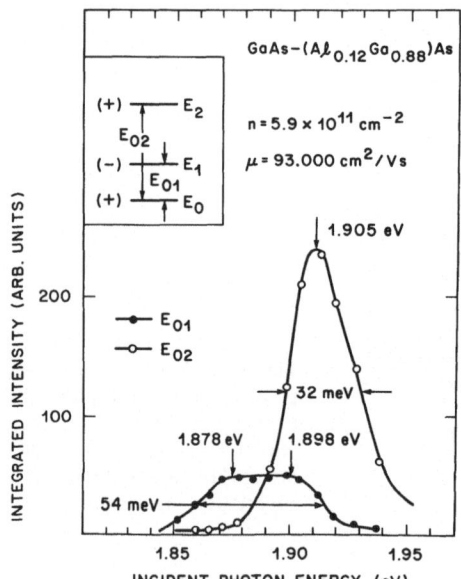

Fig. 4.20. Resonant enhancement of light scattering by spin-density inter-subband excitations of a modulation-doped multiple heterostructure

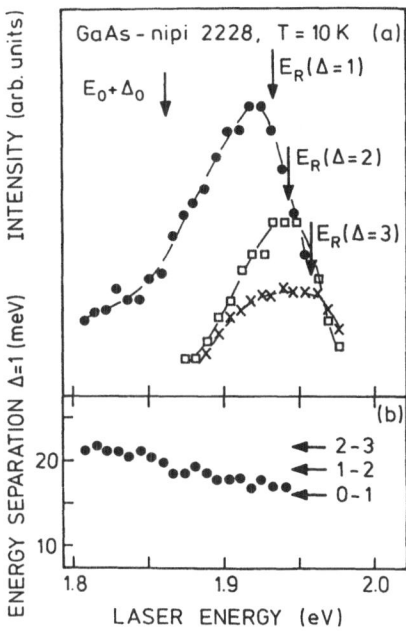

Fig. 4.21. Intensity and peak position of inter-subband transitions in doping superlattices [4.95]

rules that go beyond the dipole approximation when k_z is comparable to $1/d$. In the case of doping nipi superlattices the high concentration of impurities leads to breakdown of wavevector conservation and the parity selection rule. The resonant profile of $\Delta = 1$, $\Delta = 2$, and $\Delta = 3$ transitions in a nipi structure are shown, as an example, in Fig. 4.21. They are considered in more detail in Sect. 4.2.2e.

b) Narrow Quantum Wells

Subband energies and spacings depend critically on the quantum well widths and heights. In the case of GaAs-(AlGa)As heterostructures with conduction band offsets of the order of 300 meV the first excited subband moves close to the top of the well for GaAs layers of width ~ 50 Å. In such samples the inter-subband transitions broaden considerably. The first inelastic light scattering results in these narrow quantum wells have been reported recently [4.69]. Direct infrared dipole transitions are observed only for infrared light polarized normal to the layers [4.70, 71]. Infrared subband absorption in GaAs quantum wells was first observed with a large angle of incidence [4.71]. More recent results which were obtained in a similar geometry with high quality multiple quantum wells are shown in Fig. 4.22a (from [4.70b]). A special prism configuration, already used to study surface accumulation and inversion layers [4.72], is more convenient for these measurements. Infrared absorption spectra obtained in the prism configuration are shown in Fig. 4.22b for different quantum well thicknesses. With decreasing thickness, the $0 \rightarrow 1$ absorption shifts to higher energy and broadens

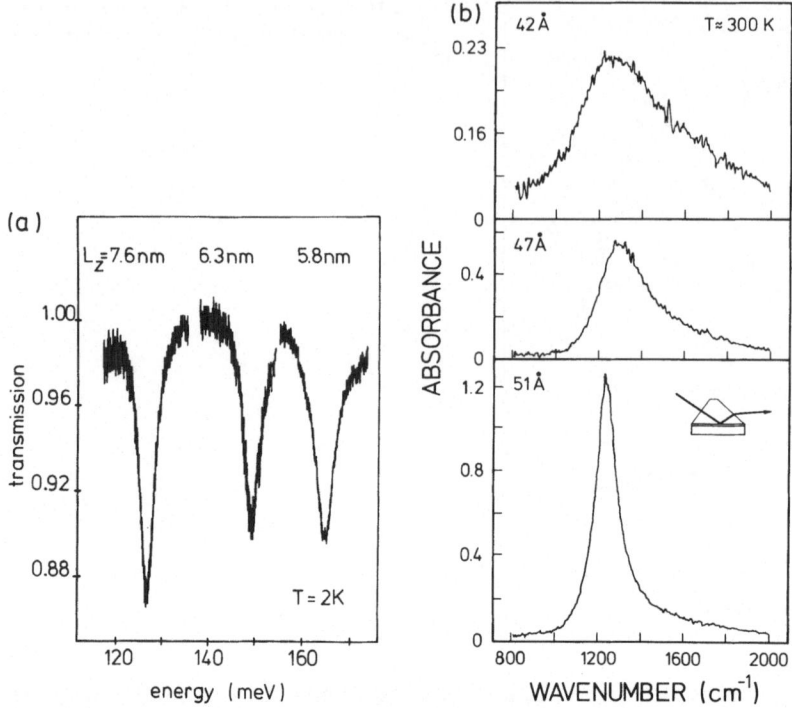

Fig. 4.22. (a) Infrared transmission of narrow modulation-doped GaAs quantum well structures (Brewster angle geometry) [4.70b]. **(b)** Absorbance of narrow modulation-doped GaAs quantum well structures (prism geometry) [4.70a]

asymmetrically towards the higher energy side. The broadening is assigned to delocalization of the excited subband that moves close to the top of the well with decreasing thickness. Other mechanisms that contribute to the broadening are fluctuations in layer thickness and interface roughness. Non-parabolicity of the conduction band is important in the subband energies, but makes a relatively small contribution to the observed linewidths.

Resonant inelastic light scattering spectra from a symmetric modulation doped multiple quantum well with $d_1 = 51$ Å are shown in Fig. 4.23. There is a small difference between peak energies in polarized and depolarized spectra, which is due to the depolarization field effect (4.36, 37). Because of resonant enhancement effects the peak positions in light scattering spectra depend slightly on photon energy. Therefore, no special significance should be given to the small differences between peak positions in infrared spectra and in polarized light scattering spectra.

Electronic excitations in such narrow quantum layers, where the excited subband is close to the top of the wells, have attracted considerable attention because of possible applications as ultrafast infrared detectors in the energy range 50–150 meV [4.73]. For such applications it is essential to have a precise

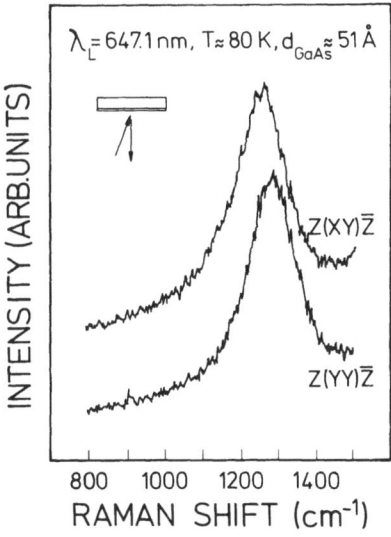

$\lambda_L = 647.1\,nm,\ T \approx 80\,K,\ d_{GaAs} \approx 51\,Å$

INTENSITY (ARB.UNITS)

$\overline{Z}(XY)\overline{Z}$

$Z(YY)\overline{Z}$

800   1000   1200   1400

RAMAN SHIFT (cm^{-1})

Fig. 4.23. Polarized and depolarized inter-subband excitation spectra of narrow GaAs quantum wells [4.69]

knowledge of inter-subband relaxation times. Time-resolved experiments show that in narrow wells the times are of the order of 10 ps [4.70, 74] with a decreasing tendency for thicker wells. For much wider quantum wells, with subband spacings smaller than E_{LO}, the relaxation times are much longer [4.75]. Infrared and light scattering experiments with time resolution are expected to provide further insight into the electron relaxation times in quantum wells.

c) Mixed Excitations

Sooryakumar et al. [4.76] have carried out an experimental investigation of the q-dependence of the single-particle and collective inter-subband excitations of symmetric modulation doped GaAs-(AlGa)As multiple quantum wells. These excitations have *mixed* inter-subband and intra-subband character and derive from the non-vertical inter-subband transitions shown in Fig. 4.2c. The light scattering results are reproduced in Fig. 4.24. The upper three pairs of spectra correspond to the scattering geometry shown in the inset, which is identical to that of Fig. 4.2a. The lowest set of spectra was obtained in the *right angle* configuration shown in Fig. 4.3b. In this geometry the scattered light propagates in the plane of the heterostructures and $k = 2.9 \times 10^5$ cm^{-1}. The sample consists of 30 periods of $d_1 = 221$ Å and the $(Al_{0.23}Ga_{0.77})$As barriers have $d_2 + 2d_3 = 597$ Å. The electron mobility is 2.5×10^5 cm^2/V s.

The depolarized $(\parallel, \perp)$ spectra show that the spin-density inter-subband excitations, at E_{01}, acquire a width $2\hbar q v_F$. The broadening is due to the non-vertical character of the transitions at finite in-plane wavevector $q = k$ (see Fig. 4.2c). On the other hand, the collective inter-subband excitation I_-, measured in the polarized $(\parallel, \parallel)$ spectrum, remains narrow and shows at the highest q a shift towards higher energies. This shift illustrates the q-dependence

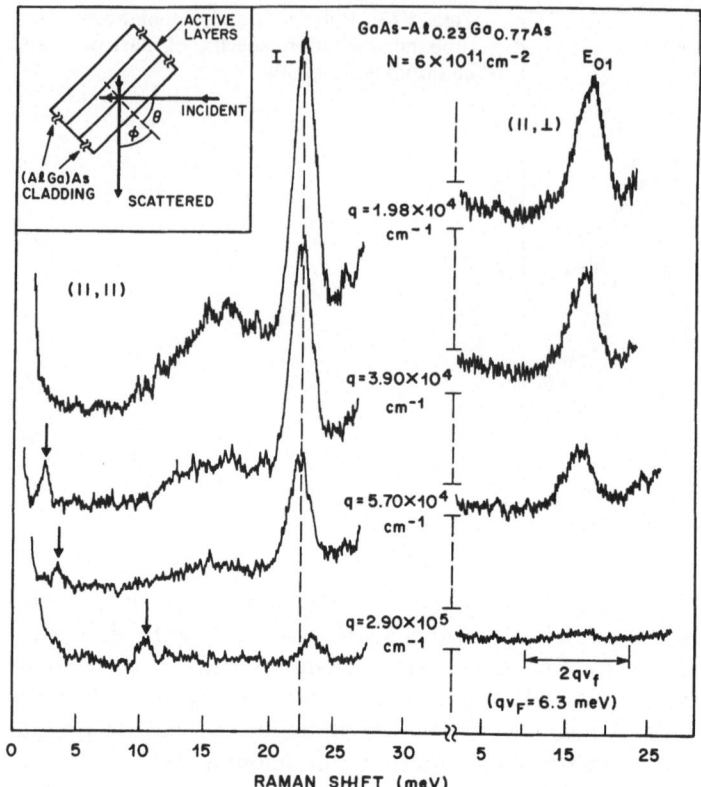

Fig. 4.24. Polarized and depolarized inter-subband excitation spectra with different in-plane wavevector [4.76]

of inter-subband excitations. The polarized spectra also show the band of the plasmon of the multilayers. For the largest value of q we have $qd = 2.37$. In this limit, as seen from (4.10a) and (4.10b), these modes have single layer behavior.

Sooryakumar et al. [4.76] have interpreted the results for the collective inter-subband excitations in terms of the solutions of (4.41–45) and neglected coupling to higher transitions. Following *Tsellis* and *Quinn* [4.19, 67], the energy of the I_- mode was written, to order q^2, as

$$E^2_-(q) = E^2_-(0) + aq + bq^2 ,$$ (4.46)

where $E_-(0)$ is the lower energy solution of (4.39). In the evaluation of a and b, the multiple quantum wells are modelled as a set of parallel uniformly spaced 2D electron gases. It is found that the largest contribution to the linear q term is *negative* because of the decrease in the Coulomb energy due to the in-plane variation of the induced charge density. The largest contribution to the q^2 term is *positive* and of the form $(3/4)\hbar^2 q^2 v_F^2$, compared with $(3/5)\hbar^2 q^2 v_F^2$ of 3D electron gases. At small q values there is a near cancellation between the linear and quadratic terms. For larger q values the quadratic term takes over, giving rise to a relatively small increase in the mode energy. The calculated energy shifts are in good agreement with the measured ones.

d) Photoexcited Systems

Resonant inelastic light scattering by photoexcited 2D electron systems was first reported in 1981 in experiments carried out in *undoped* GaAs-(AlGa)As heterostructures [4.77]. The results are reproduced in Fig. 4.25. These measurements were made with the 6471 Å line of a Kr ion laser. The photon energy ($\hbar\omega_L = 1.916$ eV) is in resonance with the $E_0 + \Delta_0$ optical gap of the GaAs quantum wells (see Fig. 4.1). To achieve carrier photoexcitation the laser beam is focussed to a spot of radius ~ 10 μm. At the relatively "low" power density of ~ 200 W/cm^2 the dominant feature is the peak of the LO phonon of GaAs. This indicates that carrier photoexcitation is low. When the power density is increased to $\sim 10^3$ W/cm^2 new bands appear in polarized and depolarized spectra. They are assigned to the collective and single particle excitations associated with the inter-subband transitions indicated in the figure. Single particle excitations of the photoexcited holes were not reported. Their scattering cross section is believed to be much smaller because it is not enhanced at the $E_0 + \Delta_0$ gap.

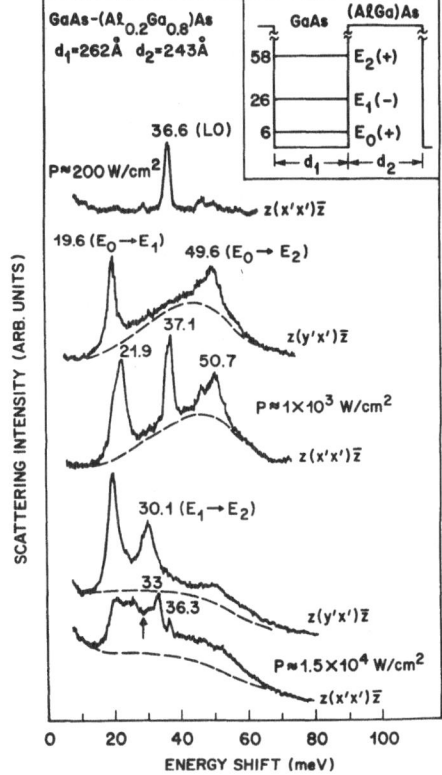

Fig. 4.25. Light scattering spectra from an undoped multiple quantum well structure with different laser power densities [4.77]

At the higher power density of $\sim 1.5 \times 10^4$ W/cm^2 there is a new band at 30.1 meV in the depolarized $z(y'x')\bar{z}$ spectrum of spin-density excitations. This has been interpreted as arising from transitions of photoexcited electrons that now also populate the next higher conduction subband. The features in the polarized $z(x'x')\bar{z}$ spectrum display large broadening because the energies of collective excitations depend on carrier density which, due to absorption of the incident light, is not the same in all the quantum wells. It is striking in these results that for a wide range of photoexcited carrier density ($10^{11} - 10^{12}$ cm^{-2}) there is no identifiable change in the single particle energies measured in the $z(y'x')\bar{z}$ spectra. This indicates that photoexcitation does not cause substantial changes in the electrostatic potential within the GaAs quantum wells and that photoexcited electrons and holes become confined in the GaAs layers with identical densities. This is because spatial gradients would create large electrostatic potentials that would modify the spacings between the subbands. The lack of sensitivity of the energies of spin-density inter-subband excitations to carrier density also indicates that final-state interactions are small, as predicted by theory [4.25].

Light scattering spectra of photoexcited electrons at much lower power levels, typically 1–50 mW/cm^2, were measured in undoped multiple GaAs-(AlGa)As heterostructures [4.78]; and also in heterostructures in which the centers of the GaAs quantum wells were doped with Si donors [4.79, 80]. These experiments allowed the measurements of the Stark effect of inter-subband transitions [4.78], of resonant states of Si donors [4.79] and of the inter-subband excitations in tilted magnetic fields [4.80].

Figure 4.26 shows resonant spectra (at the $E_0 + \varDelta_0$ optical gap) of a structure with $d_1 = 460$ Å. The spectra exhibit characteristic doublets at energies close to the expected positions of the $c_0 \rightarrow c_1$ and $c_0 \rightarrow c_2$ inter-subband transitions. The higher energy components of the doublets (E'_{01} and E'_{02}) are due to excitations from the ground state of the Si donors (associated with c_0) into $1s$-resonances derived from c_1 and c_2. The lower energy components are inter-subband transitions of the electrons photoexcited into the ground subband c_0. The transitions of these photoexcited electrons were investigated in tilted magnetic fields [4.80] where the in-plane component of the magnetic field couples the cyclotron motion of the electrons with the subband states [4.1]. Figure 4.27 shows the measured energies for two different tilt angles θ of the magnetic field with respect to the normal. Level repulsion is evident even at the small angle $\theta = 8°$. These results are interpreted with a theory for these transitions that is valid for *parabolic wells*, where analytical expressions for the energy levels are easily obtained [4.81, 82]. The agreement between measured and calculated energies is very good, particularly in the prediction of a splitting proportional to θ for $\hbar\omega_C = E_{01}$ (see inset). The reason why a square well can be approximated by a parabolic one is that the corresponding c_0 and c_1 states are very similar in these two potential wells.

The results obtained in nominally undoped quantum wells under large electric fields normal to the layer also illustrate the power of the light scattering

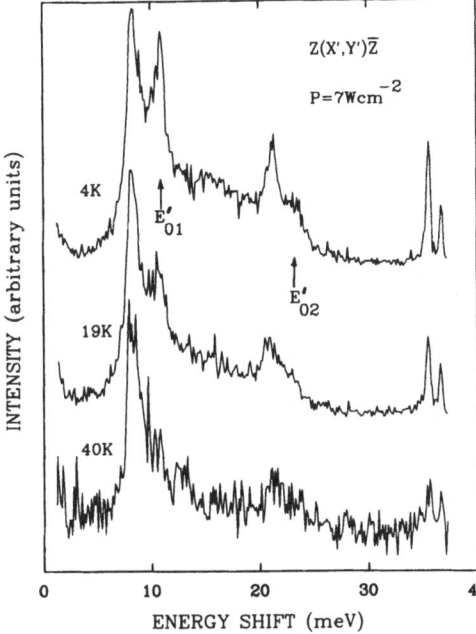

Fig. 4.26. Temperature dependence of Si-doped GaAs quantum wells [4.79]

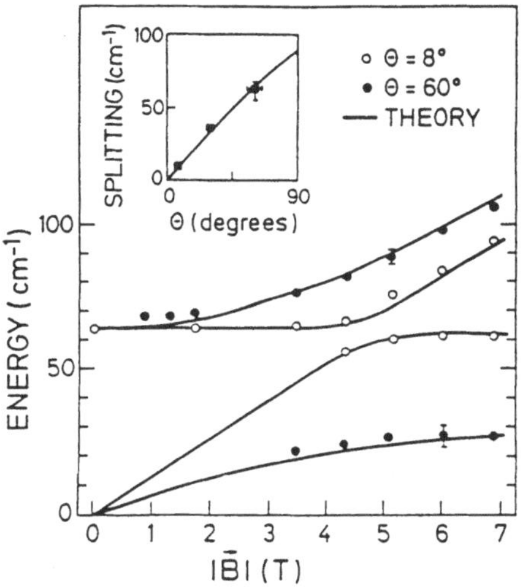

Fig. 4.27. Energy of coupled inter-subband Landau level excitations as a function of magnetic field [4.80]

method in studies of the 2D electron systems [4.78]. These spectra reveal very clearly the Stark effect of the odd parity $c_0 \rightarrow c_1$ inter-subband transitions. The observed shifts are in good agreement with theory up to fields in excess of 10 kV/cm. The intensities of light scattering by the odd parity transition increases with electric field. This was interpreted as further relaxation of parity selection rules induced by the electric field (for a discussion of parity selection rules see Sect. 4.2.2.a and [4.7]).

In 1986 *Menéndez* et al. [4.83] introduced light scattering by photoexcited electrons as a new method for the determination of conduction band offsets at semiconductor heterojunctions. The conduction band offset is defined as the ratio $Q_e = \Delta E_C / \Delta E_G$, where ΔE_G is the difference between the forbidden energy gaps of the two semiconductors and ΔE_C is the part of ΔE_G that corresponds to the conduction band. The basis of the light scattering method lies in the determination of the energy level structure in the conduction band from the measurement of inter-subband transitions in quantum wells. The conduction band discontinuities ΔE_C are obtained from an effective mass fit to the transition energies. The method has two features that make it attractive. One is the possibility of applying it selectively to the conduction band. The other is the capability to determine, also by resonant Raman scattering, the band gap of the barrier semiconductor. The method uses photoexcited carriers because, as explained above, the fact that the quantum layer remains neutral under photoexcitation ensures that the potential well is not distorted by the incorporation of free carriers. This is important since a reasonably accurate knowledge of the quantum well shape is required in the effective mass analysis of the transition energies.

The inelastic light scattering method was initially applied to the GaAs-AlAs heterojunction system, first in square wells [4.83] and later in parabolic wells [4.84]. Other heterojunction systems investigated are GaSb-AlSb [4.85] and more recently the strained-layer system InAs-GaAs [4.86]. These results and their impact on the understanding of band offsets at semiconductor heterojunctions have been considered in a recent review article [4.87]. In the case of the GaAs-AlAs system the values determined from light scattering are in agreement with those now generally accepted. The value of the offset determined in the GaSb-AlSb suggested the existence of new, weaker form of the common anion rule [4.85, 87]. The most unexpected result was obtained in the InAs-GaAs

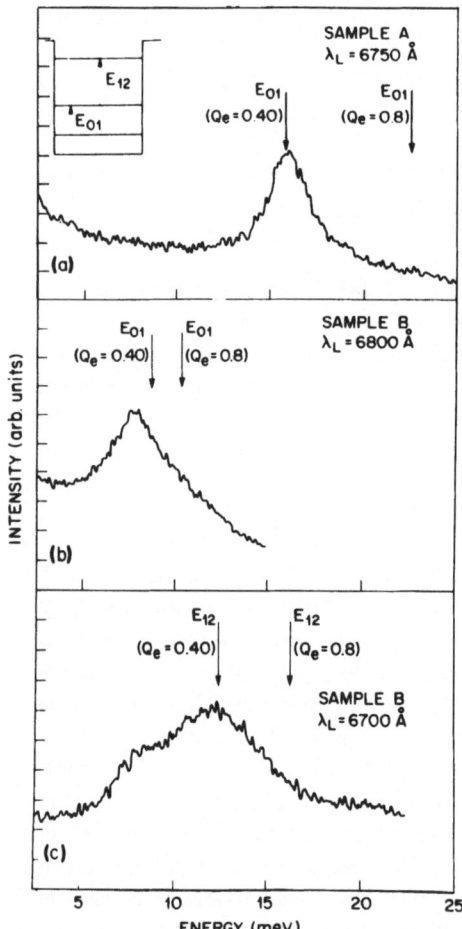

Fig. 4.28. Depolarized light scattering spectra in two InGaAs-GaAs undoped multiple quantum well samples. The inset shows the transitions of photoexcited electrons. The second excited level in the conduction band is not confined in sample A and transition E_{12} is not observed. The arrows indicate the best fit for the transition energies [4.86]

system. Figure 4.28 shows depolarized spectra obtained from two undoped $(In_{0.05}Ga_{0.95})$As-GaAs multiple quantum wells grown on GaAs (001) substrates. In sample A the $(In_{0.05}Ga_{0.95})$As wells have thickness $d_1 = 193$ Å. The relatively low energy of the peaks compared to those of larger thickness GaAs-(AlGa)As heterojunctions (Fig. 4.25) indicates that the conduction band offset is smaller. The value of Q_e is obtained by comparing the measured spacing E_{01} with that calculated from a square well model in which ΔE_C is the adjustable parameter. This procedure yields a value of $Q_e = 0.4$. This result is controversial because other experimental work indicates a larger conduction band offset of $Q_e = 0.8$. However, as indicated by the arrows in Fig. 4.28, the larger value of the offset is incompatible with the light scattering determination of the subband spacings.

We conclude this section on photoexcited systems with a short comment on time-resolved inelastic light scattering experiments. Measurements of spectra of inter-subband transitions with picosecond time resolution allow the determination of the inter-subband scattering times of the photoexcited electrons [4.75]. Non-equilibrium LO phonon populations generated by photoexcited carriers have been studied by time-resolved Raman scattering [4.88]. These results demonstrate that inelastic light scattering gives a direct insight into the dynamics of carrier relaxation phenomena and the crucial role played by electron-phonon interactions.

e) Space-Charge Induced Quantum Wells

Resonant inelastic light scattering experiments in purely space-charge induced quasi-2D systems have been reported in the cases of periodic doping superlattices (nipi structures) [4.89–91] and of δ-doped single layers [4.92]. In nipi doping superlattices the first direct evidence of carrier quantization and tunable optoelectronic properties was obtained from electronic light scattering and photoluminescence measurements [4.89]. Examples of spectra by single-particle and collective modes are shown in Fig. 4.19. The interesting properties of nipi structures have been reviewed in [4.93–95]. These superlattices have a number of exciting features such as tunable effective bandgap, long photoexcited carrier lifetimes, quantization of states in space-charge induced potential wells, and tunable absorption coefficients. Recent experiments have been carried out in more sophisticated sample geometries [4.96].

Periodic doping structures are composed of ultrathin n- and p-doped layers of an otherwise homogeneous semiconductor. Electrons are transferred from the donor impurities to the spatially separated acceptor sites. The sequence of ionized impurities causes a periodic space-charge potential along the direction perpendicular to the layers. Electron states quantized in nearly parabolic potential wells in the n-type layers. The shape of the potential well and the subband energies can be easily calculated by solving the one-dimensional Poisson equation

$$\frac{d^2U}{dz^2} = -\frac{4\pi e^2}{\varepsilon_0}\left[\varrho(z) - N_D(z) + N_A(z)\right], \tag{4.47}$$

where $N_D(z)$ and $N_A(z)$ are the distributions of ionized donors and acceptors, $\varrho(z)$ is the free carrier density, and ε_0 is the static dielectric constant. The tunability of the well depth with excitation intensity is observed by photo- or electro-luminescence [4.89, 95], due to recombination across the "indirect gap in real space". That is represented in the real space diagram of Fig. 4.29, which also

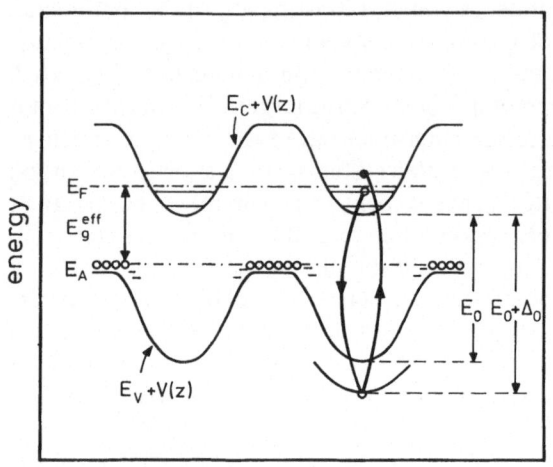

Fig. 4.29. Real space energy diagram of an excited nipi crystal. Raman processes are shown schematically

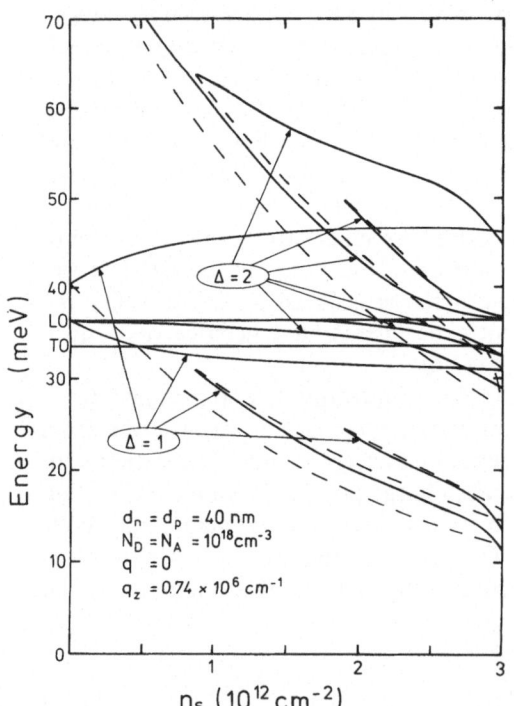

Fig. 4.30. Calculated collective and single-particle excitations of a doping superlattice. Coupling to LO-phonons is included [4.68]

shows the two optical transitions in resonant inelastic light scattering by inter-subband excitations.

In the highly excited case, many subbands are occupied and their energies are closely spaced. To obtain the energies of the collective inter-subband excitations it is necessary to solve the system of coupled modes in (4.41–45). This task has been carried out in [4.68, 97]. Figure 4.30 shows results of numerical calculations for a doping superlattice with $d_n = d_p = 40$ nm, $N_D = N_A = 10^{18}$ cm^{-3} [4.68]. The wavevectors were chosen to match the light scattering wavevectors for photon energies close to the $E_0 + \Delta_0$ gap of GaAs. The carrier density per layer, n_s, is tuned by photoexcitation. With increasing n_s the subband splittings are reduced and the number of possible transitions depends on the number of occupied subbands. The collective excitations are strongly affected by the Coulomb interaction and coupling between different transitions. At $q = 0$ the V_{ijmn} and $\tilde{V}_{ijmn}$ matrix elements are non-zero only for pairs of transitions with the same parity [4.13]. The odd parity $\Delta = 1$ transitions ($0 \rightarrow 1, 1 \rightarrow 2, 2 \rightarrow 3 \ldots$) exhibit one high energy mode close to the LO phonon energy. Therefore, it splits into two modes that show only a weak n_s dependence. This behavior is very different from that of the single particle energies (dashed line). The other $\Delta = 1$ modes have only weak depolarization fields. Their collective mode energies lie between the single particle energies and away from the LO phonon energy. The even parity $\Delta = 2$ transitions show similar behavior. In the high density case the different set of transitions merge and the system displays quasi-3 D behavior.

These calculations explain the experimental results. An example is shown in Figure 4.31. The sample parameters correspond to those used in the calculations. In the case of empty potential wells (no excitation), the solution of (4.46) predicts a well depth of 510 meV. The depth is reduced by photoexcitation. In Fig. 4.31 the laser power density is 100 W/cm^2, which results in a carrier concentration of 2.8×10^{12} cm^{-2} per layer and a well depth of ~ 100 meV as measured by photo-luminescence. Three distinct but broad peaks are measured in the depolarized spectrum of Fig. 4.31. The dashed lines indicate the hot luminescence close to the $E_0 + \Delta_0$ gap. The peaks are labelled $\Delta = 1$, $\Delta = 2$ and $\Delta = 3$, indicating transitions between first nearest, second nearest, and third nearest subbands. Self-consistent subband calculations show that three subbands are occupied, and the calculated single particle transition energies are marked by arrows. Each measured band is a superposition of different subband transitions. The peak positions shift with excitation intensity, and with increasing carrier concentration the subband spacings decrease. The mean values of the measured single particle excitations are plotted in Fig. 4.32 as function of n_s. The luminescence peak energy was used to determine the effective bandgap and n_s. There is good agreement between measured and calculated results. The solid lines correspond to the individual subband transitions and the dashed lines give the weighted average of the $\Delta = 1$, $\Delta = 2$, and $\Delta = 3$ transitions. For high carrier concentrations the single particle transitions merge and the subbands broaden into minibands. The arrow marks the density at which the Fermi energy moves above the top of the wells leading to 3 D behavior. Similar experiments and self-consistent calculations have been

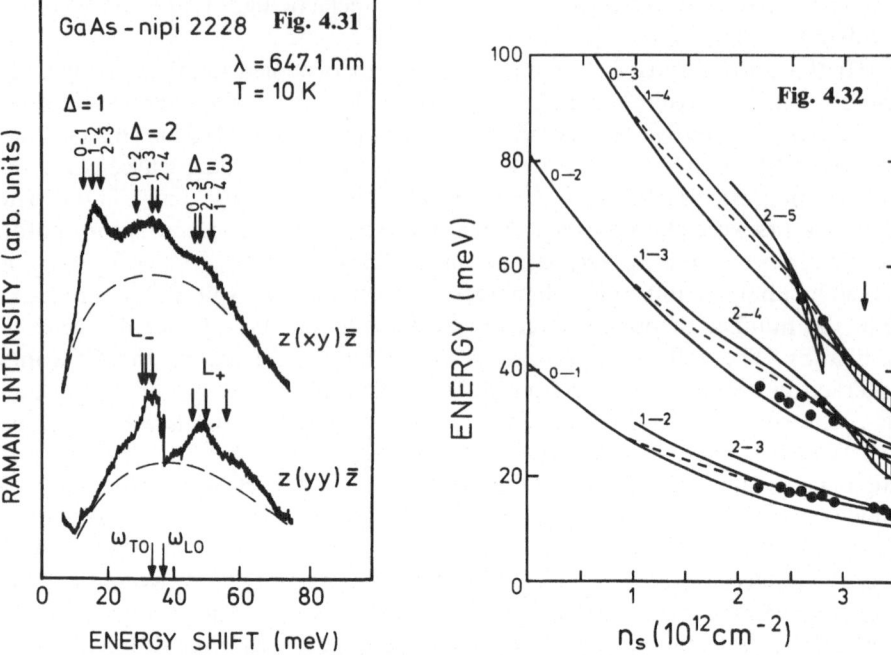

Fig. 4.31. Polarized and depolarized spectra of nipi crystal obtained with a laser power density of 100 W cm^{-2}. The arrows mark calculated transition energies [4.95]

Fig. 4.32. Comparison of experimental subband transitions to the results of self-consistent calculations for a nipi crystal with $N_D = N_A = 1 \times 10^{18}$ cm^{-2} and $d_n = d_p = 400$ Å [4.95]

performed for a wide range of parameters like thickness, doping, and photoexcited carrier concentration.

These light scattering results have been obtained with photon energies near the $E_0 + \Delta_0$ optical gap. The precise position of the resonance energy depends on subband occupation and is slightly different for different transitions [4.95]. Resonant enhancement profiles are shown in Fig. 4.21a. The calculated peak energies are marked by arrows. Each measured band consists of different transitions, depending on subband occupation, and the resonance for the individual transitions is not resolved directly. However, the main peaks in the spectra shift with photon energy. This shift is plotted in Fig. 4.21b for the $\Delta = 1$ band. Calculated subband spacings are marked by arrows. For the lower photon energy the $2 \rightarrow 3$ transition appears to dominate; and at higher energies the $0 \rightarrow 1$ is strongest. Such detailed studies of excitations in doping superlattices give information on the shapes of the potential wells, tunability of the 2D electron system and the transition from 2D to 3D behavior.

Additional information can be obtained from spectra of collective modes, also shown in Fig. 4.31 [$z(yy)\bar{z}$ configuration]. These spectra are complicated because of coupling between the several transitions and the LO phonons. The

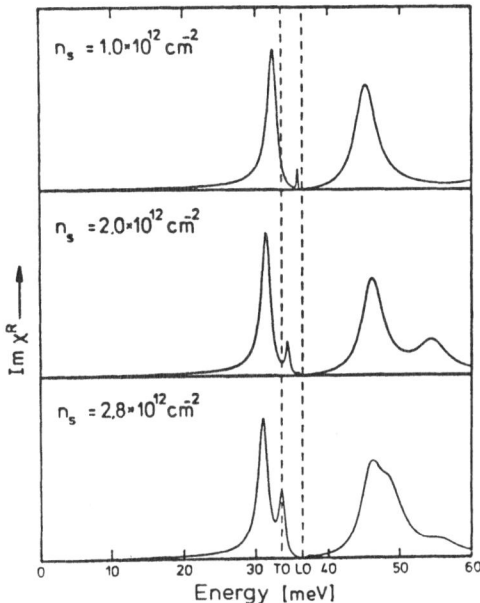

Fig. 4.33. Calculated line shapes of collective modes of a doping superlattices with different carrier concentrations [4.68]

arrows indicate the strongest lines of the calculated collective mode spectra, in reasonable agreement with experiment. *Zeller* [4.68] has carried out numerical interpretations of such complex spectra. Starting from single particle energies the collective modes were calculated with (4.41–45). It turns out that the scattering structures below the TO phonon contain contributions from the collective modes of the $\Delta = 1$ and $\Delta = 2$ transitions. The peak between the TO and LO phonons is composed of the lower mode of the $2 \to 4$ and all other high energy transitions. At higher energies several collective modes contribute. With increasing carrier concentrations it was found that the collective excitations converge to the characteristic features of the 3D electron gas. *Zeller* [4.68] calculated theoretical spectra by numerical evaluation of the charge-density response function in the RPA approximation. Some of the results are shown in Fig. 4.33. The main experimental features are reproduced by the calculations. The differences can be attributed to effects not included in the calculation, such as breakdown of wavevector conservation, that should be important at the large impurity concentrations $N_D = N_A \sim 10^{18}$ cm^{-3} [4.9]. Resonant light scattering effects also play a role in the spectral lineshapes.

The shapes of space-charge induced potential wells can be altered by changing the doping distribution. If the doped region is drastically reduced, a V-shaped potential is achieved. The case of extremely narrow doping layers, or δ-doping, has attracted considerable interest. In such structures the donor atoms are situated in one or a few atomic planes. The resulting potential well is shown schematically in the lowest part of Fig. 4.19, together with the spectra of single particle excitations. Such structures have been subjected to extensive investigation to obtain quantitative information on the donor distribution [4.92]. Deviations from the ideal δ-doping profile result in different subband energies, and spectroscopy of subband transitions is a sensitive tool to obtain direct information on the potential well and the doping distribution. The analysis of the experimental results of δ-doping layers is similar to that described above for the nipi structures.

4.3 Two-Dimensional Hole Gases

In the previous section we described the 2D electron gas confined in GaAs potential wells in terms of parabolic subbands. This is not possible in the case of the valence band states. Their much more complex structure is related to the degeneracy of the valence states of the diamond and zincblende-type semiconductors (see Fig. 4.1). This degeneracy leads to separate, but coupled, sets of heavy and light hole subbands. The first calculations of valence subband energies and dispersions have been performed by *Bangert* et al. [4.98] and by *Okhawa* and *Uemura* [4.99, 100] for Si surface space charge layers. The valence subbands of heterostructures, single heterojunctions, and multiple quantum wells of III-V semiconductors have been the subject of more recent calculations [4.56–60, 101a]. This work is based on the Luttinger hamiltonian, which describes the degenerate valence bands within $k \cdot p$ theory [4.102]. In the absence of an external magnetic field and ignoring the K linear terms, the hamiltonian can be written as

$$H_v = -\frac{1}{2m_0}\left(\gamma_1 + \frac{5}{2}\gamma_2\right)K^2 + \gamma_2(K_x^2 J_x^2 + K_y^2 J_y^2 + K_z^2 J_z^2)$$

$$+ 2\gamma_3\{K_x K_y\}\{J_x J_y\} + 2\gamma_3\{K_z K_x\}\{J_z J_x\} + 2\gamma_3\{K_y K_z\}\{J_y J_z\}, \quad (4.48)$$

where m_0 as the free electron mass, J_x, J_y, and J_z are the three components of total electron angular momentum, K_x, K_y, and K_z the three components of the electron wavevector and K^2 its squared modulus. The parentheses are defined as

$$\{K_x K_y\} = \tfrac{1}{2}(K_x K_y + K_y K_x) \quad (4.49)$$

with similar expressions for the other components of wavevector and for the components of angular momentum. γ_1, γ_2, and γ_3 are material parameters known as Luttinger parameters (for numerical values of γ_i see [4.101b]).

In effective mass theories, the energies and wave functions of the valence subbands are given by the solutions of Schrödinger equations in which the hamiltonian is the sum of the quantum well potential $V(z)$ and the Luttinger hamiltonian where the wavevector components K_z, K_y, and K_x are replaced by the operators $(1/i)\partial/\partial z$, $(1/i)\partial/\partial y$, and $(1/i)\partial/\partial x$. The six wavefunctions of the valence band states at $K=0$ serve as a basis. However, in the case of GaAs the spin-orbit splitting is relatively large ($\Delta_0 = 350$ meV) and only the four Γ_8 states need to be considered. They are the "*heavy*" states with total angular momentum $J=3/2$ and $J_z = \pm 3/2$; and the "*light*" states with $J=3/2$ and $J_z = \pm 1/2$. In Si $\Delta_0 = 44$ meV and the spin-orbit split-off states with $J=1/2$ and $J_z = \pm 1/2$ also need to be included.

When (4.48) is used to describe the valence subbands of quantum well heterostructures and of holes confined in surface space charge layers, the term in $\{K_x K_y\}$ causes coupling between light and heavy subbands. The terms in $\{K_z K_x\}$ and $\{K_y K_z\}$ lead to couplings between subbands that correspond to different confinement states. For $K \neq 0$ these couplings create considerable mixing between all the valence subband states. As a consequence of these complex interactions the dispersions of the valence subbands are highly nonparabolic and show unexpected features. Figure 4.34 shows the calculated dispersions of two heavy valence subbands (h_0 and h_1) and one light subband (l_0), in the case of a p-type GaAs-(AlGa)As symmetric quantum well heterostructure. The electron-like dispersion of the l_0 subband is a major prediction of the effective-mass theories and is due to its coupling to the lower h_1 subband. As we shall see in Sect. 4.3.2 this prediction is supported by the light scattering measurements of valence subband spacings.

For $K \neq 0$ the wavefunctions of the subbands shown in Fig. 4.34 are superpositions of the heavy and light Γ_8 states of GaAs and also of different confinement states of the square quantum well. Such complex character of the valence subband states has a profound influence on the inelastic light scattering matrix elements and selection rules. For this reason the spectra obtained with photon energies in resonance with the quantum well states of the fundamental E_0 gap are much richer than spectra excited at the $E_0 + \Delta_0$ gap. (See Fig. 4.17 for the

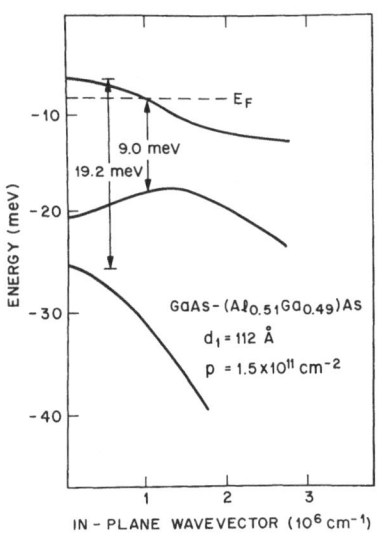

Fig. 4.34. Calculated subband dispersion of holes in a GaAs-(AlGa)As quantum well [4.56]

optical transitions in the case where the intermediate states are in the h_2 subband). The additional mixing between Landau levels introduced by an external magnetic field allows the combined subband Landau level transitions considered in Sect. 4.2.1b.

4.3.1 Si Surface Space-Charge Layers

The first spectroscopic studies of hole space-charge-layers of Si were reported in the mid-seventies [4.103, 104]. These early optical absorption measurements were carried out with far infrared lasers and were restricted to relatively small energies. Optical absorption was measured more recently at higher energies in systems of higher hole concentrations [4.105, 106]. We have seen that the inelastic light scattering method is sensitive when the photon energy is in resonance with the optical transitions that involve the free holes. For holes in Si, the Γ-point fundamental direct gap $E_0' \simeq 3.4$ eV is the most accessible. Results have been reported for hole accumulation layers in metal-oxide-semiconductor structures (MOS) [4.107, 108a]. The spectra cover a wide energy range of inter-subband excitations of holes confined at (100), (110), and (111) Si surfaces.

The MOS capacitor arrangement used in the light scattering experiments allowed changes in the carrier concentration over a wide range by changing the gate voltage. The measured spectra exhibit both broad and sharp bands that are

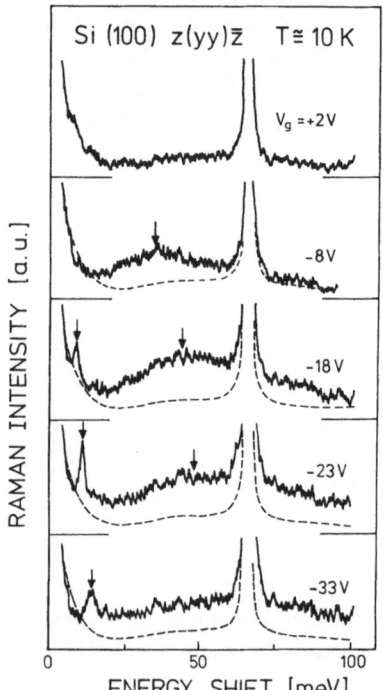

Fig. 4.35. Light scattering spectra of hole accumulation layers on (100) Si for different gate voltages [4.108a]

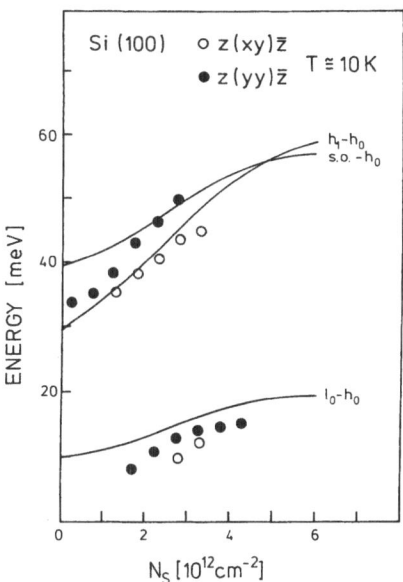

Fig. 4.36. Measured and calculated subband energy seperations for holes in (100) Si [4.108a]

related to the specific character of the structure of energy levels of the valence subbands. A typical polarized spectrum for holes at (100) surfaces of Si is shown in Fig. 4.35. The strong peak at 64.3 meV corresponds to optical phonons of the Si crystal. At the gate voltage of $+2$ V no holes are present at the surface. With increasing negative voltages, there is a broad band that shifts to higher energy as the hole concentration increases and an additional sharp peak which appears at 10 meV. For large negative voltages the broad band is observed as a background that extends to energies beyond 100 meV. The depolarized spectra show similar features but slightly shifted to lower energies.

These results have been compared with self-consistent hole subband calculations. Figure 4.36 compares measured and calculated peak energies as a function of hole concentration. The solid lines represent the energy differences between subbands at zero wavevector. The sharp peak at lower energies is close to the spacing between the l_0 and h_0 subbands. For low carrier densities the experimental points deviate from theory. This is possibly due to the fact that a finite depletion charge was assumed in the calculation, while the experiments were performed with accumulation layers for which the subband splitting vanishes with $n_s \to 0$. The peak energies of the broad structures are close to the calculated separation between the h_0 and h_1 subbands and to the energy difference between the spin-orbit split-off subband s. o. and h_0. The low density behavior that indicates a finite splitting at $n_s \to 0$ supports the interpretation that the spin-orbit split-off valence subband contributes to the spectra.

Contrary to the case of 2 D electron gases, the lineshapes of the hole subband transitions are strongly influenced by the different dispersions of the individual subbands. Figure 4.37 shows the calculated dispersion along the (010) direction

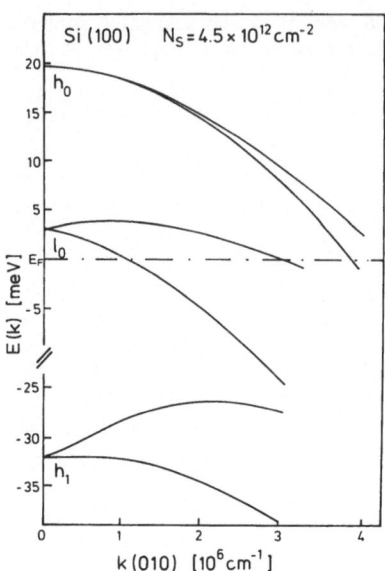

Fig. 4.37. Dispersion and surface electric field induced spin splitting of three subbands for a (100) surface of Si [4.108a]

at $n_s = 4.5 \times 10^{12}$ cm^{-2}. In bulk Si all the states have spin degenracy because of the inversion symmetry of the diamond structure. This degeneracy is lifted in the presence of a surface electric field. Each of these split branches has wavefunction admixtures of both spin orientations. However, their dispersions are different and show considerable warping (a dependence on the direction of K). The sharp and broad features in the measured spectra can be interpreted by the similarities and differences in the dispersions of the subbands. The $h_0 \rightarrow l_0$ transition appears as a relatively sharp band because the lower branch of the l_0 subband is nearly parallel to the h_0 subband. The broad structure is composed of different transitions, such as $h_0 \rightarrow h_1$ and $SO \rightarrow h_0$, in which the individual subbands exhibit very different dispersions.

Similar results have been obtained for hole accumulation layers with other surface orientations. The experimental results are in reasonable agreement with self-consistent subband calculations. However, a detailed comparison is difficult because of the broad spectral lineshapes. There are discrepancies between measured subband transitions in infrared absorptions and in light scattering. The differences are possibly due to the fact that the subband transitions exist over a relatively wide continuum and the techniques give peak intensities at different parts of the spectral range.

At high hole concentrations, light scattering from holes overlaps with the optical phonon of Si. A careful study of the lineshape indicates a coupling between the two excitations. The intensity of the phonon line is reduced and the phonon lineshape becomes asymmetric. This has been explained by quantum mechanical Fano-type interference [4.108b] similar to that extensively studied in bulk heavily doped p-type Si [4.7].

4.3.2 GaAs-(AlGa)As Quantum Wells

The first light scattering studies of high mobility 2 D hole gases confined in GaAs quantum wells were reported in p-type modulation doped GaAs-(AlGa)As heterostructures fabricated by molecular beam epitaxy [4.109, 110]. The layer sequence in the quantum wells is shown in Fig. 4.38. The spectra were excited by

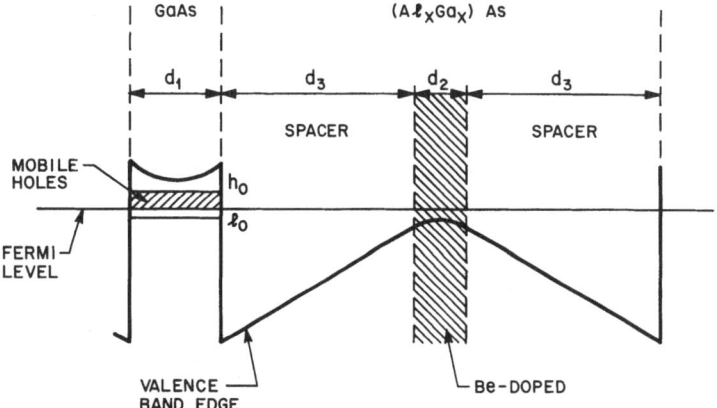

Fig. 4.38. Real space energy diagram of valence band edge in GaAs-(AlGa)As quantum well structures

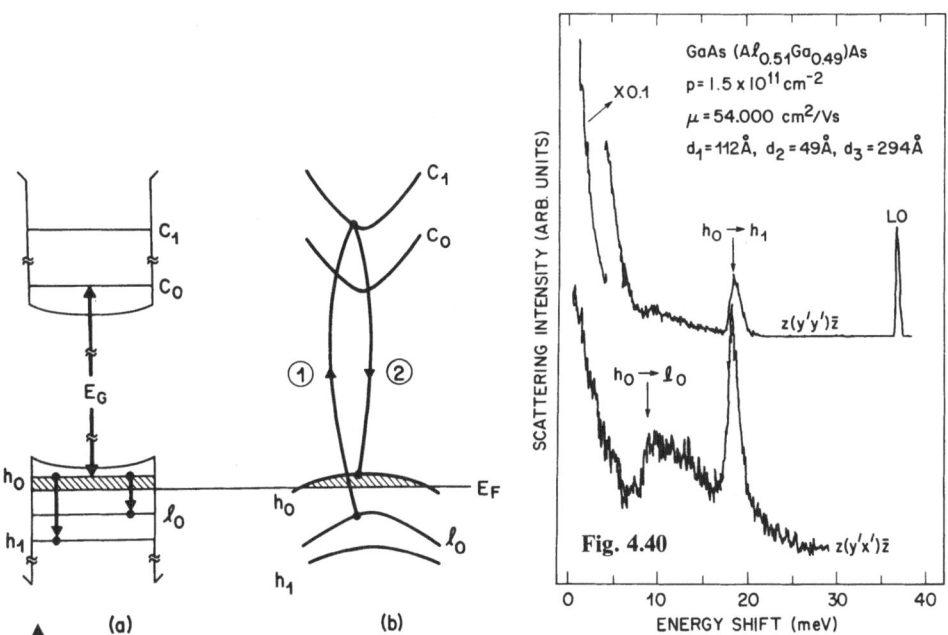

Fig. 4.39. Schematic representation of energy levels and subband dispersion in p-type GaAs-(AlGa)As quantum wells. Optical transitions active in resonant light scattering are also shown

Fig. 4.40. Polarized and depolarized spectra of 2 D holes in modulation doped GaAs quantum wells [4.109]

means of tunable infrared dye lasers with photon energies of 1.6–1.7 eV, chosen to be in resonance with optical transitions of the higher lying quantum well states derived from the Γ_8 and Γ_6 valence and conduction states. Figure 4.39a shows the two inter-subband transitions of interest and Fig. 4.39b shows the two optical transitions that participate in resonant light scattering by the $h_0 \rightarrow l_0$ excitations. This procedure for obtaining spectra by the 2D hole gases has the advantage of having large resonant enhancements and, at the same time, minimizing the background due to optical emission at the fundamental gap E_G.

Figure 4.40 shows the spectra measured in a sample with relatively low hole density and high mobility. The structures labeled $h_0 \rightarrow l_0$ and $h_0 \rightarrow h_1$ that are clearly seen in the $z(y'x')\bar{z}$ spectra and are also present in the $z(y'y')\bar{z}$ spectra, have been assigned to the excitations derived from these inter-subband transitions. The polarized $z(y'y')\bar{z}$ spectrum shows two additional features. One is the peak at the energy of the LO phonon of GaAs (at 36.6 meV). The other is the intense and broad structure at energies below those of conventional inter-subband transitions. Figure 4.41 shows that this low energy scattering actually has a maximum at an energy shift of ~ 1 meV. This feature is observed only in polarized spectra. Figure 4.42 shows spectra measured for a higher hole concentration. In these results the $h_0 \rightarrow l_0$ are much broader and the low energy scattering in the polarized spectrum (labeled LEEX, for low energy excitations) now shows a complex structure with a maximum at 4.7 meV and a cutoff at 10 meV.

The differences in spectral lineshapes of $h_0 \rightarrow l_0$ and $h_0 \rightarrow h_1$ excitations can be explained in terms of the differences in dispersion between the subbands. They are direct evidence of the complex valence subband structure predicted by extensive theoretical research [4.56–60]. The $h_0 \rightarrow h_1$ transitions appear as relatively narrow bands. This indicates that the two subbands are nearly parallel in a region of the 2D Brillouin zone. The $h_0 \rightarrow l_0$ excitations appear as broad

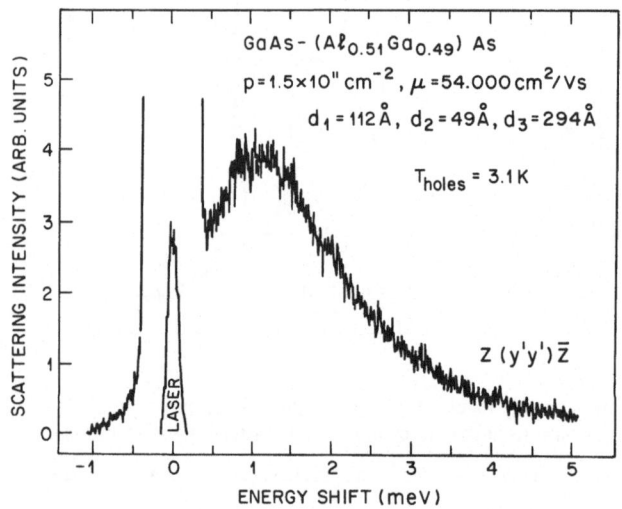

Fig. 4.41. Low energy part of a polarized spectrum like that in Fig. 4.40 [4.110]

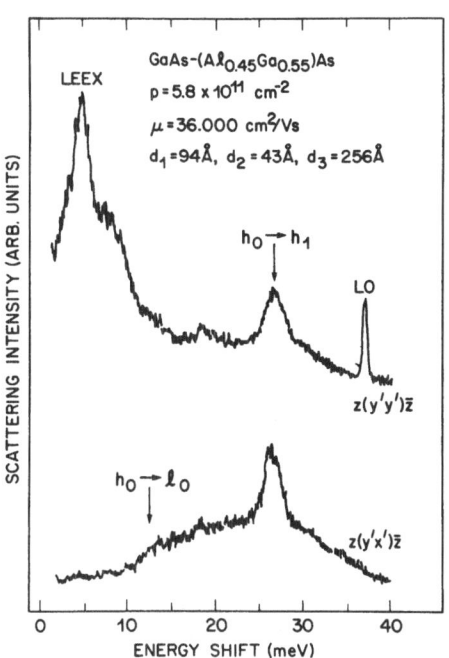

Fig. 4.42. Polarized and depolarized spectra of holes in symmetric GaAs quantum wells [4.109]

structures because the two subbands have very different disperions. The measured energies of $h_0 \rightarrow h_1$ excitations are in excellent agreement with the spacings of levels of a square quantum well model that incorporates the bandgap offsets at GaAs-(AlGa)As heterostructures [4.83–111]. For the $h_0 \rightarrow l_0$ transitions, however, this model gives energies that are higher than the measured ones. The discrepancy has been explained by the effective mass calculations shown in Fig. 4.34. In that figure we see that the onset of the continuum of $h_0 \rightarrow l_0$ excitations does not occur at $K=0$ but a the Fermi wavevector k_F. This interpretation of the light scattering results is possibly the first direct experimental evidence of electron-like dispersion of a valence subband in quantum well heterostructures.

A high magnetic field applied normal to the layers splits the $h_0 \rightarrow l_0$ and $h_0 \rightarrow h_1$ transitions into complex multiplets [4.59, 112, 113]. Figure 4.43 shows effective-mass calculations of the first three valence subbands using the parameters of the sample in Fig. 4.40. The variation of the Fermi level with B is displayed by the heavy line on the uppermost Landau levels. The wavefunction of each Landau level is a linear combination of several harmonic oscillator wavefunctions [4.57, 59, 113]. In the notation used in Fig. 4.43, the transitions allowed in light scattering by the dipole approximation have changes in Landau level quantum number $\Delta m = 0, \pm 2$; some of these transitions are indicated by vertical lines. Figure 4.44 shows results of *Heiman* et al. [4.112] for the energies of the strongest peaks in the light scattering spectra of $h_0 \rightarrow l_0$ and $h_0 \rightarrow h_1$ transitions of the low density sample. The solid lines show the energies calculated on the basis of the transitions shown in Figure 4.43. The $1 \rightarrow 1$ ($h_0 \rightarrow l_0$) transitions show good

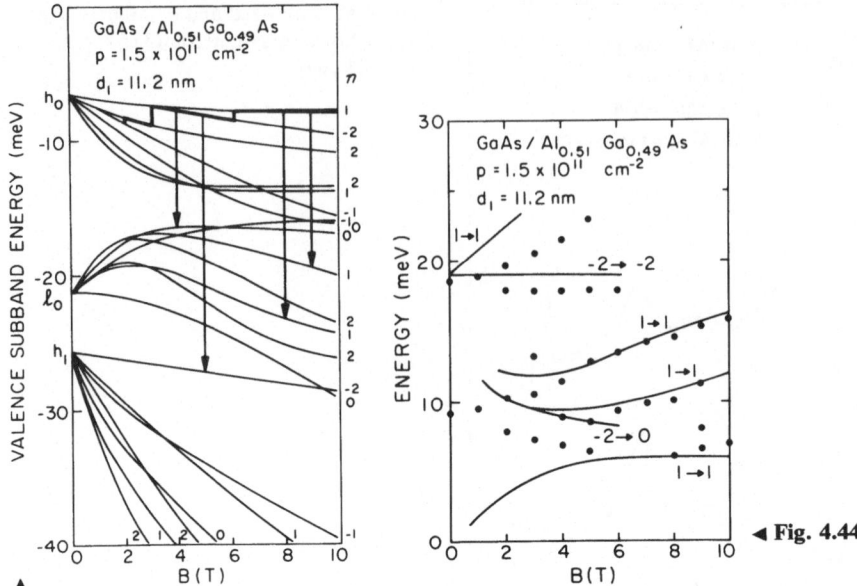

Fig. 4.43. Calculated valence Landau levels as function of magnetic field [4.112]

Fig. 4.44. Peak energies in inelastic light scattering spectra of holes in symmetric GaAs quantum wells as function of magnetic field. Points are experiment and lines are theory [4.112]

agreement above 5 Tesla, but poor coincidence at low fields. The $-2 \to -2$ ($h_0 \to h_1$) and $-2 \to 0$ ($h_0 \to l_0$) transitions are expected to disappear relatively abruptly at $B = 6.2$ Tesla, when the $m = 2$ level is no longer occupied. Two peaks disappear between 6 and 7 Tesla. The upper peak at 18 meV is field-independent. This, as seen in Fig. 4.43, is characteristic of the $-2 \to -2$ transiton. The peak at lower energy, labeled $-2 \to 0$, also disappears for $B \gtrsim 5$ Tesla. Although there are discrepancies of $\sim 1-2$ meV between measured and calculated energies, the experiments confirm rather specific predictions of theory, including the electron-like behavior of some of the Landau levels of the l_0 subbands.

The intensities of these light scattering spectra display resonant enhancements of the type represented in Fig. 4.39b. The measured profiles of the resonant enhancements are shown in Fig. 4.45 for the case of polarized spectra of the high density sample. The profiles of three excitations, measured in spectra like those in Fig. 4.42, as well as that of the LO phonon, not shown in Fig. 4.45, have maxima at the energies of optical transitons from the valence subbands to the c_1 states of the conduction band, as described by the process shown in Fig. 4.39b. For this mechanism of resonant light scattering, the predicted width of the resonant enhancement profile is $\Delta E \approx \hbar^2 k_F^2 / 2m_e^*$, where m_e^* is the conduction band effective mass and k_F is the Fermi wavevector. For a free hole density of 5.8×10^{11} cm^{-2}, $\Delta E \approx 20$ meV, in reasonable agreement with the measured width.

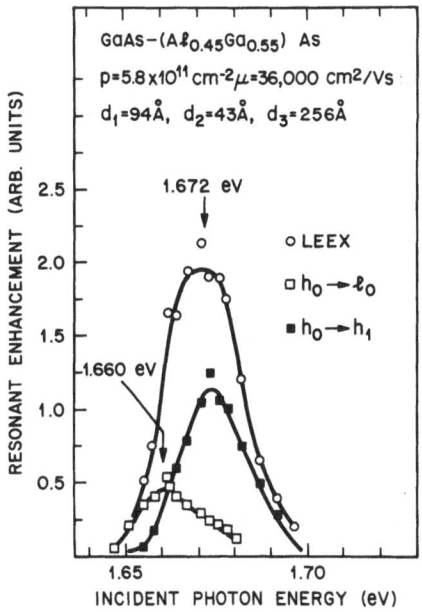

RESONANT ENHANCEMENT (ARB. UNITS)

GaAs$-($A$\ell_{0.45}$Ga$_{0.55})$ As

p=5.8x10^{11} cm^{-2} μ=36,000 cm^2/Vs

d$_1$=94Å, d$_2$=43Å, d$_3$=256Å

1.672 eV

○ LEEX

□ h$_0 \rightarrow \ell_0$

■ h$_0 \rightarrow$ h$_1$

1.660 eV

INCIDENT PHOTON ENERGY (eV)

Fig. 4.45. Resonant enhancement of inelastic light scattering in p-type symmetric GaAs quantum well [4.110]

The spectral lineshapes and peak positions of the $h_0 \rightarrow h_1$ excitations were found to change with incident photon energy [4.109, 110]. This was interpreted in terms of the resonant enhancement mechanism of Fig. 4.39b as due to the large width of the $h_0 \rightarrow h_1$ excitations because of the non-parabolicity and warping of the subbands. *Goldberg* et al. [4.118] have recently reported that the

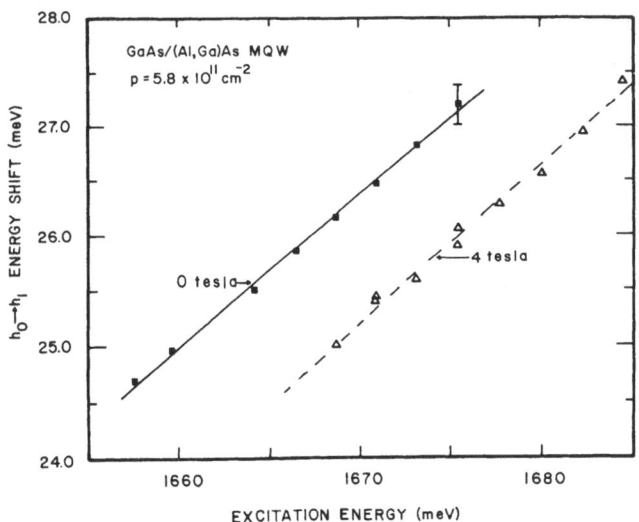

h$_0 \rightarrow$ h$_1$ ENERGY SHIFT (meV)

GaAs/(Al,Ga)As MQW

p = 5.8 x 10^{11} cm^{-2}

0 tesla

4 tesla

EXCITATION ENERGY (meV)

Fig. 4.46. Energy of the $h_0 \Rightarrow h_1$ inter-subband transition as function of excitation energy at two magnetic fields [4.114]

shift in the energy of the $h_0 \rightarrow h_1$ excitations with changes in photon energy persists in the presence of a magnetic field. The results for the sample with high free hole density are shown in Fig. 4.46 at $B=0$ and $B=4$ Tesla. They indicate that the shifts of the peak energies do not depend strongly on magnetic field. This is an unexpected result that cannot be easily explained in terms of non-parabolicity because of the quantization of the in-plane motion in a high perpendicular magnetic field.

The shift of the $h_0 \rightarrow h_1$ inter-subband transition can be understood in terms of the inhomogeneous broading caused by fluctuations in the well thickness. This effect can be evaluated within the simple infinite square well model. In this case, the change in the bandgap energy δE_g due to a fluctuation δd_1 in well thickness is

$$\delta E_g = A_g \frac{1}{\mu^*},$$ (4.50)

where $(1/\mu^*) = (1/m_e^* + 1/m_h^*)$ is the reduced effective mass, and m_h^* is the heavy hole effective mass. The change in the $h_0 \rightarrow h_1$ transition energy is

$$\delta E_{01}^h = A_{01}^h \frac{1}{m_h^*}.$$ (4.51)

It is straightforward to show that in the infinite square-well model $(A_{01}^h/A_g) = 3/4$. If different well widths resonate at different energies $E_g + \delta E_g$ the slope of the lines in Fig. 4.46 is given by

$$\frac{\delta E_{01}^h}{\delta E_g} = \frac{3}{4} \frac{\mu^*}{m_h^*}$$ (4.52)

taking $m_c^* = 0.085$ (the mass of the c_1 subband) and $m_h^* = 0.37$ a slope of 0.14 is calculated, in excellent agreement with the slopes of 0.14 for the data shown in Fig. 4.46. The total variation in the energy of the $h_0 \rightarrow h_1$ transition is 2 meV. It can be accounted for by a difference of two atomic layers, a width fluctuation that is not unusual for the GaAs-(AlGa)As quantum well heterostructures.

In Fig. 4.47 we see that a relatively small temperature increase from 3 to 20 K causes a large broadening of the scattering assigned to $h_0 \rightarrow h_1$ excitations. This behavior is in striking contrast to that of the inter-subband excitations of 2D electron gases, in which the light scattering spectra show substantial lineshape changes only for temperatures $T \gtrsim 100$ K. A possible interpretation of the broadening of these hole transitions is based on the different dispersions and non-parabolicity of the subbands. Given the large mass of the holes, a small temperature increase results in occupation of states over a larger fraction of the Brillouin zone, thus increasing the energy range covered by the $h_0 \rightarrow l_0$ and $h_0 \rightarrow h_1$ transitions. However, further investigation of this temperature broadening is required. For example, effects related to localization of holes in the quantum well

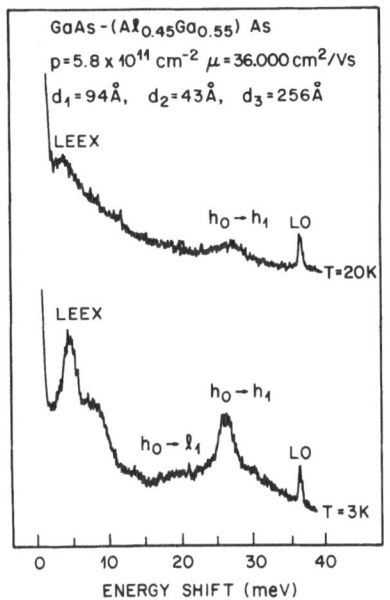

GaAs-$(Al_{0.45}Ga_{0.55})$ As

p=5.8 x 10^{11} cm^{-2} μ=36.000 cm^2/Vs

d_1=94Å, d_2=43Å, d_3=256Å

LEEX

$h_0 \rightarrow h_1$ LO

T=20K

LEEX

$h_0 \rightarrow h_1$

$h_0 \rightarrow l_1$

LO

T=3K

0   10   20   30   40

ENERGY SHIFT (meV)

Fig. 4.47. Temperature dependence of the $z(y'y')\bar{z}$ polarized spectra shown in Fig. 4.42 [4.110]

fluctuations described above could be partly responsible for the temperature dependence of the spectra.

The low energy scattering shown in Figs. 4.41 and 4.42 is below the energy of conventional excitations and is observed only in polarized spectra. Such a selection rule is taken as evidence that the excitations are associated with charge density fluctuations. Although the positions and lineshapes of the low energy scattering have a large dependence on hole density, they do not show a significant dependence on scattering wavevector. This indicates that they are not multilayer plasmons like those considered in Sect. 4.2.1a. On the other hand, as seen in Fig. 4.47, scattering by the low energy excitations shows temperature broadening similar to that of inter-subband excitations. This behavior led to speculation [4.109] that the existence of strong low energy scattering might be evidence of a lifting of the Kramers degeneracy of the ground valence subband, similar to that observed in single heterojunctions [4.115]. However, this interpretation conflicts with magneto-transport data in these symmetric quantum wells [4.116].

A final interpretation of the low energy collective modes is not available at the present time. The fact that they have the temperature dependence of the inter-subband excitations is proof that they originate in the 2D hole gas. Because their energies are low, they can be used to determine the hole temperatures from Stokes-anti-Stokes intensity ratios. In fact, the spectrum of Fig. 4.41 shows that the hole temperature can be kept below 3 K in these light scattering experiments. Light scattering by low energy excitations has been used to determine the temperature of holes in order to obtain energy loss rates of hot holes via acoustic phonon-hole interactions [4.117].

Only minor differences are observed in the positions of $h_0 \rightarrow h_1$ bands in $z(y'y')\bar{z}$ and $z(x'y')\bar{z}$ spectra. This behavior is very different from that found in n-type quantum wells, discussed in Sect. 4.2.2, where the spectral differences are due to depolarization field effects of the inter-subband excitations. *Ando* [4.60], in a self-consistent Hartree calculation of light scattering spectra of holes in GaAs quantum wells, predicts larger depolarization field effects than have been measured. The absence in the spectra of a phonon-like mode coupled to free holes is also evidence of weak depolarization field effects.

Resonant Raman scattering by longitudinal optical phonons in the p-type quantum wells has recently been reported by two groups [4.118, 119]. The resonant enhancement profiles in the two cases are very different. The interpretation of the results reported in [4.118] involves breakdown of wavevector conservation induced by ionized impurity scattering. *Zucker* et al. [4.119], on the other hand, have interpreted their results in higher mobility p-type and undoped quantum wells in terms of intrinsic Raman processes, and suggested that breakdown of wave vector conservation could be induced by fluctuations in the layer width that are intrinsic to quantum wells. If this is the case, then there would be only minor correlation between the shapes of resonant enhancement profiles and the carrier mobility [4.119].

Resonant inelastic light scattering by photoexcited holes has also been reported in Be-doped multiple quantum wells [4.120] and, more recently, in GaAs-(AlGa)As heterostructures grown in the [111] direction [4.121]. The work in the doped system will be considered in the next section. The results obtained in quantum wells with the superlattice along the [111] direction give further information on the Luttinger parameters of GaAs.

4.4 Shallow Impurities in Quantum Wells

Impurities with shallow energy states confined in GaAs quantum wells were first investigated by the photoluminescence of transitions across the energy gap [4.122–124]. Inelastic light scattering [4.79, 120, 125, 126] and infrared optical absorption [4.127, 128] give a direct insight into the structure of impurity energy states. This brief section presents highlights of the light scattering work.

Theoretical investigations [4.127, 129–132] have shown that in GaAs-(AlGa)As heterostructures, quantum confinement leads to significant shifts and splittings of impurity levels when the well width becomes comparable to the effective Bohr radius of the impurity state. The magnitude of confinement effects depends on the quantum well parameters and also on the position of the impurity within the well. The latter is a source of inhomogeneous broadening. For large well depths confinement leads to an enhancement of the binding energy. Another important aspect of confinement is the appearance of resonances, i.e., impurity states derived from the higher subbands that overlap with the 2D continuum of the lower subbands. The donor states derive from the Γ_6 states of the conduction band. They show, within the effective mass approximation, a hydrogen-like spectrum with Bohr radius $a_0 \approx 100$ Å. The main effect of confinement is a substantial increase of the binding energy [4.129, 130] and the appearance of a ladder of impurity resonances [4.132]. The acceptor states derive mainly from the four-fold degenerate Γ_8 states of the valence band, which split into two Kramer doublets of Γ_6 and Γ_7 symmetry [4.135].

Inelastic light scattering by donors in GaAs-(Al_xGa_{1-x})As quantum well heterostructures has been reported by *Perry* et al. [4.79, 125, 126]. They observed a line assigned to $1s \rightarrow 2s$ transitions and additional peaks attributed to donor-resonance excitations. The investigations were carried out in symmetric quantum wells having well thicknesses in the range 88 Å–460 Å. Si donors were incorporated at the centers of the GaAs wells with concentrations in the range $1 - 5 \times 10^{15}$ cm^{-3}. Data were obtained with incident photon energies in resonance with the $E_0 + \Delta_0$ optical gap of GaAs. Figure 4.48 shows spectra from 88 Å quantum wells. The broad feature in the lowest spectrum is due to the $1s \rightarrow 2s$ transitions, which could be measured only in a very narrow range of incident photon energies. With increasing photon energy the scattering transforms into $E_0 + \Delta_0 (D^0 - h)$ photoluminescence [4.125]. The positions of the $1s \rightarrow 2s$ tran-

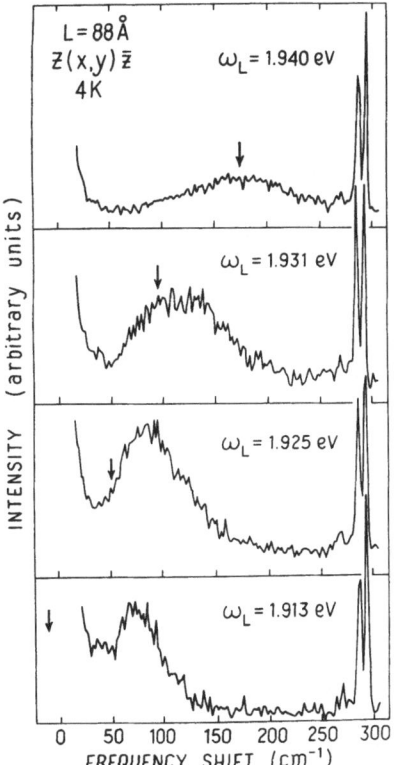

Fig. 4.48. Raman spectra Si-doped GaAs quantum wells [4.125]

sition measured in the light scattering spectra is in good agreement with the calculated values [4.129].

Light scattering by impurity resonances has been shown in Fig. 4.26 for the case of a quantum well of width 460 Å. The higher energy components of the doublets are due to excitations from the ground state of the Si donors into $1s$-resonances associated with the first excited conduction subband c_1. The lower energy components are, as we have already mentioned, the inter-subband transitions of the photoexcited electrons. The doublet splittings reflect the lower binding energies of the resonances relative to the ground state of the impurity. The measured values are in good agreement with the calculated results [4.132]. The rapid quenching of the high energy components with increasing temperatures is a result of donor ionization.

Light scattering by Be-acceptors in GaAs-$(Al_xGa_{1-x})As$ quantum wells was studied by *Gammon* et al. [4.120]. The quantum wells were doped with Be in the concentration range $7 \times 10^{15} - 3 \times 10^{16}$ cm^{-3}. The center- or edge-doped quantum wells have thicknesses in the range 92–165 Å. The incident photon energies were in resonance with higher excited states of the fundamental gap. Figure 4.49 displays the effects of confinement on acceptor levels. The three acceptor transitions *A*, *B*, and *C* involve levels associated with the $1s$ and $2s$ $J = 3/2$ states of the bulk. The measured positions of the *C* line agree well with theory in the case of a center-doped well, but disagree in the case of edge-doped wells [4.131]. The *A* and *B* lines are shown in the spectra of Fig. 4.49 for center-doped samples. Their positions and widths are close to theoretical predictions [4.131].

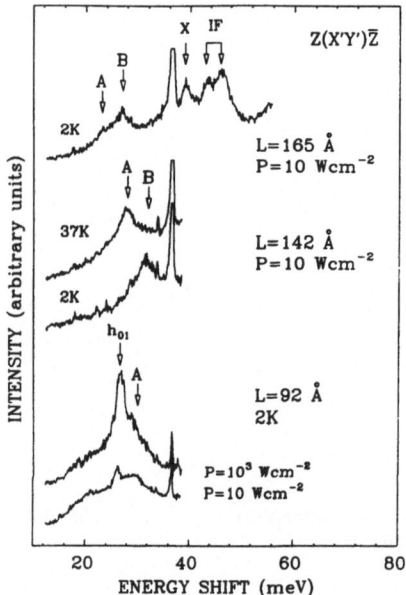

Fig. 4.49. Well-width dependence of Raman spectra of Be-doped GaAs quantum wells [4.120]

4.5 Concluding Remarks

We have presented a review of recent results that highlight applications of the light scattering method in studies of quasi-two-dimensional electron systems. We have seen that the method can be used to measure the excitations associated with the free electron motion in the plane as well as that of the restricted motion normal to the plane. These features, in conjunction with the very advantageous option to measure spectra of collective and single particle excitations, make light scattering a very versatile spectroscopic tool. With the use of optical multi-channel detection, and taking advantage of large resonant enhancements, it is possible to carry out experiments at very low laser power densities. This creates the possibility to study remarkable low temperature many-body phenomena, like the fractional quantization of the Hall effect, that are at the frontier of condensed-matter physics. We expect important applications in the area of time-resolved spectroscopy, where the method could reveal dynamical behavior of hot electrons. There is increasing interest in systems where the electrons have one-dimensional behavior (quantum wires) and also in zero-dimensional systems (quantum dots). We expect that inelastic light scattering, especially with the new techniques of micro-Raman spectroscopy, will play a prominent role in the elucidation of the intriguing properties of these novel semiconductor microstructures.

References

4.1   T. Ando, A.B. Fowler, F. Stern: Rev. Mod. Phys. **54**, 437 (1982)
4.2   T. Ando (ed.): Proc. 6th Int. Conf. Electronic Properties of Two-Dimensional Systems, Surf. Sci. Vol. **170** (1986)
4.3   J.M. Worlock (ed.): Proc. 7th Int. Conf. Electronic Properties of Two-Dimensional Systems, Surface Sci. Vol. **196** (1988)
4.4   D.S. Chemla, A. Pinczuk (eds.): Special Issue on Semiconductor Quantum Wells and Superlattices: Quantum Wells and Applications, IEEE J. Quantum Electron. QE-**22**, 1609-1921 (1986)
4.5   A. Mooradian: In *Festkörperprobleme IX*, ed. by O. Madelung (Pergamon-Vieweg, Braunschweig 1969) p. 74
4.6   M.V. Klein: In *Light Scattering in Solids* ed. by M. Cardona, Topics Appl. Phys., Vol. 8, 2nd ed. (Springer, Berlin, Heidelberg 1983) p. 147
4.7   G. Abstreiter, M. Cardona, A. Pinczuk: In *Light Scattering in Solids IV*, ed. by M. Cardona, G. Güntherodt, Topics Appl. Phys., Vol. 54 (Springer, Berlin, Heidelberg 1984) p. 5
4.8   A. Pinczuk, L. Brillson, E. Anastassakis, E. Burstein: Phys. Rev. Lett. **27**, 317 (1971)
4.9   A. Pinczuk, G. Abstreiter, R. Trommer, M. Cardona: Solid State Commun. **30**, 429 (1979)
4.10   E. Burstein, A. Pinczuk, S. Buchner: In *Physics of Semiconductors*, ed. by B.L.H. Wilson (The Institute of Physics, London, 1979) p. 1231
4.11   G. Abstreiter, K. Ploog: Phys. Rev. Lett. **42**, 1308 (1979)

4.12   A. Pinczuk, H.L. Störmer, R. Dingle, J.M. Worlock, W. Wiegmann, A.C. Gossard: Solid State Commun. **32**, 1001 (1979)
4.13   A. Pinczuk, J.M. Worlock: Surface Sci. **113**, 69 (1982)
4.14   G. Abstreiter: In *Molecular Beam Epitaxy and Heterostructures*, ed. by L.L. Chang, K. Ploog (Nijhoff, Dordrecht,1985) p. 425
4.15   G. Abstreiter, R. Merlin, A. Pinczuk: In [4.4], p. 1771
4.16   E. Burstein, A. Pinczuk, D.L. Mills: Surface Sci. **98**, 451 (1980)
4.17   D. Hamilton, A.L. McWorter: In *Light Scattering Spectra of Solids*, ed. by G.B. Wright (Springer, New York 1969) p. 309
4.18   S. Katayama, T. Ando: J. Phys. Soc. Jap. **54**, 1615 (1985)
4.19   A.C. Tsellis, J.J. Quinn: Phys. Rev. B **29**, 3318 (1984)
4.20   S. Das Sarma: App. Surf. Sci. **11/12**, 535 (1982)
4.21   J.K. Jain, P.B. Allen: Phys. Rev. Lett. **54**, 947 (1985); Phys. Rev. B **32**, 997 (1985)
4.22   M. Babiker, N.C. Constantinou, M.G. Cottam: Solid State Commun. **57**, 877 (1986)
4.23   H. Chiu, J.J. Quinn: Phys. Rev. B **9**, 4724 (1974)
4.24   C. Kallin, B.I. Halperin: Phys. Rev. B **30**, 5655 (1984)
4.25   T. Ando: J. Phys. Soc. Jap. **51**, 3893 (1982)
4.26   D. Olego, A. Pinczuk, A.C. Gossard, W. Wiegmann: Bull. Am. Phys. Soc. **28**, 447 (1983)
4.27   A. Pinczuk, H.L. Störmer, A.C. Gossard, W. Wiegmann: In Proc. 17th Int. Conf. Physics of Semiconductors, ed. by J.D. Chadhi, W.A. Harrison (Springer, New York 1985) p. 329
4.28   A. Pinczuk, J.P. Valladares, C.W. Tu, A.C. Gossard, J.H. English: Bull. Am. Phys. Soc. **32**, 756 (1987)
4.29   D. Olego, A. Pinczuk, A.C. Gossard, W. Wiegmann: Phys. Rev. B **25**, 7867 (1982)
4.30   A. Pinczuk, J.M. Worlock, H.L. Störmer, A.C. Gossard, W. Wiegmann: J. Vac. Sci. Technol. **19**, 561 (1981)
4.31   A. Pinczuk, J.P. Valladares, C.W. Tu, D. Heiman, A.C. Gossard, J.H. English: Bull. Am. Phys. Soc. **33**, 573 (1988)
4.32   A.L. Fetter: Ann. Phys. **88**, 1, 1976
4.33   S. Das Sarma, J.J. Quinn: Phys. Rev. B **25**, 7603 (1982)
4.34   W.L. Bloss, E.M. Brody: Solid State Commun. **43**, 523 (1982)
4.35   A.H. MacDonald: J. Phys. B **18**, 1003 (1985)
4.36   W. Kohn: Phys. Rev. **123**, 1242 (1961)
4.37   A. Pinczuk, D. Heiman, A.C. Gossard, J.H. English: In Proc. 18th Int. Conf. Physics of Semiconductors, ed. by O. Engström (World Scientific, Singapore 1987) p. 557
4.38   G.F. Giuliani, J.J. Quinn: Phys. Rev. Lett. **51**, 919 (1983)
4.39   J.W. Wu, P. Hawrylak, J.J. Quinn: Phys. Rev. Lett. **55**, 879 (1985)
4.40   J.K. Jain, P.B. Allen: Phys. Rev. Lett. **54**, 2437 (1985)
4.41   R. Merlin, K. Bajema, R. Clashe, F.Y. Juang, P.K. Bhattacharya: Phys. Rev. Lett. **55**, 1768 (1985)
4.42   P. Hawrylak, J.J. Quinn: Phys. Rev. Lett. **57**, 380 (1986)
4.43   S. Das Sarma, A. Kobayashi, R.E. Prange: Phys. Rev. B **34**, 5309 (1986)
4.44   S. Das Sarma, A. Kobayashi, R.E. Prange: Phys. Rev. Lett. **56**, 1280 (1986)
4.45   A. Pinczuk, M.G. Lamont, A.C. Gossard: Phys. Rev. Lett. **56**, 2092 (1985)
4.46a G. Fasol, N. Mestres, H.P. Hughes, A. Fischer, K. Ploog: Phys. Rev. Lett. **56**, 2517 (1986)
4.46b T. Egeler, S. Beeck, G. Abstreiter, G. Weimann, W. Schlapp: Superlattices and Microstructures (1988) to be published
4.47   F. Stern, Phys. Rev. Lett. **18**, 546 (1967)
4.48   G. Fasol, N. Mestres, M. Dobers, A. Fischer, K. Ploog: Phys. Rev. B **36**, 1536 (1987)
4.49   N. Mermin: Phys. Rev. B **1**, 2362 (1970)
4.50   J.M. Worlock, A. Pinczuk, Z.J. Tien, C.H. Perry, H. Störmer, R. Dingle, A.C. Gossard, W. Wiegmann, R.L. Aggarwal: Solid State Commun. **40**, 867 (1981)

4.51   Z.J. Tien, J.M. Worlock, C.H. Perry, A. Pinczuk, R.L. Aggarwal, H.L. Störmer, A.C. Gossard, W. Wiegmann: Surf. Sci. **113**, 89 (1982)

4.52   J.M. Worlock, A.C. Maciel, C.H. Perry, Z.J. Tien, R.L. Aggarwal, A.C. Gossard, W. Wiegmann: In *Application of High Magnetic Fields in Semiconductor Physics*, ed. by G. Landwehr (Springer, Heidelberg 1983) p. 186

4.53   S. Das Sarma: Phys. Rev. B **29**, 2334 (1984)

4.54   J.K. Jain, S. Das Sarma: Phys. Rev. B **36**, 5949 (1987)

4.55   W. Beinvogl, F. Koch: Phys. Rev. Lett. **40**, 1736 (1978)

4.56   M. Altarelli, U. Ekenberg, A. Fasolino: Phys. Rev. B **32**, 5138 (1985)

4.57   G. Landwehr, E. Bangert: Superlattices and Microstructures **1**, 363 (1985)

4.58   G.D. Sanders, Y.C. Chang: Phys. Rev. B **31**, 6892 (1985)

4.59   D.A. Broido, L.J. Sham: Phys. Rev. B **31**, 888 (1985);
        S. Eric-Yang, D.A. Broido, L.J. Sham: Phys. Rev. B **32**, 6630 (1985)

4.60   T. Ando: J. Phys. Soc. Japan **54**, 1528 (1985)

4.61   A. Pinczuk, J.P. Valladares, D. Heiman, A.C. Gossard, J.H. English, C.W. Tu, L.N. Pfeiffer, K. West: Phys. Rev. Lett. **61**, 2701 (1988)

4.62   A. Pinczuk, J.M. Worlock, H.L. Störmer, R. Dingle, W. Wiegmann, A.C. Gossard: Solid State Commun. **36**, 43 (1980)

4.63   M.V. Klein: In [4.4] p. 1760

4.64   W.P. Chen, Y. Chen, E. Burstein: Surf. Sci. **58**, 263 (1976)

4.65   S.J. Allen, Jr., D.C. Tsui, B. Vinter: Solid State Commun. **20**, 425 (1976)

4.66   D.A. Dahl, L.J. Sham: Phys. Rev. B **16**, 651 (1977)

4.67   A.C. Tsellis, J.J. Quinn: Solid State Commun. **46**, 779 (1983)

4.68   C. Zeller: Ph.D. Thesis, Tech. Univ. Munich (1984)

4.69   G. Abstreiter, T. Egeler, S. Beeck, A. Seilmeier, H.J. Huebner, G. Weimann, W. Schlapp: In [4.3] p. 613

4.70a   A. Seilmeier, H.J. Huebner, G. Abstreiter, G. Weimann, W. Schlapp, Phys. Rev. Lett. **59**, 1345 (1987)

4.70b   R. Zachai, M. Besson, T. Egeler, M. Zachau, G. Abstreiter, G. Weimann, W. Schlapp: to be published

4.71   L.C. West, S.J. Eglash: Appl. Phys. Lett. **46**, 1156 (1985)

4.72   F. Koch: In *Interfaces, Quantum Wells and Superlattices*, NATO ASI Series, ed. R. Taylor (1988) p. 67

4.73   B.F. Levine, R.J. Malik, J. Walker, C.G. Bethea, D.A. Kleinman, J.M. Vandenberg: Appl. Phys. Lett. **50**, 273 (1987)

4.74   A. Seilmeier, H.J. Hübner, M. Wörner, G. Abstreiter, G. Weimann, W. Schlapp: Solid State Electronics **31**, 767 (1988)

4.75   D.Y. Oberli, D.R. Wake, M.V. Klein, J. Klem, T. Hendersen, H. Morkoc: Phys. Rev. Lett. **59**, 696 (1987)

4.76   R. Sooryakumar, A. Pinczuk, A.C. Gossard, W. Wiegmann: Phys. Rev. B **31**, 2578 (1985)

4.77   A. Pinczuk, J. Shah, A.C. Gossard, W. Wiegmann: Phys. Rev. Lett. **46**, 1341 (1981)

4.78   K. Bajema, R. Merlin, F.Y. Juang, S.C. Hong, J. Singh, P.K. Bhattacharya: Phys. Rev. B **36**, 1300 (1987)

4.79   T.A. Perry, R. Merlin, B.V. Shanabrook, J. Comas: Phys. Rev. Lett. **54**, 2623 (1985)

4.80   R. Borroff, R. Merlin, R.L. Greene, J. Comas: In [4.3] p. 626

4.81   R. Merlin: Solid State Commun. **64**, 99 (1987)

4.82   J.C. Maan: In *Two-Dimensional Systems, Heterostructures and Superlattices*, ed. by G. Bauer, F. Kucher, H. Heinrich (Springer, Berlin, Heidelberg 1984) p. 183

4.83   J. Menéndez, A. Pinczuk, D.J. Werder, A.C. Gossard, J.H. English: Phys. Rev. B **33**, 8863 (1986)

4.84   J. Menéndez, A. Pinczuk, A.C. Gossard, M.G. Lamont, F. Cerdeira: Solid State Commun. **61**, 601 (1987)

4.85   J. Menéndez, A. Pinczuk, D.J. Werder, J.P. Valladares, T.H. Chiu, W.T. Tsang: Solid State Commun. **61**, 703 (1987)

4.86   J. Menéndez, A. Pinczuk, D.J. Werder, S.K. Sputz, R.C. Miller, D.L. Sivco, A.Y. Cho: Phys. Rev. B**36**, 8165 (1987)
4.87   J. Menendez, A. Pinczuk: to be published in IEEE J. Quantum Electron. (1988)
4.88   J.A. Kash, J.C. Tsang, J.M. Hvam: Phys. Rev. B**54**, 2151 (1985)
4.89   G.H. Döhler, H. Kuenzel, D. Olego, K. Ploog, P. Ruden, H.J. Stolz, G. Abstreiter: Phys. Rev. Lett. **47**, 864 (1981)
4.90   C. Zeller, B. Vinter, G. Abstreiter, K. Ploog: Phys. Rev. B**26**, 2124 (1982)
4.91   G. Fasol, P. Ruden, K. Ploog: J. Phys. C**17**, 1395 (1984)
4.92   N. Krischke, A. Zrenner, G. Abstreiter, K. Ploog (unpublished), see [4.15]; N. Krischke: Diploma Thesis, Tech. Univ. Munich (1986)
4.93   K. Ploog, G.H. Döhler: Advances in Phys. **32**, 285 (1983)
4.94   G.H. Döhler: In *Two-Dimensional Systems: Physics and New Devices*, ed. by G. Bauer, F. Kuchar, H. Heinrich (Springer, Berlin, Heidelberg 1986) p. 270
4.95   G. Abstreiter: In *Two-Dimensional Systems, Heterostructures and Superlattices*, ed. by G. Bauer, F. Kuchar, H. Heinrich (Springer, Berlin, Heidelberg 1984) p. 232
4.96   C.J. Chang-Hasmain, G. Hasmain, G.H. Döhler, N.M. Johnson, J.N. Miller, J.R. Whinnery, A. Dienes: SPIE Vol. 892, *Quantum Wells and Superlattices* (1987) p. 45
4.97   P. Ruden, G.H. Döhler: Phys. Rev. B**27**, 2538 (1983); ibid B**27**, 3547 (1983)
4.98   E. Bangert, K. von Klitzing, G. Landwehr: In Proc. 12th Int. Conf. on Physics of Semiconductors, ed. by M.H. Pilkuhn, Teubner, Stuttgart (1974) p. 714
4.99   F.J. Okhawa, J. Uemura: Prog. Theor. Phys. Suppl. **57**, 164 (1975)
4.100   F.J. Okhawa: J. Phys. Soc. Japan **41**, 122 (1976)
4.101a G. Bastard, J.A. Brum: In [4.4]p. 1625
4.101b O. Madelung, M. Schulz, H. Weiss (eds.): Landolt-Börnstein Tables, Vol. 17a (Springer, Berlin, Heidelberg 1982)
4.102   J.M. Luttinger, W. Kohn: Phys. Rev. **97**, 869 (1955)
4.103   P. Kneschaurek, A. Kamgar, J.F. Koch: Phys. Rev. B**14**, 1610 (1976)
4.104   A. Kamgar: Solid State Commun. **21**, 823 (1977)
4.105   U. Claessen, F. Martelli, C. Mazure, F. Koch: (1983) unpublished
4.106   A.D. Wieck, E. Batke, D. Heitmann, J.P. Kotthaus: Phys. Rev. B**30**, 4653 (1984)
4.107   G. Abstreiter, U. Claessen, G. Tränkle: Solid State Commun. **44**, 673 (1982)
4.108a M. Baumgartner, G. Abstreiter, E. Bangert: J. Phys. C**17**, 1617 (1984)
4.108b M. Baumgartner, G. Abstreiter: Surf. Sci. **142**, 357 (1984)
4.109   A. Pinczuk, H.L. Störmer, A.C. Gossard, W. Wiegmann: In *Proc. of the 17th Int. Conf. on the Physics of Semiconductors*, ed. by J.D. Chadi, W.A. Harrison (Springer, New York 1985) p. 329
4.110   A. Pinczuk, D. Heiman, R. Sooryakumar, A.C. Gossard, W. Wiegmann: In [4.2] p. 573
4.111   R.C. Miller, D.A. Kleinman, A.C. Gossard: Phys. Rev. B**29**, 7083 (1985)
4.112   D. Heiman, A. Pinczuk, A.C. Gossard, J.H. English, A. Fasolino, M. Altarelli: In *Proc. of the 18th Int. Conference on the Physics of Semiconductors*, ed. by O. Engström (World Scientific, Singapore 1987) p. 617
4.113   M. Altarelli: In *Festkörperprobleme* XXV (Advances in Solid State Physics), ed. by P. Grosse (Vieweg, Braunschweig 1985) p. 381
4.114   B.B. Goldberg, D. Heiman, A. Pinczuk, A.C. Gossard: In [4.3], p. 619
4.115   H.L. Störmer, A. Chang, Z. Schlesinger, D.C. Tsui, A.C. Gossard, W. Wiegmann: Phys. Rev. Lett. **51**, 126 (1983)
4.116   J.P. Eisenstein, H.L. Störmer, V. Narayanamurti, A.C. Gossard, W. Wiegmann: Phys. Rev. Lett. **53**, 2579 (1984)
4.117   A. Pinczuk, J. Shah, A.C. Gossard: Solid State Electron. **31**, 477 (1988)
4.118   T. Suemoto, G. Fasol, K. Ploog: Phys. Rev. B**34**, 6034 (1986)
4.119   J.E. Zucker, A. Pinczuk, D.S. Chemla: Phys. Rev. B, in press
4.120   D. Gammon, R. Merlin, W.T. Masselink, H. Morkoc: Phys. Rev. B**33**, 2919 (1986)
4.121   B.V. Shanabrook, O.J. Glembocki, D.A. Broido, W.I. Wang: To be published

4.122   R.C. Miller, A.C. Gossard, W.T. Tsang, O. Munteanu: Phys. Rev. B**25**, 3871 (1982)
4.123   P.W. Yu, S. Chaudhuri, D.C. Reynolds, K.K. Bajaj, C.W. Litton, W.T. Masselink, R. Fischer, H. Morkoc: Solid State Commun. **54**, 159 (1985)
4.124   B.V. Shanabrook, J. Comas: Surf. Sci. **142**, 504 (1984)
4.125   B.V. Shanabrook, T. Comas, T.A. Perry, R. Merlin: Phys. Rev. B**29**, 7096 (1984)
4.126   T.A. Perry, R. Merlin, B.V. Shanabrook, J. Comas: J. Vac. Sci. Technol. B**3**, 636 (1985)
4.127   G. Bastard, E.E. Mendez, L.L. Chang, L. Esaki: Solid State Commun. **45**, 367 (1983)
4.128   N.C. Jarosik, B.D. McCombe, B.V. Shanabrook, J. Comas, J. Ralston, G. Wicks: Phys. Rev. Lett. **54**, 1283 (1985)
4.129   C. Mailhiot, Y.C. Chang, T.C. McGill: Phys. Rev. B**26**, 4449 (1982)
4.130   R.L. Greene, K.K. Bajaj: Phys. Rev. B**31**, 913 (1985)
4.131   W. Masselink, Y.C. Chang, H. Morkoc: Phys. Rev. B**32**, 5190 (1985)
4.132   C. Priester, G. Allan, M. Lanoo: Phys. Rev. B**29**, 3408 (1984)

LEONARDI FIBONACCI (* 1175 – † 1250)

5. Raman Studies of Fibonacci, Thue-Morse, and Random Superlattices

Roberto Merlin

With 12 Figures

> "A certain man put a pair of rabbits in a place surrounded on all sides by a wall. How many pairs of rabbits can be produced from that pair in a year if it is supposed that every month, each pair begets a new pair which from the second month on becomes productive?"
>
> Leonardo Fibonacci, "Liber Abaci" (Pisa, 1202)

Many of the interesting properties of superlattices (SL) are well described by one-dimensonal (1D) wave equations [5.1]. The sequence determining the 1D modulation is controlled by the experimenter. Layer growth normally follows a periodic pattern, but it is clear that periodicity is not the only possible choice. Studies of non-periodic superlattices are motivated, in a large part, by theoretical work on random and quasiperiodic 1D Hamiltonians revealing spectra and eigenstates that are quite unlike those of periodic systems [5.2–12]. The group of Fibonacci, Thue-Morse, and random SLs covers a wide range of 1D wave behavior. Fibonacci SLs [5.13–54] are quasiperiodic structures related to *quasicrystals* [5.55]. Their spectrum is characterized by critical (power-law decaying) eigenstates and a self-similar hierarchy of gaps [5.24–25, 30–32, 37–39]. The interest in random SLs [5.23, 56–59] focuses on the problem of Anderson localization [5.2]. Thue-Morse SLs [5.60–62] belong to the class of structures based on *automatic* sequences; their behavior is, in some sense, intermediate between quasiperiodicity and randomness.

The properties of Fibonacci, Thue-Morse, and random SLs are briefly discussed in Sect. 5.1. Raman studies [5.13, 15–19, 23, 51, 58, 59, 63] are reviewed in Sect. 5.2. Only a few reports, including experiments on resonant scattering by acoustic phonons [5.15, 18] and plasmons [5.63] in Fibonacci SLs, bear directly on the spectrum of elementary excitations. Most of the available data concentrate on non-resonant scattering by acoustic phonons, mainly probing the structure of the SL along the direction of growth [5.15–19, 23, 51, 58, 59].

5.1 Basic Properties of Non-periodic Superlattices

5.1.1 Structural Properties

The complex wave properties of non-periodic systems reflect their particular symmetry (or the lack of it) in much the same way that Bloch states are the result of periodicity. The SLs considered here are defined by a sequence of two building blocks A and B of thicknesses d_A and d_B; each block can be composed by one or more layers of different materials of arbitrary thicknesses [5.13–15, 23]. A possible, although incomplete classification of non-periodic structures is based on the value of the exponent γ determining the dependence of $I(k) = |S(k)|^2$ on L, the total thickness of the sample [$S(k)$ is the structure factor of the SL for wavevectors of magnitude k parallel to the direction of growth]. Such a classification is probably incomplete because sequences may exist for which the L-dependence is not in the form of a power-law [5.64]. The periodic sequence $ABABABAB...$ gives $I(k) \propto L^2$ for $k_n = 2\pi n/d_0$ (d_0 is the period and n is an integer) and $I(k) \to 0$ for $k \neq k_n$; $\gamma = 2$ is the signature of coherent scattering. In random SLs, the scattering is incoherent and $I(k)$ is a continuous function proportional to L for large L; i.e., $\gamma = 1$. (However, a trivial set of k may still exist for which $I \propto L^2$; see [5.65].) The structure factor of a quasiperiodic (incommensurate) lattice contains, by definition [5.8], two or more periods whose ratios are given by irrational numbers. Accordingly, as in the periodic case, $\gamma = 2$. But the k values for which $I(k) \neq 0$ are now indexed by more than one integer. This set of k is dense, i.e., given an arbitrary wavevector, one can find a δ-function peak that is arbitrarily close to it. Fibonacci SLs are quasiperiodic; they show two irreducible periods (the k vectors are labelled by two integer indices) in a ratio given by the golden mean $\tau = (1 + \sqrt{5})/2 = 1.61803...$ [5.13, 16, 28]. The cases $\gamma = 1$ (random) and $\gamma = 2$ (quasiperiodic) clearly lead to two distinctive classes of non-periodic SLs. A third and last group is the one to which Thue-Morse structures belong [5.60, 66]. Here, $S(k)$ shows an *infinite* number of incommensurate periods and $I(k) \propto L^{\gamma(k)}$ where $\gamma(k) < 2$ [5.60]. The following subsections discuss these properties in some detail.

a) Fibonacci Superlattices

The Fibonacci sequence can be described as the limit of generations F_r which obey the rule $F_r = F_{r-1} \oplus F_{r-2}$ with $F_1 = \{A\}$ and $F_2 = \{AB\}$. This gives, e.g., $F_5 = \{ABAABABA\}$. To obtain a Fibonacci SL, one simply replaces A and B by two arbitray blocks of layers [5.13–15]. As mentioned above, this arrangement leads to a quasiperiodic structure with two basic periods whose ratio is τ [5.13–16, 28]. In order to prove this, consider the "Bravais" lattice given by the set of points $z_{j+1} = z_j + d_j$, where d_j is taken from the Fibonacci sequence of intervals $\{d_\alpha d_\beta d_\alpha d_\alpha d_\beta ... d_j ...\}$ (note that, in general, $d_\alpha \neq d_A$ and $d_\beta \neq d_B$; see, e.g., [5.13, 28, 50] and Sect. 5.2.1). The structure factor of the Fibonacci Bravais lattice, $S(k) = \sum_j \exp(ikz_j)$, satisfies the recursion relation:

$$S_r = S_{r-1} + \exp(ikL_{r-1})S_{r-2} , \tag{5.1}$$

where $L_r = \xi_{r-1}d_\alpha + \xi_{r-2}d_\beta$ is the length of the rth generation, and the ξ's are Fibonacci numbers. Assuming that $S_r \propto L_r$ for $r \to \infty$, it follows that $S(k) \to 0$

216   R. Merlin

unless $\exp(ikL_{r-1}) \to 1$. Noting that $\xi_r/\xi_{r-1} \to \tau$, it is easy to show that the k vectors are of the form $k_{mn} = 2\pi(m+n\tau)/d_F$, with $d_F = (\tau d_\alpha + d_\beta)$ [5.13]. It is only for these values of k that kL_{r-1} becomes arbitrarily close to an integer when $r \to \infty$. While this proof has only heuristic value, the result is correct. It can be rigorously derived using the *projection* method: because the indexing of δ-function peaks requires two integers, the Fibonacci chain can be obtained by projecting a window of a *periodic* 2D lattice along a line [5.13, 16, 28, 50, 67, 68]. The procedure allows one to calculate the quasiperiodic $S(k)$ in terms of the structure factor of the periodic system, with the result [5.16, 28, 50, 68]:

$$S(k) = \frac{L\tau^2}{d_F} \sum_{m,n} \exp\left(-i\phi_{mn}\tau^{-3}\right) \frac{\sin\phi_{mn}}{\phi_{mn}} \delta_{kk_{mm}} , \qquad (5.2)$$

where $k_{mn} = 2\pi(m+n\tau)/d_F$ and $\phi_{mn} = \pi\tau^2(md_\alpha - nd_\beta)/d_F$.

The largest peaks of $S(k)$ are for $\phi_{mn} \cong 0$. In the case $d_\alpha/d_\beta = \tau$ (1D quasicrystals [5.55]), the condition $m\tau - n \cong 0$ leads to pairs (m,n) which are consecutive Fibonacci numbers (ξ_{p-1}, ξ_p) [5.69]. The corresponding k vectors are then given by $k_{mn} = 2\pi d_F^{-1}(\xi_{p-1} + \tau\xi_p) = 2\pi\tau^p/d_F$, with integer p. The geometric progression τ^p, implying periodicity in a logarithmic scale, reflects the self-similarity of the quasicrystalline lattice [5.55]. It is not a general feature of Fibonacci SLs. If, for instance, $d_\alpha/d_\beta = 2$, the strongest peaks are for

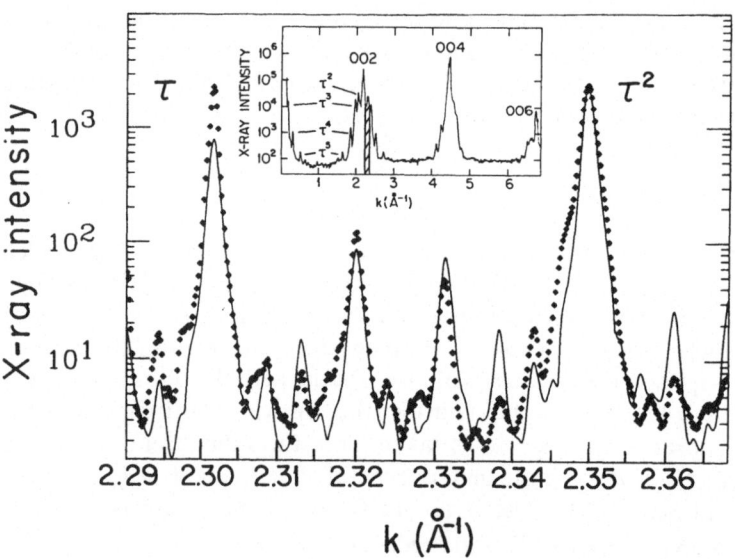

Fig. 5.1. X-ray diffraction data obtained with synchrotron radiation (●●●) and calculated diffraction profile for the ideal Fibonacci structure (——). The building blocks of the superlattice containing 13 generations are $A \equiv (17\ \text{Å–AlAs}/42\ \text{Å–GaAs})$ and $B \equiv (17\ \text{Å–AlAs}/20\ \text{Å–GaAs})$. The inset shows indexing of strong peaks as powers of τ. The shaded region indicates the range of the high resolution scan shown in the main figure. From [5.14]

$k = 2\pi m/d_\beta$ (Sect. 5.2.1). Fibonacci Bravais lattices, we emphasize, are always quasiperiodic (but not necessarily self-similar) regardless of the value of $d_\alpha/d_\beta \neq 1$.

The x-ray diffraction results [5.14] shown in Fig. 5.1 demonstrate many of the unusual properties of Fibonacci ordering. The parameters of the structure are indicated in the caption. They were chosen so that AlAs slabs of the same thickness (17 Å) are separated by either 42 Å or 20 Å slabs of GaAs, giving $d_\alpha = 59$ Å and $d_\beta = 37$ Å. The ratio d_α/d_β is approximately τ and, therefore, the corresponding Bravais lattice approximates that of a 1D quasicrystal. In agreement with the discussion above, the low resolution scan of the inset reveals dominant superlattice reflections and satellites at $k = 2\pi d_F^{-1}\tau^p$. The data obtained with synchrotron radiation indicate that, at least up to the instrumental resolution (≈ 0.0015 Å^{-1}), the diffraction peaks do indeed form a dense set. The measurements agree remarkably well with the calculated profile for the ideal Fibonacci structure (solid line in Fig. 5.1). This, and also numerical simulations, show that the ordering is largely insensitive to the unavoidable random fluctuations in the growth parameters [5.14, 21, 34].

b) Thue-Morse Superlattices

Thue-Morse generations are defined by $T_r = T_{r-1} \oplus T_{r-1}^\dagger$, where $T^\dagger$ is the complement of T: $A^\dagger \equiv B$ and $B^\dagger \equiv A$. The first three generations are $T_1 = \{AB\}$, $T_2 = \{ABBA\}$, and $T_3 = \{ABBABAAB\}$. This sequence is not quasiperiodic: the spacial periods cannot be related to a *finite* basis as, e.g., in the Fibonacci case. The recursion relation for the structure factor of the Thue-Morse Bravais lattice is [5.60]:

$$S_r = S_{r-1}[1 - \exp(ikL_{r-1})], \tag{5.3}$$

where $L_r = d2^{r-1}$ is the length of the rth generation and $d = (d_\alpha + d_\beta)$. From (5.3), the following expression for $I(k)$ can be easily derived:

$$I_r(k) = I_1(k)2^{2r-2} \prod_{j=1}^{r-1} \sin^2(kL_j/2) . \tag{5.4}$$

Assuming that $I(k)$ scales with the length like $L^{\gamma(k)}$, the equation for $\gamma(k)$ reads

$$\gamma(k) = 2 + \frac{1}{\ln 2} \lim_{r \to \infty} \left\{ \frac{1}{r} \sum_{j=1}^{r} \ln[\sin^2(kd2^{j-2})] \right\} . \tag{5.5}$$

The function γ is well behaved for all vectors k such that kd/π is a rational number. The largest value of γ, $\gamma = \ln 3/\ln 2 \cong 1.58$, is found for $kd = n\pi/3$ [5.60]. If kd/π is irrational, the limit may not exist [5.60]. It also follows from (5.5) that $\gamma = -\infty$, i.e., $I(k) = 0$, for $kd = 2^{-n}\pi$ ($n \geq 0$). Finally, the trivial set $kd = 2\pi n$ gives $I \propto L^2$, as in periodic systems.

The Thue-Morse sequence was first discovered by Thue [5.70] and rediscovered by Morse [5.71]. A third independent discovery was made by Arshon [5.72].

c) Random Superlattices

The simplest random sequence is obtained by flipping a coin. This gives equal probabilities for A and B and zero short range correlations. The corresponding random Bravais lattice shows a k-independent incoherent term $I(k) = L/[2(d_\alpha + d_\beta)]$ for $L \to \infty$ [5.65]. In addition, wavevectors for which $\exp(ikd_{\alpha,\beta}) = 1$ exhibit coherent scattering proportional to L^2 [5.65a]. The introduction of correlations leads to incoherent scattering that depends on k [5.65]. In finite samples, fluctuations can be very important leading to spectra which differ significantly from the results for the $L \to \infty$ limit [5.65b]. An example is shown in Sect. 5.2.1.

5.1.2 Electronic Properties

This section focusses on studies of the solutions to the Schrödinger equation in 1D Fibonacci lattices. The electronic spectrum of Thue-Morse chains is considered in [5.60, 62]. Disordered systems, including 1D random potentials, are reviewed in [5.2]. We note that 1D random structures show exponentially localized eigenstates for arbitrarily weak disorder [5.2]. The localization length, relevant to experiments on random SLs, is of the order of the backscattering mean free path [5.2].

The tight-binding approximation is the approach most frequently used in theoretical studies of 1D Fibonacci-type potentials [5.9–12, 24–26, 32–37, 39, 41–42, 52–54]. The model Hamiltonian is defined by:

$$u_{i,i+1}\psi_{i+1} + u_{i-1,i}\psi_{i-1} + \varepsilon_i\psi_i = E\psi_i , \tag{5.6}$$

where ε_i is the site energy, $u_{i,i+1}$ is the transfer energy associated with an electron hopping between sites i and $i+1$, ψ_i is the amplitude of the wave function at the ith site and E is the eigenenergy. In Fibonacci lattices, the sets $\{u_{i,i+1}\}$ and $\{\varepsilon_i\}$ are Fibonacci sequences of elements u_a, u_b and ε_a, ε_b ($u_a = u_b$ in the on-site or diagonal version of the problem, while $\varepsilon_a = \varepsilon_b$ in the off-diagonal case).

Different methods have been applied to solve (5.6), including brute-force diagonalization of the Hamiltonian [5.25–26, 34, 42, 44], perturbation theory [5.25, 35, 44], and renormalization group techniques [5.9–12, 24, 32, 37, 47, 52]. Results of these studies can be summarized as follows. The eigenvalues of (5.6) build a Cantor set with a scaling index that depends on the energy [5.9–11, 32, 37]; such an object is referred to as a multifractal [5.37, 50, 52]. The measure of the set of allowed energies is zero and the spectrum shows an infinite number of gaps. The eigenstates are either chaotic or self-similar (critical); the decay for the latter follows a power-law [5.9–11, 32, 37]. These findings are

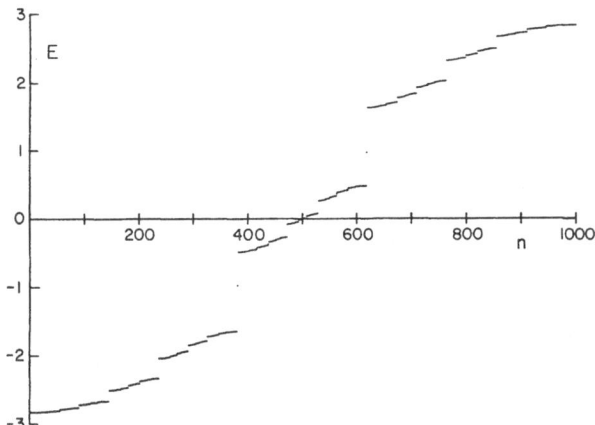

Fig. 5.2. Calculated tight-binding energy spectrum of a Fibonacci lattice. The site energies are $\varepsilon_a = \varepsilon_b = 0$, and the transfer energies are $u_a = 1$ and $u_b = 2$; n denotes the number of nodes of the corresponding eigenstate. The gap states at $E \cong \pm 1$ are localized at the edges of the sample. From [5.25]

consistent with calculations of the Landauer resistance of finite chains showing power-law behavior as a function of the length of the sample [5.36, 39, 47]. The continuous model, i.e., the Schrödinger equation, [5.27, 47, 49, 50], also exhibits critical wave functions and a Cantor set of eigenvalues, but in addition, it can sustain a discrete (trivial) set of extended eigenstates [5.49] (certain 3D tight-binding models [5.44] show analogous features). In this respect, we should note that 1D continuous models can be mapped onto discrete, i.e., tight-binding, models using an exact transformation based on the method of transfer matrices [5.47]. This closes the summary. A few specific topics on the electronic spectrum are amplified in the next paragraphs.

Figure 5.2 reproduces the tight-binding spectrum of a Fibonacci chain [5.25] corresponding to $\varepsilon_a = \varepsilon_b = 0$ in (5.6). The number of nodes in an eigenstate, n, is a good quantum number, allowing a generalization of the concept of Bloch wavevector K ($n = 2\pi K/L$ in periodic systems). The self-similarity of the spectrum is apparent in the figure and it is characterized by trifurcations [5.32]. This property originates in the fact that Fibonacci lattices have only two types of bonds, i.e., either one or two A-blocks, separating B-elements [5.32]. In the lowest order of perturbation theory, there are three eigenstates associated with isolated B-blocks (i.e., $AABAA$ sequences) and with the bonding and anti-bonding states of $AABABAA$ sequences. Higher perturbation orders lead to a hierarchy of bonds and sequences of elements with corresponding trifurcations of the spectrum [5.32]. This approach, used in renormalization group studies [5.32], establishes a classification of the eigenstates in terms of their bonding, anti-bonding, or isolated identity at each stage of the analysis.

An alternative renormalization group approach is based on the relationship between (5.6) and a nonlinear dynamical map [5.9–11, 37, 52]. Equation (5.6) can be written as

$$\begin{pmatrix} \psi_{i+1} \\ \psi_i \end{pmatrix} = \mathbb{M}(i) \begin{pmatrix} \psi_i \\ \psi_{i-1} \end{pmatrix} , \tag{5.7}$$

where $\mathbb{M}(i)$ is the corresponding transfer matrix. If i is a Fibonacci number, i.e., $i = \xi_s$, and $\mathbb{N}(s) = \mathbb{M}(\xi_s)\mathbb{M}(\xi_{s-1})\ldots\mathbb{M}(1)$, it follows from the structure of the Fibonacci sequence that [5.9–11, 37, 52]:

$$\mathbb{N}(s+1) = \mathbb{N}(s-1)\mathbb{N}(s) , \tag{5.8}$$

where $s \geq 1$ ($\xi_0 = \xi_1 = 1$). In the diagonal case ($u_a = u_b$), the $\mathbb{N}$ are unimodular, i.e., $\det(\mathbb{N}) = 1$. Defining $x_s = \mathrm{Tr}\{\mathbb{N}(s)\}/2$, one can show that [5.9–11, 37, 52]:

$$x_{s+1} = 2x_s x_{s-1} - x_{s-2} . \tag{5.9}$$

If $u_a \neq u_b$, then $\det(\mathbb{N}) \neq 1$ and this expression does not apply. However, a new set of unimodular transfer matrices satisfying (5.8) and, therefore, (5.9) can always be derived (see, e.g., [5.37]). Equation (5.9) represents a dynamical map with the invariant [5.9–11, 37, 52]

$$I \equiv x_{s+1}^2 + x_s^2 + x_{s-1}^2 - 2x_s x_{s-1} - 1 . \tag{5.10}$$

In the on-site problem, $I = (\varepsilon_a - \varepsilon_b)^2/4u^2$, and $I = (u_a - u_b)^2/4u_a^2 u_b^2$ for the off-diagonal case [5.37]. If the system is periodic, $I \equiv 0$. Equations (5.9) and (5.10) also apply to continous models where I becomes a function of E. As indicated above, these models may admit a discrete set of extended eigenstates with $I(E) \equiv 0$ [5.49–50]. In passing, we note that the expression corresponding to (5.9) in the case of Thue-Morse lattices is [5.61]:

$$(x_{s+1} - 1) = 4x_{s-1}^2 (x_s - 1) . \tag{5.11}$$

The properties of this map have not yet been elucidated.

The x's in (5.9) depend on E. A state is allowed if $|x_s(E)| \leq 1$ for large s (the eigenvalueas of the corresponding transfer matrix are $[x_s \pm (x_s^2 - 1)^{1/2}]$). It follows that the problem of solving (5.6) is reduced to that of fiding initial conditions (i.e., values of x_0, x_1, and x_2) for which the $|x_s|$'s remain bound. In principle, this problem can be solved numerically, but it is unstable: because of the Cantor-set nature of the spectrum, almost any initial set of values will lead to $|x_s| \to \infty$ for $s \to \infty$ [5.9–11, 37, 52]. An example of energy spectra derived from (5.9) is shown in Fig. 5.3 [5.37]. Sets of E which satisfy $|x_s(E)| \leq 1$ are shown for $s = 3, 4, 5$, and 6. Each set represents the energy bands of periodic systems with periods given by the sth generation of the Fibonacci lattice (note that the corresponding spectrum consists of ξ_s bands). As s gets larger, more gaps appear and dominate the spectrum. The sum of the bandwidths approaches zero for $s \to \infty$ [5.9–11, 37, 52]. Several stable cycles associated with (5.9)

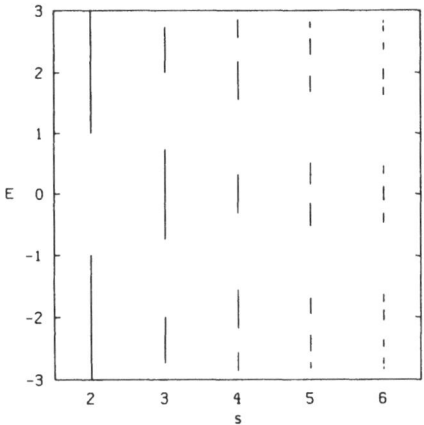

Fig. 5.3. Allowed energies [$|x_s| \leqq 1$ in (5.9)] of periodic lattices with cells given by the sth generation of the Fibonacci chain ($s = 3, 4, 5, 6$). The calculations correspon to $u_a = 1$, $u_b = 2$, and $\varepsilon_a = \varepsilon_b$ in (5.6). From [5.37]

have been identified [5.37]. For instance, $x_1 = x_2 = 0$ results in the 6-cycle $(x_0, 0, 0, -x_0, 0, 0)$, and $x_0 = x_1/(2x_1 - 1)$ gives a 2-cycle [5.37]. The latter determines the scaling properties of the spectrum near of clusters of eigenstates, while the 6-cycle relates to states near the centers of clusters [5.37]. Note that cycles are associated with power-law decaying eigenstates: if the period of the cycle is m and χ is an eigenvalue of the given transfer matrix, the associated eigenvector satisfies

$$\psi(\tau^m z) = \chi \psi(z) , \tag{5.12}$$

where we have used $\zeta_s \cong \tau^m \zeta_{s-m}$. It is worthwhile to point out that a given cycle does not necessarily correspond to an eigenstate of (5.6) because the initial conditions may not match. For example, $E = 0$ is a 6-cycle eigenstate of the off-diagonal problem but there are no 6-cycle solutions of the diagonal one. Nevertheless, it is believed that independently of the initial conditions, states close to centers of clusters approach limiting 6-cycles as $L \to \infty$. A more general statement is that the scaling properties of the spectrum depend only on I [5.37].

5.1.3 Phonons, Plasmons and Other Elementary Excitations

The spectrum of elementary excitations in non-periodic structures is expected to follow a pattern similar to that shown by electrons. This is because wave behavior is largely determined by the symmetry of the lattice and not by details of the wave equation. In particular, phonons and other excitations in random SLs are always localized for motion normal to the layers [5.2]. We also note that a formal solution of quasiperiodic Schrödinger equations can be obtained using projections from higher dimensions and radon transforms [5.45]. It is apparent that the same procedure can be applied to other excitations.

Most of the published work relevant to this section deals with Fibonacci lattices (phonons in Thue-Morse chains are considered in [5.61]). The properties

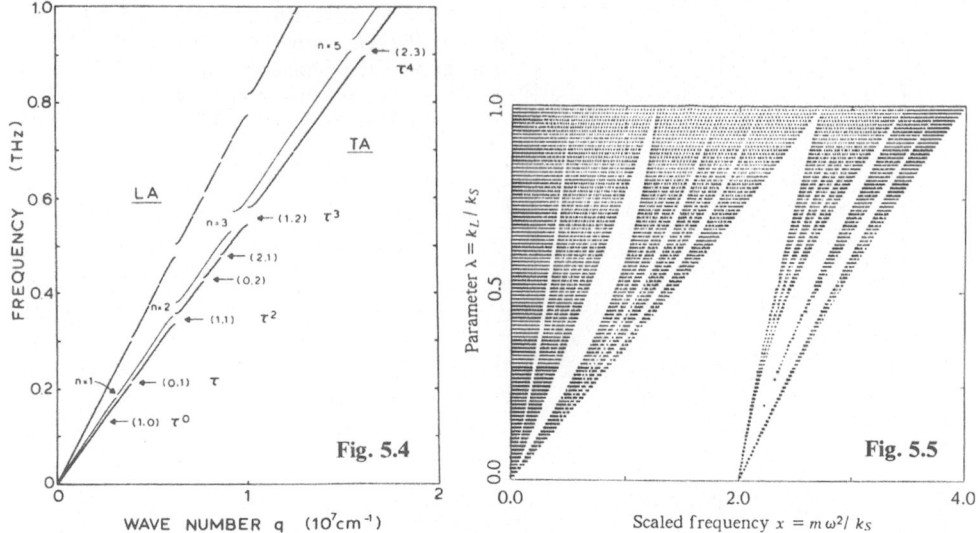

Fig. 5.4. Calculated dispersion of transverse (TA) and longitudinal (LA) acoustic phonons in a GaAs-AlAs Fibonacci superlattice (blocks A and B are the same as in Fig. 5.1). Major gaps are indexed according to the associated Bragg peak (5.2). Thin lines show the dispersion of TA modes in the periodic superlattice $ABABAB\ldots$. For the latter, the horizontal scale has been multiplied by a factor of 0.95 in order to avoid an overlap of the curves. From [5.43]

Fig. 5.5. Phase diagram of the vibrational spectrum of a Fibonacci chain with equal masses and alternating spring constants k_L and k_s. The vertical axis is $\lambda = k_L/k_s$ and ω is the phonon frequency. White areas correspond to gaps. From [5.24]

of phonons are discussed in [5.24, 25, 29, 43], polaritons in [5.40], plasmons in [5.30–31], and excitons in [5.46]. In all cases, the eigenstates are critical and the spectrum of eigenvalues is in the form of a Cantor set. The calculated dispersion of acoustic modes in a Fibonacci GaAs-AlAs SL is shown in Fig. 5.4 [5.43]. As for electrons [5.32], the phonon spectrum exhibits a self-similar structure of gaps characterized by trifurcations. The size of the gaps decreases with decreasing frequency, but there is never a crossover into a regime of extended states (the structure factor (5.2) always shows some weight at arbitrarly small wavevectors). An alternative picture of the phonon spectrum is shown in Fig. 5.5 [5.24]. The results are for a chain of identical objects connected by springs k_s and k_L that alternate in a Fibonacci sequence. The periodic case corresponds to $k_s = k_L$ ($\lambda = 1$). For $k_L = 0$, there are only two eigenstates associated with isolated objects and with two objects connected by k_s. Data obtained using decimation techniques give diagrams similar to that in Fig. 5.5 and confirm the Cantor-like nature of the spectrum [5.24].

5.2 Raman Scattering in Non-periodic Superlattices

5.2.1 Scattering by Acoustic Phonons

Raman scattering has been extensively applied to the study of acoustic phonons in periodic SLs, particularly those based on III–V semiconductors and $Ge_x Si_{1-x}$ grown on (001) substrates ([5.69] and Chap. 2 of this volume). Work on non-periodic structures has been limited so far to the latter systems [5.13, 15–19, 23, 51, 58–59]. Acoustic mode spectra provide mainly information on the structural properties of the SLs, and, to a lesser extent, on the frequency spectrum of sound waves [5.23, 73]. The experiments are usually performed using configurations where the scattering wavevector is nearly perpendicular to the layers. In zinc blende and diamond structure (001) SLs, this geometry allows scattering only by longitudinal acoustic (LA) modes propagating along $\hat{z} \equiv [001]$ (note that the polarization vectors of the incident and scattered light must be parallel [5.73]). These considerations follow from the photoelastic continuum model ([5.73], and Chap. 2 of this volume), and we should emphasize that they apply to periodic as well as to non-periodic SLs [5.23]. Within the continuum model, the intensity for Raman scattering by [001] LA-phonons is given by (Sect. 3.3.4):

$$I(\Omega) \propto \left\langle \left| \int e^{-iqz} P(z) \frac{\partial U}{\partial z} \, dz \right|^2 \right\rangle_\Omega , \tag{5.13}$$

where $U(z)$ is the amplitude of the mode with frequency Ω, $P(z)$ is the layer-dependent photoelastic coefficient P^{12} and q is the scattering wavevector. In $Ge_x Si_{1-x}$ and many III–V SLs, the P^{12} modulation dominates over the relatively weak modulation of the LA sound velocity ([5.16, 73] and Sect. 3.3.4). Thus, the phonons can be approximated by plane waves and (5.13) reduces to ($k_B T \gg \hbar\Omega$):

$$I[\Omega(K)] \propto |P_{q-K}|^2 , \tag{5.14}$$

where P_k is the Fourier transform of $P(z)$. This equation establishes the simplest link between Raman scattering and structural studies, i.e., the scattering probes $P(z)$, For a given P_k, (5.14) describes scattering by phonon *doublets* with Bloch wavevectors $K = |k \pm q|$ [5.73]. In periodic SLs, $P_k \propto L$ for $k_n = 2\pi n/d_0$ and is zero otherwise. This leads to equally spaced LA doublets at $\Omega \cong c|k_n \pm q|$ [c is an average sound velocity, see (3.11)]. Non-periodic structures generally show a dense set of doublets with weights determined by P_k [5.23]. Specific cases are discussed in the following.

In Fig. 5.6, we compare Raman spectra of Fibonacci (generation 13), Thue-Morse, and random samples with calculations using (5.13). The SLs consist, respectively, of a total of 377, 256, and 377 blocks with $A = 20$ Å-GaAs and $B = 20$ Å-AlAs [5.23] and were grown on (001) GaAs substrates. The random structure is the disordered counterpart of the Fibonacci SL; the layer deposition

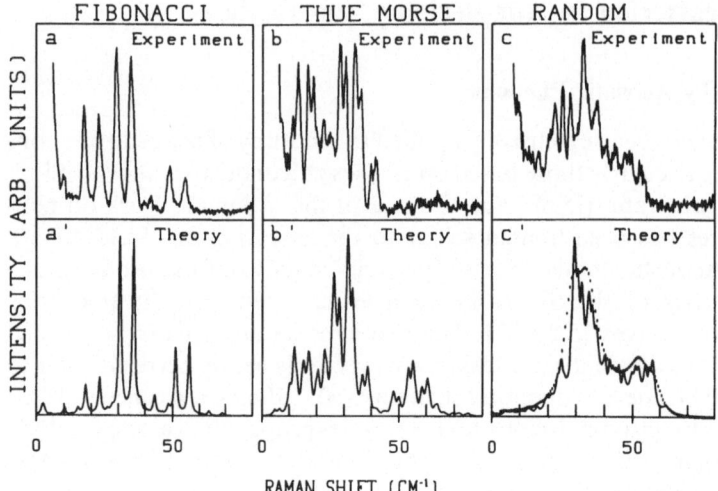

RAMAN SHIFT (CM⁻¹)

Fig. 5.6a–c. Comparison between calculated and measured LA-phonon spectra of the Fibonacci (generation 13), Thue-Morse, and random superlattices with building blocks A (20 Å–GaAs) and B (20 Å–AlAs). The dashed curve in (c') corresponds to $L = \infty$. From [5.23]

follows the Markovian process considered in [5.74]. To analyze the results in Fig. 5.6, it is convenient to introduce the sequence $\{\alpha_j\}$ where $\alpha_j = 0$ if the jth layer is A and $\alpha_j = 1$ if it is B. The expression for P_k in terms of the α_j's is, for $k \neq 0$, [5.23]:

$$P_k = \Delta P [1 - \exp(-ikl)](-ikl)^{-1} S(k) , \tag{5.15}$$

where $\Delta P = (P_{\text{GaAs}}^{12} - P_{\text{AlAs}}^{12})$ and $l = d_A = d_B = 20$ Å; the structure factor is given by

$$S(k) = \sum_j \alpha_j e^{i(klj)} . \tag{5.16}$$

For the Fibonacci SL, $S(k)$ can be obtained using the projection method mentioned in Sect. 5.1.1a. The result, analogous to (5.2), is [5.16, 23]:

$$S(k) = \frac{L}{2l\tau^2} \frac{\sin[\pi\tau^{-1}(m-n)]}{(m-n)} \delta_{k, k_{mn}} , \tag{5.17}$$

with $k_{mn} l/(2\pi) = (m(\tau - 1) + n)/\tau$. For a more general case, i.e., $d_A \neq d_B$, see [5.16]. In (5.17), $S(k)$ is largest for $s \equiv (m - n) = 0$. But P_k in (5.15) vanishes at the corresponding k values, hence, no scattering results. The integers $s = \pm 1, \pm 2$ lead to $kl/(2\pi) = \tau^{-1}, \tau^{-2}, \tau^{-3}$ for $|kl/(2\pi)| \leq 1$. The major LA doublets in Fig. 5.6a, a' are associated with these values of k. The fact that these doublets follow a power-law (τ^p) behavior is consistent with the discussion in Sect. 5.1.1a: the ratio $r = d_\alpha/d_\beta$ for the Bravais lattice of this sample is $60/40 \cong \tau$ (the separation between centers of neighboring GaAs slabs is either 60 or 40 Å). The recursion relation for the structure factor of the Thue-Morse SL is given by (5.3) with $L_{r-1} = 2^{r-1}l$. The strongest lines in Fig. 5.6b, b' can be identified in terms of a small set of wavevectors giving the largest γ's of $S(k)$ (see Sect. 5.1.1b); the highest exponent ($\gamma = \ln 3/\ln 4$) corresponds to $kl/\pi = 1/3$, $2/3$ [5.23]. The analytical expression of $S(k)$ for the random SL (in the limit $L \to \infty$) is given in [5.23]; $S(k)$ shows maxima at $kl/(2\pi) \cong \tau^{-1}$, τ^{-2}, in reasonable agreement with the experimental data. The sharp features in the spectra of Figs. 5.6c and c' are fluctuations (noise) due to finite size effects [5.23]; this is evident in the comparison with the $L \to \infty$ limit shown in Fig. 5.6c' (see, however, [S. 65a]). The Raman spectrum of the random GaAs-AlAs SL reported in [5.59] also shows narrow peaks reflecting fluctuations.

Finite-size effects in Fibonacci SLs have been studied in [5.19, 51]. Computer simulations indicate that, in most cases, finite-size effects become too small to be detected experimentally beyond the ninth or tenth generation. The work reported in [5.51] deals with periodic SLs for wich the unit cell is a (low-index) Fibonacci generation. Data showing LA scattering in Fibonacci GaSb-AlSb structures corresponding to the fourth (5 layers) and sixth (13 layers) generation of the sequence are reproduced in Fig. 5.7 [5.19]. The features of interest are the large differences between the spectra, not only in regard to the widths but also the positions of the peaks (in periodic SLs only the widths are substantially affected by the reduced size [5.75]). The generation-dependence of the scattering is not as yet well understood. Nevertheless, it is remarkable that the behavior of the Raman spectra resembles, in some sense, that of the electronic spectrum Fig. 5.3. This suggests the possibility of a link between (5.1), giving $S_r(k)$, and (5.9).

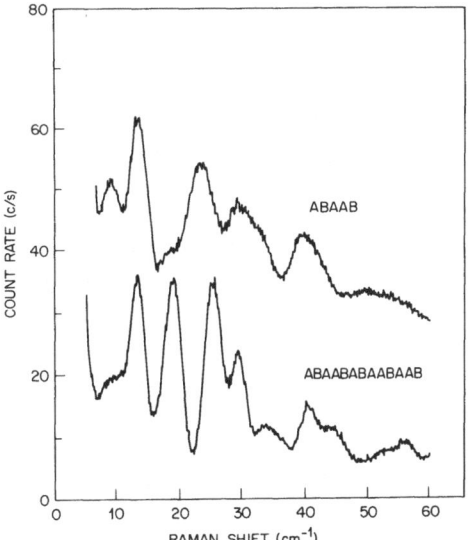

Fig. 5.7. Acoustic phonon spectra of structures with $A \equiv (25$ Å–GaSb) and $B \equiv (40$ Å–AlSb), corresponding to four (*top trace*) and six (*bottom trace*) generations of the Fibonacci sequence. From [5.19]

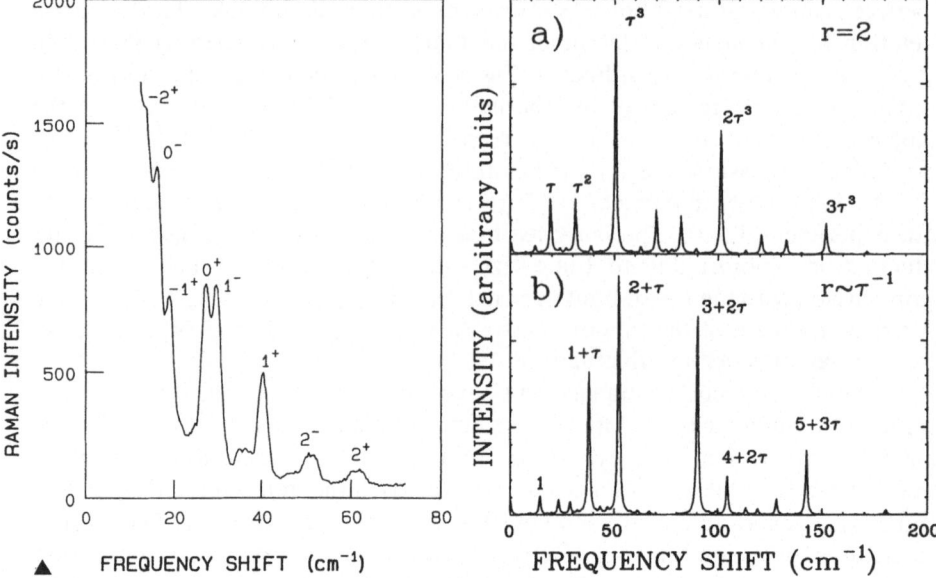

Fig. 5.8. Raman spectrum of a Fibonacci superlattice (generation 11) consisting of 54 Å–Si (*A* blocks) and 42 Å–Ge$_{0.48}$Si$_{0.52}$ (*B* blocks). The peaks are labeled according to $\Omega_p^\pm = |2\pi c\tau^p/d_F \pm cq| (p = -2, -1, 0, 1, 2)$. The variable c is the average LA sound velocity and q is the scattering wavevector. From [5.16]

Fig. 5.9. Calculated forward-scattering spectra of 12th generation GaAs-AlAs Fibonacci superlattices with (*a*) $r=2$ and (*b*) $r \approx \tau^{-1}$ ($r = d_A/d_B$). The results show LA phonon scattering and correspond to $k_B T \gg \hbar\Omega$

The results in Fig. 5.6 indicate that (5.13) describes well the positions of the Raman peaks, but not their relative intensities. This problem, also noticed in periodic SLs (see Chap. 2), is likely to be due to the breakdown of the local approximation for very thin layers. The agreement between theory and experiments improves with increasing layer thickness. The intensities of the LA-doublets in the spectrum of the (generation 11) Fibonacci Si-Ge$_x$Si$_{1-x}$ SL, shown in Fig. 5.8, are well accounted for by (5.13) [5.16].

The spectra of the Fibonacci SLs in Figs. 5.6a and 5.8 show τ^p-geometric progressions of doublets [5.16, 23]; a feature that is often considered as the signature of Fibonacci ordering. Actually, power-law behavior is *not* an intrinsic property of these structures (Sect. 5.1.1a). The complete geometric progression is expected for $r = d_\alpha/d_\beta \equiv \tau$. If this condition is only approximately met, a few τ^p-terms dominate the spectra but the progression is incomplete. This applies to the samples in Fig. 5.6a ($r=3/2$) and Fig. 5.8 ($r=25/16$). In cases where r differs appreciably from τ, τ^p-peaks do not necessarily dominate the spectra [5.15]. This is shown in Fig. 5.9. The forward-scattering calculations ($q=0$) are for Fibonacci SLs of generation 12 with $A \equiv (8$ Å–AlAs/60 Å–GaAs), $B \equiv (8$ Å–AlAs/26 Å–GaAs) ($r=2$) and $A \equiv (8$ Å–AlAs/26 Å–GaAs), $B \equiv$

(8 Å–AlAs/47 Å–GaAs) ($r \cong \tau^{-1}$), and $k_B T \gg \hbar\Omega$. The $r \cong \tau^{-1}$ structure shows no hint of power-law behavior, whereas there are a few peaks following a geometric progression in the spectrum of the SL with $r = 2$ (labels τ, τ^2, τ^3 in Fig. 5.9a). The latter case can be understood as follows. The strongest Raman lines are generally associated with the dominant Fourier components of the 1D modulation, For $r = 2$, the structure factor, given by (5.2), exhibits large δ-function peaks for small integer indices leading to the first powers of τ (note that $\tau^2 = 1 + \tau$ and $\tau^3 = 2\tau + 1$). However, one can easily show using (5.2) that the largest peaks of $S(k)$ are at $k = 2\pi m/d_\beta$. Moreover, $S(k)$ is *periodic* for $r = 2$ (the SL is, of course, quasiperiodic) and this is reflected in the calculated spectrum of Fig. 5.9a. If $k_B T$ is not $\gg \hbar\Omega$, as for the results reported in [5.16] for a Si-Ge$_x$Si$_{1-x}$ SL with $r = 2$, the periodic behavior can be obscured by strong thermal-factor weighting at low frequencies.

The results discussed in the previous paragraphs were obtained under non-resonant conditions. A comparison between off-resonance and resonant spectra for a Fibonacci SL is shown in Fig. 5.10 [5.18]. The reflectivity data of the inset indicates that $\omega_L = 2.409$ eV falls in a relatively featureless region of the spectrum while $\omega_L = 1.916$ eV is close to a critical point for optical transitions (at ~ 1.89 eV). The parameters of the structure are $A \equiv (17$ Å–AlAs/42 Å–GaAs) and $B \equiv (17$ Å–AlAs/20 Å–GaAs); x-ray patterns of the same sample are depicted in Fig. 5.1. The differences between the spectra are striking. The off-resonance results exhibit the characteristic τ^p-doublets, in agreement with the x-ray measurements and the fact that $r = 59/37 \cong \tau$ [5.18]. The resonant spectrum cannot be accounted for by (5.13). Its complex lineshape has been ascribed to a weighted density of states of [001] LA-phonons [5.13, 18]. This assignment is based on the observation that the measured positions of the dips correlate with major gaps in the calculated 1D phonon spectrum. The correlation is shown in

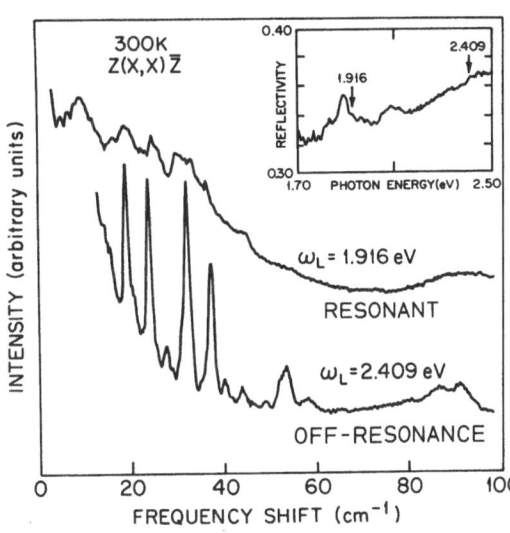

Fig. 5.10. Resonant ($\omega_L = 1.916$ eV) and nonresonant ($\omega_L = 2.409$ eV) Raman spectra of a GaAs-AlAs Fibonacci superlattice (generation 13) showing LA scattering. The parameters are the same as in Fig. 5.1. *Inset:* normal reflectivity data. The scattering resonance is due to the electronic transition associated with the peak at ≈ 1.89 eV. From [5.18]

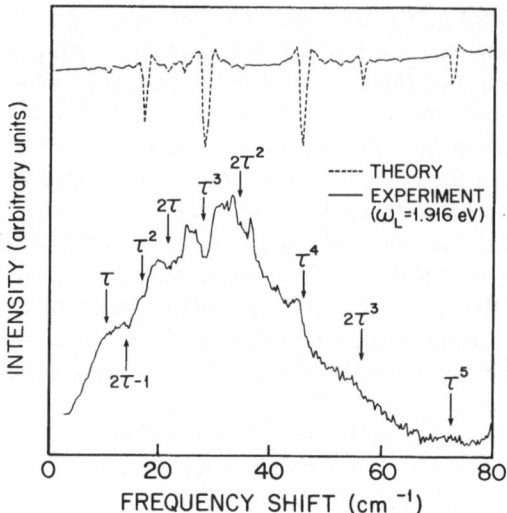

Fig. 5.11. The resonant spectrum of Fig. 5.10 after dividing by $[n(\Omega)+1]$; $n(\Omega)$ is the Bose factor. The dashed curve is the calculated density of states of [001] LA modes. Arrows denote expected midfrequencies of main gaps in units of $\pi c d_F^{-1}$. From [5.15]

Fig. 5.11 [5.18]. The relevant electronic resonance at ~ 1.89 eV was tentatively assigned in [5.18] to an *intrinsic* gap excitation localized at the surface (see Fig. 5.2, and [5.25]). However, this matter, and consequently, the origin of the resonant scattering are not yet settled.

5.2.2 Scattering by Plasmons

The Raman intensity for plasmon scattering in a layered electron gas is given by [5.30–31, 56]:

$$I(\Omega, q, k) \cong \sum_v |\Phi_v(k)|^2 \delta(\Omega^2 \pm \Omega_v^2) , \quad \text{with} \tag{5.18}$$

$$\Phi_v(k) = \sum_l e^{ikz_l} \Phi_v(l) , \tag{5.19}$$

where k is the component of the scattering wavevector normal to the layers, z_l is the position of the lth layer, and $\Phi_v(l)$ and Ω_v are the eigenvectors and eigenvalues of the plasmons. Hence, the scattering directly probes the plasmon wave function.

Equations (5.18–19) have been shown to account extremely well for the experimental data on modulation-doped periodic SLs (see Chap. 6). Theoretical work on plasmons in non-periodic structures is described in [5.30–31, 56]. In particular, calculations for Fibonacci SLs reveal rich spectral features reflecting the critical nature of the eigenstates [5.30–31]. Unfortunately, broadening effects severely limit the possibility of resolving the expected fine structure in the experiments [5.30, 63]. Raman spectra of plasmons in a modulation-doped GaAs-Al$_x$Ga$_{1-x}$As Fibonacci SL are shown in Fig. 5.12a [5.63]. The

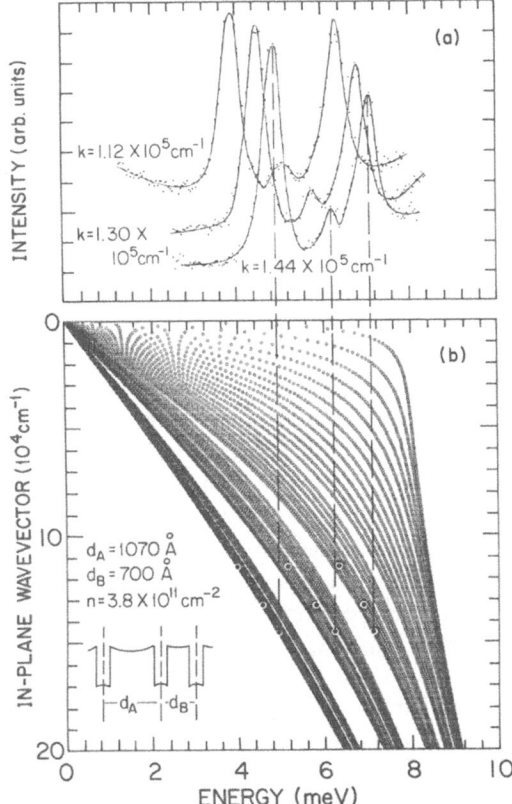

Fig. 5.12. (a) Raman spectra showing scattering by plasmons in a modulation-doped Fibonacci structure, k is the component of the scattering wavevector perpendicular to the layers. (b) Calculated dispersion of plasma modes. The parameters of the sample are indicated in the *inset*. From [5.63]

sample consists of a set of 55 quasi-2D electron systems localized in the GaAs wells ($n \cong 3.8 \times 10^{11}$ cm^{-2}); the separation between neighboring wells is either $d_\alpha = 1070$ Å or $d_\beta = 700$ Å according to the Fibonacci sequence (see the inset in Fig. 5.12b). The plasmon spectrum, shown in Fig. 5.12b, was calculated from [5.30–31]

$$\left(\frac{\Omega^2}{\Omega_p^2} - 1\right)\Phi(l) = \sum_{l' \neq l} \Phi(l') \exp\left(-k|z_l - z_{l'}|\right) , \qquad (5.20)$$

where $\Omega_p = (2\pi e^2 nq/\varepsilon m^*)^{1/2}$; q is the in-plane scattering wavevector, ε is the background dielectric constant and m^* is the effective mass.

As shown in Fig. 5.12b, the positions of the Raman peaks correlate with calculated plasmon eigenenergies. Numerical simulations using (5.18) indicate that the three main lines consist of a large number of much narrower features separated by $\sim 0.1 - 0.3$ meV [5.63]. The narrow lines merge into the observed three structures for a broadening of ~ 0.3 meV. The calculations agree reasonably well with the measured q-dependence of the Raman shifts, but do not account for the relative intensities of the three structures [5.63]. It is not clear whether this problem reflects difficulties with (5.18) or an experimental artifact due to departures from nominal values of the sample parameters.

5.3 Concluding Remarks

The experimental results reviewed in this article are, to a large degree, preliminary. The only exception is the work on non-resonant scattering by LA phonons where the few remaining questions, particularly in regard to the intensity of the doublets, do not (necessarily) call for studies of non-periodic structures. Resonant scattering by LA modes is a different matter, as the results on Fibonacci SLs [5.18] do not appear to have a periodic counterpart. It is clear that further studies are required to clarify the origin of the resonant LA spectra. The work on plasmons is also of preliminary nature, but it already shows limitations imposed by broadening effects which may be difficult to avoid. Other areas remain largely unexplored. For instance, interface phonons and polaritons in non-periodic SLs have not as yet been investigated using Raman scattering. In addition, resonant scattering (as a probe of the electronic spectrum of the SL), may be a useful technique to elucidate some of the expected features which rely on the non-periodic modulation. The available optical data, mainly on Fibonacci SLs [5.15], are inconclusive, and most theoretical predictions [5.9–12, 24–26, 32–37, 39, 41, 42, 52–54] still await experimental confirmation.

Acknowledgements. I am grateful to a number of colleagues, in particular K. Bajema, P.K. Bhattacharya, Z. Cheng, R. Clarke, M. Garriga, F. Juang, J. Nagle, K. Ploog, L. Sander, R. Savit, and J. Todd, for their contributions to the work described in this article. G. Schwartz made available to me results of experiments prior to publication. Stimulating discussions with M. Kohmoto and A. H. MacDonald are gratefully acknowledged. This work was supported by the U.S. Army Research Office under Contract DAAG 29-85-K-0175, and by the National Science Foundation Grant DMR-8602675.

References

5.1   L. Esaki: In *Proceedings of the 17th International Conference on the Physics of Semiconductors*, ed. by J.D. Chadi, W.A. Harrison (Springer, Berlin, Heidelberg 1984) p 473
5.2   P.W. Anderson: Phys. Rev. **109**, 1492 (1958)
       For a review, see P.A. Lee, T.V. Ramakrishnan: Rec. Mod. Phys. **57**, 287 (1985)
5.3   B. Simon: Adv. Appl. Math. **3**, 463 (1982)
5.4   J.B. Sokoloff: Phys. Rep. **126**, 189 (1985)
5.5   C.M. Soukoulis, E.N. Economou: Phys. Rev. Lett. **48**, 1043 (1982)
5.6   J. Bellissard, D. Bessis, P. Moussa: Phys. Rev. Lett. **49**, 701 (1982)
5.7   C. de Lange, T. Janssen: Phys. Rev. B**28**, 195 (1983)
5.8   S. Ostlund, R. Pandit: Phys. Rev. B**29**, 1394 (1984)
5.9   M. Kohmoto, L.P. Kadanoff, C. Tang: Phys. Rev. Lett. **50**, 1870 (1983)

5.10 S. Ostlund, R. Pandit, D. Rand, H.J. Schellnhuber, E.D. Siggia: Phys. Rev. Lett. **50**, 1873 (1983)
5.11 M. Kohmoto, Y. Oono: Phys. Lett. A**102**, 145 (1984)
5.12 C. Tang, M. Kohmoto: Phys. Rev. B**34**, 2041 (1986)
5.13 R. Merlin, K. Bajema, R. Clarke, F.-Y. Juang, P.K. Bhattacharya: Phys. Rev. Lett. **55**, 1768 (1985)
5.14 J. Todd, R. Merlin, R. Clarke, K.M. Mohanty, J.D. Axe: Phys. Rev. Lett. **57**, 1157 (1986)
5.15 R. Merlin, K. Bajema, R. Clarke and J. Todd: In *Proceedings of the 18th International Conference on the Physics of Semiconductors*, ed. by O. Engströn (World Scientific, Singapore 1987) p. 675
5.16 M.W.C. Dharma-wardana, A.H. MacDonald, D.J. Lockwood, J.-M. Baribeau, D.C. Houghton: Phys. Rev. Lett. **58**, 1761 (1987)
5.17 M. Nakayama, H. Kato, S. Nakashima: Phys. Rev. B**36**, 3472 (1987)
5.18 K. Bajema, R. Merlin: Phys. Rev. B**36**, 4555 (187)
5.19 A.T. Macrander, G.P. Schwartz, J. Bevk: Phys. Rev. B**37**, 8459 (1988)
5.20 D.C. Hurley, S. Tamura, J.P. Wolfe, K. Ploog: Phys. Rev. B**37**, 8829 (1988)
5.21 R. Clarke, R. Merlin: Phys. Rev. Lett. **59**, 2237 (1987)
5.22 F. Laruelle, V. Thierry-Mieg, M.C. Joncour, B. Etienne: In *Proceedings of the Third International Conference on Modulated Semiconductor Structures*, J. de Phys. **48**–C5, 529 (1987)
5.23 R. Merlin, K. Bajema, J. Nagle, K. Ploog: In *Proceedings of the Third International Conference on Modulated Semiconductor Structures*, J. de Phys. **48**–C5, 503 (1987)
5.24 J.P. Lu, T. Odagaki, J.L. Birman: Phys. Rev. B**33**, 4809 (1986)
5.25 F. Nori, J.P. Rodriguez: Phys. Rev. B**34**, 2207 (1986)
5.26 T. Odagaki, L. Friedman: Solid State Comm. **57**, 915 (1986)
5.27 J. Kollâr, A. Sütõ: Phys. Lett. A**117**, 203 (1986)
5.28 M.C. Valsakumar, V. Kumar: Pramana **26**, 215 (1986)
5.29 K. Machida, M. Fujita: J. Phys. Soc. Jpn. **55**, 1799 (1986)
5.30 S. Das Sarma, A. Kobayashi, R.E. Prange: Phys. Rev. B**34**, 5309 (1986)
5.31 P. Hawrylak, J.J. Quinn: Phys. Rev. Lett. **57**, 380 (1986)
5.32 Q. Niu, F. Nori: Phys. Rev. Lett. **57**, 2057 (1986)
5.33 L. Chen, G. Hu, R. Tao: Phys. Lett. A**117**, 120 (1986)
5.34 Y. Liu, R. Riklund: Phys. Rev. B**35**, 6034 (1987)
5.35 M.C. Valsakumar, G. Ananthakrishna: J. Phys. C**20**, 9 (1987)
5.36 M. Goda: J. Phys. Soc. Jpn. **56**, 1924 (1987)
5.37 M. Kohmoto, B. Sutherland, C. Tang: Phys. Rev. B**35**, 1020 (1987)
5.38 J.B. Sokoloff: Phys. Rev. Lett. **58**, 2267 (1987)
5.39 T. Schneider, A. Politi, D. Würtz: Z. Phys. B**66**, 469 (1987)
5.40 H.-R. Ma, C.-H. Tsai: Phys. Rev. B**35**, 9295 (1987)
5.41 H. Hiramoto, S. Abe: Jpn. J. Appl. Phys. **26**, Suppl. 3, 665 (1987)
5.42 M. Fujita, K. Machida: J. Phys. Soc. Jpn. **56**, 1470 (1987)
5.43 S. Tamura, J.P. Wolfe: Phys. Rev. B**36**, 3491 (1987)
5.44 V. Kumar, G. Ananthakrishna: Phys. Rev. Lett. **59**, 1476 (1987)
5.45 J.P. Lu, J.L. Birman: Phys. Rev. B**36**, 4471 (1987)
5.46 S.-R. Yang, S. Das Sarma: Phys. Rev. B**37**, 4007 (1988)
5.47 M. Kohmoto: Phys. Rev. B**34**, 5043 (1986)
5.48 K.W.-K. Shung, L. Sander, R. Merlin: Phys. Rev. Lett. **61**, 455 (1988)
5.49 A.H. MacDonald, G.C. Aers: Phys. Rev. B**36**, 9142 (1987)
5.50 A.H. MacDonald: In *Interfaces, Quantum Wells and Superlattices*, ed. by C.R. Leavens and R. Taylor (Plenum, 1988) p. 347
5.51 B. Jusserand, F. Mollot, M.C. Joncour, B. Etienne: In *Proceedings of the Third International Conference on Modulated Semiconductor Structures*, J. de Phys. **48**–C5, 577 (1987)

5.52   M. Kohmoto: Int. J. Mod. Phys. B1, 31 (1987)
5.53   J.A. Vergés, L. Brey, E. Louis, C. Tejedor: Phys. Rev. B35, 5270 (1987)
5.54a H. He: Phys. Stat. Sol. b144, 631 (1987)
5.54b R. Merlin: IEEE J. Quantum. Electron. 24, 1791 (1988)
5.55   D. Levine, P.J. Steinhardt: Phys. Rev. Lett. 53, 2477 (1984)
5.56   S. Das Sarma, A. Kobayashi, R.E. Prange: Phys. Rev. Lett. 56, 1280 (1986)
5.57   A. Chomette, B. Deveaud, A. Regreny, G. Bastard, Phys. Rev. Lett. 57, 1464 (1986)
5.58   R. Clarke, T. Moustakas, K. Bajema, D. Grier, W. Dos Passos, R. Merlin: In
       *Proceedings of the Third International Conference on Superlattices, Microstructures and
       Microdevices*, Superlattices and Microstructures 4, 371 (1988)
5.59   G.P. Schwartz, G.J. Gualtieri, W.A. Sunder: J. Vac. Sci. Technol. A5, 1500 (1987)
5.60   Z.C. Chen, R. Savit, R. Merlin: Phys. Rev. B37, 4375 (1988)
5.61   F. Axel, J.P. Allouche, M. Kleman, M. Mendes-France, J. Peyriere: J. Phys. 47, C3, 181
       (1986)
5.62   R. Ricklund, M. Severin, Y. Liu: Int. J. Mod. Phys. B1, 121 (1987)
5.63   J.P. Valladares, A. Pinczuk, R. Merlin, A.C. Gossard, J.H. English: to be published
       (1989)
5.64   S. Aubry, C. Godrèche, F. Vallet: J. Physique 48, 327 (1987)
5.65a S. Hendricks, E. Teller: J. Chem. Phys. 10, 147 (1942)
5.65b For a class of random layered systems, A. Garg and D. Levine [Phys. Rev. Lett. 60, 2160
       (1988)] have shown that the fluctuations of $I(k)$ do not vanish for $l \to \infty$, as in the
       phenomenon for laser speckle
5.66   S. Aubry, C. Godrèche, J.M. Luck: Europhys. Lett. 4, 639 (1987)
5.67   V. Elser: Phys. Rev. B32, 4892 (1985)
5.68   R.K.P. Zia, W.J. Dallas: J. Phys. A18, L341 (1985)
5.69   The reader will find the relations $1 + \tau = \tau^{-2}$ and $\tau - 1 = \tau^{-1}$ useful for operating with the
       golden mean.
5.70   A. Thue: Norske Vid. Selsk. Skr. I Mat.-Natur. Kl. No. 1, 1–67 (1912)
5.71   M. Morse: Trans. Am. Math. Soc. 22, 84–100 (1921)
5.72   S. Arshon: Mat. Sb. 2, 769–779 (1937)
5.73   M.V. Klein: IEEE J. QE-22, 1760 (1986)
5.74   M.R. Schröder: In *Number Theory in Science and Communication*, Springer Ser. Inf.
       Sci., Vol. 7 (Springer, Berlin, Heidelberg 1985) p. 315
5.75   M. Nakayama, K. Kubota, H. Kato, N. Sano: J. Appl. Phys. 60, 3289 (1986)

6. Multichannel Detection and Raman Spectroscopy of Surface Layers and Interfaces

James C. Tsang

With 26 Figures

6.1 Overview

Raman spectroscopy has always been associated with low light levels [6.1]. The recent interest in the use of Raman spectroscopy for the study of surfaces, interfaces, and ultra thin layers dictates an interest in the detection of light levels considerably lower than those commonly encountered in Raman studies of bulk materials. The signals from the small number of atoms or molecules at a buried interface or in a monolayer adsorbed at an interface can be orders of magnitude weaker than the signals obtained from bulk samples. Even in the case of strongly absorbing materials where the bulk Raman signals come from only a few tens of nanometers near the surface of the sample (i.e. for values of the absorption coefficient $\alpha > 10^5$ cm^{-1}), the surface contribution to the Raman scattering will usually be less than one percent of the bulk signals. Since the signals produced in Raman experiments on opaque solids are almost always small enough to require photon-counting detection, the still smaller signals obtained in surface and buried interface experiments pose a substantial experimental challenge.

Recent advances in optical detector technology have produced significant improvements in our ability to detect extremely low light level signals [6.2]. Two to three orders of magnitude, or greater, improvements in signal-to-noise ratios over those of conventional single channel, photon counting photomultipliers have been obtained by using these new detectors [6.3]. These improvements make it possible to study a variety of phenomena previously inaccessible to Raman spectroscopy. In this article, we describe how these new detectors allow Raman spectroscopy to be applied to semiconductor and metal surfaces and interfaces. We review the parameters that determine the performance of these detectors, demonstrate their use, and provide examples of the kinds of information that can be obtained. Although the enhancements in sensitivity provided by these detectors have implications for the use of Raman spectroscopy in a wide range of applications, here the focus will be on applications involving surfaces and interfaces.

The use of multichannel detection to improve the signal-to-noise ratio in a Raman scattering experiment is not the only way that the small signals from surfaces and interfaces can be enhanced. Another method is to have multiple interfaces in single sample. However, there are many systems of considerable experimental and theoretical interest that are not amenable to this method, for example, (1) those that cannot presently be grown as multilayer structures, such as many semiconductor-metal and semiconductor-oxide interfaces, (2) those that involve chemistry at an exposed interface, or (3) those involving questions concerning the evolution of the properties of a single layer into those of a multilayer.

A second alternative is the use of resonant Raman scattering to enhance the scattering cross section for the interfacial layer [6.4]. This requires optical excitation sources at the appropriate photon energies and power levels, which may not always be available. In addition, the strengths of the resonances on surfaces and at interfaces can be considerably weaker than in the bulk, since the surface or interface resonances are often broadened, producing smaller enhancements than usual. Furthermore, the use of resonant Raman scattering to enhance Raman signals often includes the mapping out of the actual resonance. This requires the observation of both the resonant and the non-resonant Raman scattering. The new detectors would be essential for the latter part of a resonance Raman experiment. Because resonance Raman scattering requires the exploration of a two-dimensional parameter space, including both the excitation energy and the Raman shift, the new detectors facilitate these measurements simply because they acquire the data at a much faster rate.

We do not consider the special case of surface enhanced Raman scattering (SERS) in this chapter. This subject was reviewed by *Otto* [6.5] in the previous volume of this series. Many of the substrates discussed in this chapter, in particular the semiconductors and transition metals, do not show a significant surface enhanced Raman effect. Furthermore, SERS produces large changes in the Raman selection rules and the relative intensities of various features in the Raman spectra of the adsorbed species. This makes it difficult to use the Raman spectra for the characterization of adsorbates, since observed changes in the Raman spectra may be due to SERS itself.

This article builds on the article by *Chang* and *Long* [6.2] (CL) in Volume 2 of this series. It goes beyond CL in describing the behavior of and considering results obtained by using two new types of detectors, the microchannel plate photomultiplier with a position sensitive resistive anode (mepsicron) and the low noise silicon charge coupled device (CCD) detector, that were not widely available at the time that Chang and Long prepared their article. An effort has been made to limit the overlap with the material discussed in CL. The interested reader may wish to review the CL article for background on the previous generation of multichannel detectors.

The present article is organized as follows. A number of general considerations are discussed in the second section. Given the difficulty of obtaining a detectable signal in these experiments, we first consider the kinds of problems that could be addressed by Raman scattering. This survey leads to an estimate of the level of sensitivity required in this type of experiment. A discussion of some efforts to apply Raman spectroscopy to the characterization of surfaces and interfaces prior to 1983 provides independent confirmation of the estimates of the required improvements in experimental sensitivity and of the reasons for attempting such studies.

In the third section, the new experimental approaches to this problem are discussed. Both proximity focussed mepsicrons and low noise Si CCD's are covered, especially with regard to the parameters that govern their performance in low light level applications. The advantages and disadvantages of both are discussed and their performance compared. There is a brief discussion of the means by which the Raman scattered light is collected and analyzed.

Section 6.4 shows how multichannel detectors of various types have recently been used in Raman studies of surfaces and interfaces. The focus in this section is less on the surface and interface science that has been done than on the kinds of experimental questions that can now be asked. Because this chapter is concerned with the use of multichannel detectors, we do not review recent measurements on semiconductor surfaces which have used single channel photomultipliers. These include studies of band bending induced changes in selection rules [6.6], the oxidation of III-V semiconductors [6.7], the metallization of III-V semiconductors [6.8], and the damage introduced into a semiconductor surface by polishing [6.9].

First, Raman studies of the adsorption of molecules onto metal and semiconductor surfaces are described. Both UHV work [6.10], demonstrating the ability of Raman spectroscopy to overlap with traditional vacuum-solid surface studies, and work on Langmuir-Blodgett films demonstrating its ability to characterize adsorbed monolayers in a non-UHV environment [6.11], are reviewed. Next, a number of experiments aimed at characterizing buried surfaces and reacted layers are described. These include (1) the study of new phases at interfaces and surfaces associated with the reaction of oxides [6.12] and deposited metal layers [6.13], (2) the modification of the long range order at a surface by chemical reactions at the surface [6.14], and (3) the lattice dynamics of thin continuous films, as exemplified by ultra-thin epitaxially grown semiconductor layers showing both interface and bulk vibrational modess [6.15]. These results emphasize the value of a surface sensitive probe that can be applied to buried interfaces.

The final section reviews the characteristics of the problems that can be addressed by Raman scattering using the detectors described in Sect. 6.2. A number of other ways in which these new detectors enhance the range of problems that can be studied using Raman spectroscopy are mentioned.

6.2 Raman Spectroscopy and Interface Science

6.2.1 Why Raman Scattering from Surfaces?

Most of our detailed knowledge of solids and solid surfaces has been obtained by the spectroscopic study of various types of excitations of these systems. The energies, lifetimes, and symmetry properties of these excitations help to define the electronic, optical, magnetic, and thermal properties of any solid and provide critical tests of our microscopic understanding of these properties. This chapter is concerned with the vibrational excitations of surfaces, interfaces, and thin films. A review of the table of contents of this volume will show that considerable interest exists for the use of Raman spectroscopy to study single-particle and collective electronic and magnetic excitations of thin layers and interfaces. However, the scattering cross sections for these excitations are often much smaller than for phonon excitations, making their detection on a surface or at an interface an even more formidable problem [6.16].

Vibrational Raman scattering allows the study of the chemical composition of a system, since the vibrational energies can identify the bonds between different atoms in the solid. While the chemical composition of most bulk solids can be easily determined, this is not always the case for the phases at an interface between two materials, since the amount of material involved can be minuscule. The bonding between two solids can vary from the relatively weak van der Waals'-type interaction to covalent bonds between the atoms on the two sides of the interface. The vibrational modes of these bonds can range in energy from a few tens of wavenumbers to over a thousand wavenumbers, a range easily spanned by Raman scattering. The interaction between the different materials at the surface can also perturb the pre-existing bonds in the two materials, producing shifts in their vibrational frequencies. An example of this can be seen when two materials with different lattice constants such as Ge and Si, are in contact at an interface. There can be large strain fields which produce shifts in the Raman frequencies [6.17] of the bulk phonons. Even for values of strain near the yield limits, the shifts are of the order of a few wavenumbers, below the current limits of resolution of surface electron spectroscopies but well within the resolution of inelastic light scattering. While the vibrational modes introduced by bonding across an interface can be observed using electron spectroscopies when the interface is within a nanometer of the surface, because of the sub-nanometer interaction depth of charged particles with solids, optical techniques must be used for overlayer thicknesses on the order of tens of nanometers or greater.

In addition to the changes that can occur at the interface between two materials, the interactions between two solids, or a solid and an ambient atmosphere, can result in the formation of a new thin-film phase, distinct from either of the original phases. Examples of this are the oxidation of a surface [6.12] and the reaction of a deposited transition metal layer on Si to form a silicide [6.13]. These layers are often too thin to be studied by conventional Raman scattering but too thick and remote from the surface to be studied by electron spectroscopy.

The bonding of atoms at an interface provides only a partial description of the interface. The geometry of the atomic sites at the interface can also play a significant role in determining its physical properties. While vibrational Raman spectroscopy cannot provide an image of the atoms at a surface, its sensitivity to the symmetry properties of the vibrational excitations can be used to provide information about the symmetry of the interfacial layer. In addition to the chemical changes that occur when two materials are brought together, there can be important physical changes. The etching of a thermal oxide on crystalline Si by a reactive plasma can generate sub-surface lattice damage that has a substantial effect on the electronic properties of devices fabricated on the etched Si [6.18] At the solid-vacuum interface, many surfaces display complex reconstructions resulting in surface structures whose point- and space-group symmetries are drastically different from the bulk. Vibrational Raman scattering has been shown to be sensitive to similar structural transitions in a number of bulk quasi-two-dimensional systems, such as the transition metal dichalcogenide charge density wave systems [6.19]. At the present time, little is known about the existence of these reconstructions at buried interfaces, although it has recently been demonstrated that the Si (111) 7×7 reconstruction is stable under the evaporation of amorphous Si [6.20]. One reason for this lack of knowledge is the paucity of experimental probes which can look through a covering overlayer and still be sensitive to the atoms at the interface and their local environment.

In addition to the identification of the phases at the interface, the study of these systems raises questions concerning the abruptness of the interfaces. The intermixing of different phases will produce a variety of observable effects in the lattice dynamics of the systems. Examples of these have been seen in Raman studies of the reaction of Al on GaAs [6.8].

Finally, Raman spectroscopy can be used to study the interaction between the electronic properties of a system and the interfacial modes. The coupling of the light from a visible optical excitation source to the vibrational modes in a Raman experiment is through the electronic polarizability of the system [6.16]. Resonant Raman scattering has been widely used to study both the electronic properties of solids and the electron-phonon interaction [6.21]. The forces responsible for the distortions of the bulk crystal structure at the surfaces of many materials are still not understood, but in many cases it is clear that the electron-phonon interaction can play a major role in stabilizing the distorted phase [6.22].

6.2.2 Experimental Requirements

The requirements on a vibrational spectroscopy of surfaces and interfaces begin with the specification that it be able to detect the signal from the surface or interface layer. The sensitivity to the signal from the surface layer must be accompanied by the ability to discriminate against signals from surrounding phases both above and below the interface. To see what these statements mean quantitatively, we first consider the magnitude of the Raman scattering signals generated by various types of monolayers and ultra-thin films and then consider the problem of detecting these signals in the presence of optical emission from different types of substrates and overlayers.

The Raman scattering cross sections, or scattering efficiencies, of many materials have been measured and can be used to provide an estimate of the magnitude of the signals from a monolayer or an ultra thin film of the material.

Table 6.1. Raman parameters for some molecules and solids

Material and laser wavelength	Raman energy [cm^{-1}]	Scattering efficiency [10^{-7}cm^{-1}sr^{-1}]	Absorption coefficient α_{ar} [cm^{-1}]	Index of refraction
Diamond 5145 Å	1332	4.9–6.06	0	2.42
Si 5145 Å	522	519–706	1.4×10^4	$4.2 + 0.6i$
Ge 5145 Å	302	8670	6.0×10^5	$4.6 + 2.5i$
C_6H_6 4880 Å	992	2.2	0	1.50
$C_6H_5NO_2$ 4880 Å	1345	6.0	0	1.56

For the case of a molecule whose electronic and vibrational states are localized, this estimate, based on solution or condensed phase values, should be quite accurate. This need not be the case for crystalline solids, however, since the states that contribute to the electronic polarizability and the Raman-active phonons are all delocalized, so that surface effects are a significant perturbation when dealing with an ultra-thin layer. The scattering efficiencies of the dominant vibrational modes for several materials are given in Table 6.1 for an excitation energy $\hbar\omega_i = 2.4$ or 2.54 eV [6.23, 24]. Also given in Table 6.1 are the absorption coefficient and the index of refraction of each material. The internal intensity, I_s^i of the Raman scattered light is given by

$$I_s^i = I_i^i \frac{S}{(S + \alpha_i + \alpha_s)} \{1 - \exp[-L(S + \alpha_i + \alpha_s)]\} , \qquad (6.1)$$

where I_i^i is the intensity of the incident light in the sample, S is the scattering efficiency, α_i and α_s are the absorption coefficients of light at the incident and scattered frequencies and L is the length of the sample. S is defined in terms of the Raman effciency A by $S = A \sum_{j=1}^{l} |e_s R_j e_i|^2$ where e_i and e_s are the polarization unit vectors for the incident and the scattered light, R_j is the Raman tensor for the vibrational mode of interest and l is its degeneracy. The scattering cross section per molecule or unit cell $d\sigma/d\Omega$ is related to the scattering efficiency by $A = (d\sigma/d\Omega)\varrho_0$ where ϱ_0 is the density of molecules or unit cells.

The magnitudes of the maximum intensities of the Raman scattering from thin layers of these materials can be obtained from Table 6.1. The experimentally detected intensity will depend on the scattering efficiency, size of the sample, and a number of parameters which describe (1) the experimental system used to obtain the Raman spectrum and (2) the optical constants of the sample which determine the relationship between the intensity of the light inside the sample and radiation outside the sample.

High incident power levels can produce large signals, but can also damage or destroy a sample. Use of spectrometer-type analyzers with narrow entrance slits or interference-type rejection filters requiring well collimated beams limits the cross-sectional area of the sample being illuminated if the scattered light is to be collected with reasonable efficiency. Since adsorbed molecular mono-layers can be readily desorbed, chemical reactions can be optically activated and laser light can enhance mixing of different phases, the constraints on the level of the incident optical power for surface and interface Raman scattering can be more severe than for bulk studies. In the absence of experimental evidence explicitly showing that laser induced heating and sample damage is not important, incident powers, I_i, of less than 100 mW are generally the rule. In this section we therefore use the value $I_i = 100$ mW in estimating the magnitude of the raman signal from a molecular monolayer or an ultra thin film.

The lenses used to focus the exciting radiation on the samples and collect the scattered light introduce transmission losses, v_i and v_s, at their interfaces reducing the scattered signal by $v_i v_s$. The product of v_i and v_s is assumed to be 0.7 when a simple single glass lens is used to focus the incident light and a multiple lens system to collect the scattered light. In addition, since underfilling the entrance slits of the monochromator can result in experimental problems, the entrance slits are slightly overfilled so that the throughput of the entrance slits, v_e is also assumed to be 0.7.

The solid angles of collection of scattered light, Ω, are typically on the order of 0.5–1.0 sr. For the estimates of the scattered signals in this section, we take $\Omega = 0.6$ sr. Use of larger collection angles can conflict with the requirements that (1) the image of the illuminated sample on the entrance slits be not significantly larger than the entrance slits so that the scattered light can be efficiently transmitted, and (2) the scattered light collection geometry be properly matched to the optical speed of the analyzer or monochromator, which is usually slower than f5.

Almost all Raman scattering experiments are currently performed using multistage scanning monochromators [6.1]. Typically, these monochromators have 2 or 3 diffraction gratings and between 6 and 10 mirrors. While grating efficiencies can be as high as 90% over limited ranges of wavelengths [6.25] and mirror reflectivities can be greater than 98% using Ag or dielectric mirrors, for most broad band applications, grating efficiencies of 70% and mirror reflectivi-ties of 90% are typical. As a result, the monochromator efficiency, v_m is taken to be 10%.

We calculate signal levels for a monolayer or interfacial layer scatterer with respect to the performance of a conventional single channel photon counting photomultiplier with a GaAs photocathode. Because these detectors must be cooled, the quantum efficiency can be slightly degraded by the windows isolating the cooler from the monochromator. It will vary significantly with excitation wavelength. We use a nominal value for the quantum efficiency of the detector, $v_d = 15\%$.

The samples considered in this article are often buried layers. While in some cases the overlayer is transparent, in many cases it is opaque with an optical absorption coefficient, $\alpha(\hbar\omega_i)$ as high as 10^6 cm^{-1} [6.26]. For such values of the absorption coefficient, any finite thickness of a covering layer will produce a large decrease in the intensity of the Raman scattering from the interface. For $\hbar\omega_i = 2.4$ eV, the Raman spectrum from an interfacial layer buried under 20 nm of Ge where $\alpha = 5 \times 10^5$ cm^{-1} will be reduced in intensity by more than an order of magnitude. The reduction in the intensity of the scattering by the presence of an overlayer, v_o, depends on the composition of the overlayer.

The strengths of the exciting fields in the sample and the scattered fields outside the sample depend on the dielectric mismatch between the sample and the external environment. The transmission of light across the surface at normal incidence is $4n/[(n+1)^2 + k^2]$ where $n + ik$ is the complex index of refraction. For non-normal angles of incidence, the situation is described by similar but more complicated expressions [6.27], which depend on the angles of incidence and scattering and the polarization of the light. For the materials listed in Table 6.1, the reflections from the sample-air interface of both the incident, R_i, and scattered, R_s, light in near-normal geometries, will result in reductions in the strengths of the detected Raman signals from their internal sample values of less than 10% for the bulk transparent organic compounds to over 70% for the bulk opaque semiconductors.

The measured intensity of the Raman scattered light from an interface buried under a material with a large index of refraction will also be reduced by the collapse of the solid angle, Ω_s, within the sample over which the scattered light is collected for a given solid angle, Ω, outside of the sample. This lowers the efficiency with which the Raman scattered light is collected by a factor $v_\Omega = \Omega_s/\Omega$. Snell's law brings any external ray incident on or leaving the surface of the sample towards the normal to the surface inside the sample. While this has no significant effect on the excitation, it greatly reduces the collection angle within the sample. This effect has been treated explicity by *Anastasskis* et al. [6.28]. It can result in a factor of 4–5 decrease in the internal collection angle for the dielectric constants of Si and Ge in Table 6.1 and a factor of 16–25 decrease in the signal due to v_Ω.

For the above values of the performance of a conventional single channel Raman scattering system and the parameters describing the coupling of light in and out of the sample to the collection optics, Table 6.2 gives the signal levels of the Raman scattering intensities for thin layers of the materials listed in Table 6.1. The signals from diamond, Si, and Ge are calculated for 0.3 nm of material. The signals from benzene and nitrobenzene are calculated on the basis of a single molecular monolayer using surface densities of 1.6×10^{14} and 10^{14} cm^{-2} respectively. Table 6.2 also provides an "internal" signal level. The internal signal level assumes 100% efficiency for the coupling of the exciting light into the sample and is the intensity in the sample of the scattered light based just on (6.1). The "external" intensity takes the internal value and corrects it for all of the system parameters listed above i.e., it involves the product for a given system

Table 6.2 Raman scattering signals from various materials. The external signal includes the effect of collection angle, monochromater throughput, and detector efficiency.

Material	Thickness	Internal signal [cps sr^{-1}]	External signal [cps]
Diamond	0.3 nm	3900	0.036
Si	0.3 nm	4.0×10^5	3.6
Ge	0.3 nm	6.8×10^6	63
C_6H_6	1.0 ml	1350	2.4
$C_6H_5NO_2$	1.0 ml	2760	4.4

of the internal intensity and the factors, v_i, $1 - R_i$, $1 - R_s$, Ω, v_Ω, v_s, v_o, v_e, v_m, and v_d. While none of these factors are especially small, together they generate a substantial reduction in the observed signal level. For the three tetrahedrally bonded group IV materials in Table 6.1, we take $v_o = 0.1$. The reflection and geometric optics corrections to the collection efficiency used in Table 6.2 are appropriate for an overlayer with the optical constants of Si. The molecules are assumed to sit on an optically inert surface and the reflection and geometric optic corrections to the scattering intensity are for the dielectric constants of the solutions of the molecules. The values listed in Table 6.2 are for the total integrated scattering intensity in the specified line in Table 6.1. In order to spectrally resolve the line, the intensity of the scattering has to be measured over the line width. The peak intensities of the lines will be considerably smaller than their integrated intensities. While the magnitudes of the internal signals for the opaque semiconductors listed in Table 6.1 are substantial, the difficulty of coupling light efficiently into and out of the samples and the need to resolve the lines will reduce the peak signal levels to below 10 cps in all cases and below 1 cps for all except Ge. The observation of the other characteristic vibrational modes of these materials, for example, the remaining first order modes of benzene and nitrobenzene which are needed for a complete description of these molecules, and the second order Raman scattering in the semiconductors which provides direct information about the phonons away from $k = 0$, will require the ability to detect signals yet another order of magnitude smaller. With the possible exception of Ge, the resolution of the lines whose integrated scattering levels are given in Table 6.2 are well beyond the capabilities of single channel detectors, since their dark counts range between 2 and 10 cps. Reasonable signal-to-noise ratios would demand integration times per channel on the order of 100 to 1000 s. The accumulation of 500 channel spectrum would require between 1 day and 1 week, a prohibitive length of time.

An adsorbate layer must rest on a substrate. If the substrate is a highly reflecting metal such as Al, the intensity of the Raman spectrum will depend on the exact scattering geometry, including the polarizations of the incident and scattered light. The reversal in phase on reflection of light polarized in the plane of the conducting surface means that the electric field at the surface of the metal is zero, producing no driving field. On the other hand, for p polarized light, a layer sitting above a reflecting substrate will interact with the constructive sum of the incident and the reflected electric fields. The addition of the exciting fields can produce up to a factor of 4 increase in the scattering intensity. Similar considerations for the scattered fields produce an analogous decrease in the intensity of s polarized scattered light and an additional increase in the intensity of p polarized scattered light. The combined effects of the optical constants of a substrate can produce enhancements of the signal levels of over an order of magnitude (or alternatively, a considerable decrease) as shown by *Greenler* and *Slager* [6.29].

The presence of an overlayer and a substrate can also introduce background emission which obscures the Raman spectrum of the interface and makes it more difficult to identify. This background includes both luminescence and Raman scattering from the material above and below the interface. Any detector in such an experiment must not only be very sensitive, it must also be highly linear in its response with a wide dynamic range. This is necessary if the background contributions are to be correctly labeled and appropriately subtracted from the experimentally obtained spectra. Furthermore, given our interest in multi-channel detectors, the cross talk between channels of the detector must be sufficiently low that the presence of a strong substrate feature does not make it impossible to observe a weak neighboring line due to an interface mode.

A best case example is adsorption on an Al substrate. Al is typical of many simple metals. Its fcc crystal structure means there is no first order Raman scattering. Because of the small penetration depth of light, the second order Raman scattering is presently undetectable. Even in the case of a roughened surface where the surface plasmon polaritons can couple to light [6.30], optically excited hot luminescence in the visible range has not been reported and poses no problem for the observation of low level signals due to adsorbed species. Therefore, a Raman experiment using an Al substrate is currently limited only by the sensitivity and dark count of the detector.

A worse case is one in which the background is much stronger than the desired Raman signal. A semiconducting substrate can produce efficient broad band photoluminescence at energies from just above to well below the bandgap. The photoluminescence signal levels can easily exceed 10^4 cps cm^{-1} even above the band gap. The detection of a 0.01 to 1 cps signal 4 to 6 orders of magnitude smaller than a real background signal from the sample is beyond the scope of this article. Fortunately, for most semiconductors, it is possible to find spectral regions where photoluminescence is not a major problem. While many metals including the noble metals, transition metals and transition-metal compounds where there are observable optical transitions between the Fermi level and the d bands [6.31] also show weak luminescence, it is often possible to chose an excitation energy where the Raman spectrum is shifted far away from the luminescence.

Another source of background signals is Raman scattering from the sample substrate and any overlayer. All of the semiconductors in Table 6.1 produce detectable broad band, higher order Raman scattering whose integrated intensity is several orders of magnitude smaller than that of the first order Raman scattering [6.32]. The strength of the scattering is strongly affected by the wavelength dependence of both the scattering efficiency and the absorption coefficient of the sample [6.21, 33]. Another important parameter is the scattering geometry. Because the first and second order Raman lines of crystalline Si, for example, have well defined symmetry selection rules, it is possible to choose a scattering geometry where the dominant Γ_1 and $\Gamma_{25'}$ contributions to the first and second order scattering are symmetry forbidden, thus reducing the intensity of each by up to two orders of magnitude. If the surface induced scattering is symmetry allowed in a geometry where the substrate scattering is symmetry forbidden, then the intensity of the background scattering can be significantly reduced, greatly simplifying the identification of the surface scattering. Finally, although the higher order scattering is broad band, it does have structure and there are spectral regions which show only weak second order scattering. For example, there is very little higher order Raman scattering from Si for Raman shifts below 200 cm^{-1} [6.32].

These facts mean that for Raman experiments involving semiconductor surfaces, it can be necessary to obtain spectra where a first order substrate contribution can be more than three orders of magnitude stronger than the surface contribution. The surface contribution can be superimposed on a second order background which can be one or two orders of magnitude larger with considerable structure of its own. The isolation of the surface induced Raman spectrum will require an optical detector with a dynamic range two to four orders of magnitude larger than the surface signals, with linearity of better than 1%. It also means that if a multichannel detector is used, the isolation between channels must be sufficient so that when combined with the spectral resolution of the monochromator, the experimental system can resolve a weak surface feature against a strong bulk line at a neighboring energy.

An example of these Raman background effects and the Raman signal levels from a surface layer is shown in Fig. 6.1 which displays the Raman spectrum of 2 monolayers of Ge grown epitaxially on Si(100) and covered by 10 nm of Si [6.15]. Also shown are the Raman spectra of the Si substrate and the difference spectrum obtained by subtracting the Raman spectrum of Si from the Ge-Si sample spectrum. The Raman spectra were obtained in the scattering geometry where the $\Gamma_1 + \Gamma_{12} + \Gamma_{25'}$ symmetry vibrations of the diamond structure are allowed so that both the first and second order scattering were observed. The line at 522 cm^{-1} is the first order Si mode while that near 302 cm^{-1} arises from Ge-Ge vibrations of the Ge layer [6.17]. The 410 cm^{-1} mode is due to Si-Ge vibrations at the Si-Ge interface [6.34]. The broad band centered about 300 cm^{-1} seen in both the Si substrate and Ge-Si sample spectra arises from second order acoustic phonon scattering in Si [6.32]. Under 4765 Å excitation, the Ge induced modes are weaker than the symmetry allowed second order scattering from the Si substrate and the Si overlayer capping the sample. The first order Si peak is several orders of magnitude stronger than the Ge lines. Since the Si substrate spectrum shows considerable structure near 300 cm^{-1}, the sub-

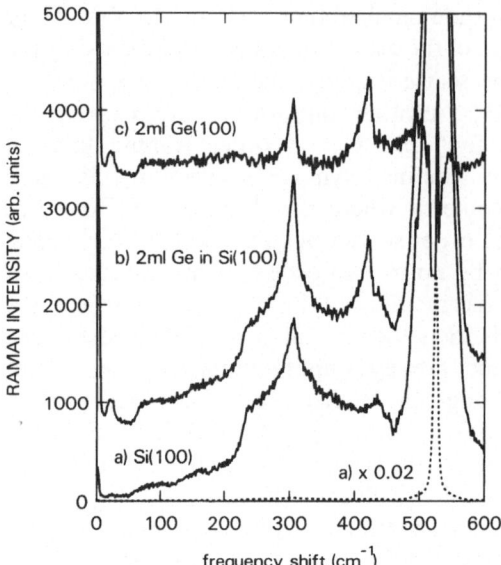

Fig. 6.1. Raman spectrum of 2 mono-layers of Ge epitaxially grown on Si (100) and covered by 10 nm of epitaxial Si. Both the first order $\Gamma_{25'}$ symmetry optic phonon scattering and the Γ_1 symmetry second order scattering are allowed for the geometry used to obtain these results. (a) Pure Si (100) (——) [x0.02 (–––)]. (b) Ge in Si. (c) Difference of (a) and (b)

traction of the spectrum obtained from a pure Si(100) sample requires a high degree of linearity on the part of the detector as well as considerable sensitivity for the observation of the Ge-Ge vibrations in the Ge layer.

6.2.3 State of the Art ca. 1983

Section 6.2.2 showed that the signal levels obtained from nanometer thick films of opaque semiconductors in Raman scattering were within two orders of magnitude of what could be detected with a single channel photomultiplier. *Connell* et al. [6.35] showed that enhancements of more than a factor of 20 of the Raman scattering intensities from thin film samples were achievable if the samples were made part of a dielectric stack which optimized the electric fields in the sample plane. Using this structure in conjunction with a single-channel detector Raman system, they demonstrated the capabilities of Raman scattering to provide information about solid-solid interfaces and reactions.

A sample of thickness t was made the front surface of an anti-reflection coating on an Al mirror and the thickness d of the dielectric spacer layer adjusted to maximize the electric field in the sample plane. The peak in the intensity of the optical field at the sample plane arises from satisfying the same conditions for the minimization of the reflectivity from this composite system i.e., the appropriate interference of the waves reflected from the sample and the Al behind the sample when the dielectric thickness is about $\lambda_0/4$ [6.36]. The net enhancement of the intensity of the scattering from a film of thickness t in an appropriately designed multi-layer system over that from a semi-infinite sample (including the effects on both the incident and the scattered fields) is $8\alpha_i t/(1-R)^2$ where R is the

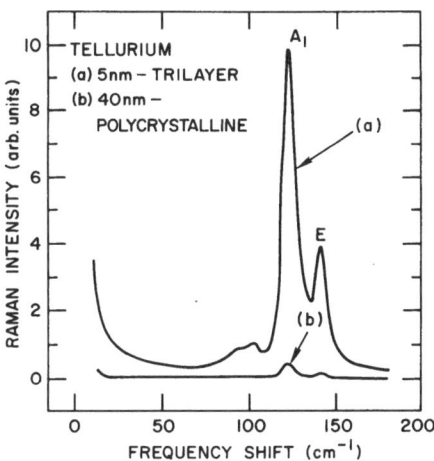

Fig. 6.2. Raman spectrum of 5 nm of Te in an IERS configuration compared to the unenhanced spectrum of 40 nm of Te [6.35]

reflectivity of the sample and α_i the sample absorption. Figure 6.2 shows the results of *Connell* et al. for the Raman spectrum of a deposited Te layer with and without the interference enhancement of the scattered signal produced by the multilayer structure [6.35]. The lines between 75 and 150 cm^{-1} are the characteristic vibrational modes of crystalline Te. The enhancement of the Raman scattering intensity is clearly seen in the multilayer system. This enhancement is obtained in a scattering geometry where the incident and scattered **E** fields are in the plane of the surface, which means that the polarization selection rules can be explored in detail. This is in contrast to the reflection enhancement of *Greenler* and *Slager* [6.29] which was restricted to incident and scattered light polarized normal to the surface.

Nemanich et al. applied this technique to a variety of solid-gas and solid-solid reactions. Evaporation of thin Ti layers onto the spacer layers resulted in the observation of the 137 cm^{-1} Raman active mode of hcp Ti. Their results [6.37] on evaporated Ti also showed a broad background extending to beyond 400 cm^{-1} (Fig. 6.3) which was correlated with the presence of oxygen in the evaporator. By comparison with the Raman spectra of deliberately oxidized Ti films and of crystalline Ti_2O_3 and TiO_2, they were able to show that the native oxide on a Ti film was an amorphous form of Ti_2O_3.

Nemanich et al. also used their interference enhanced Raman scattering technique to study the formation of silicide phases on amorphous Si [6.38]. Ten nm of amorphous Si was deposited on SiO_2-Al structures with the appropriate oxide thickness. Varying thicknesses of Pd were then laid down on the freshly deposited amorphous Si. Nemanich et al. were able to observe the initial reaction of the deposited metal on the amorphous Si (Fig. 6.4a). The sharp lines between 100 and 220 cm^{-1} in Fig. 6.4a are the characteristic vibrational modes of Pd_2Si. These lines are superimposed on Raman scattering from acoustic phonons in the amorphous Si layer. The broad band below 500 cm^{-1} is due to optic phonon scattering from the amorphous Si layer. The Raman spectra showed that the first

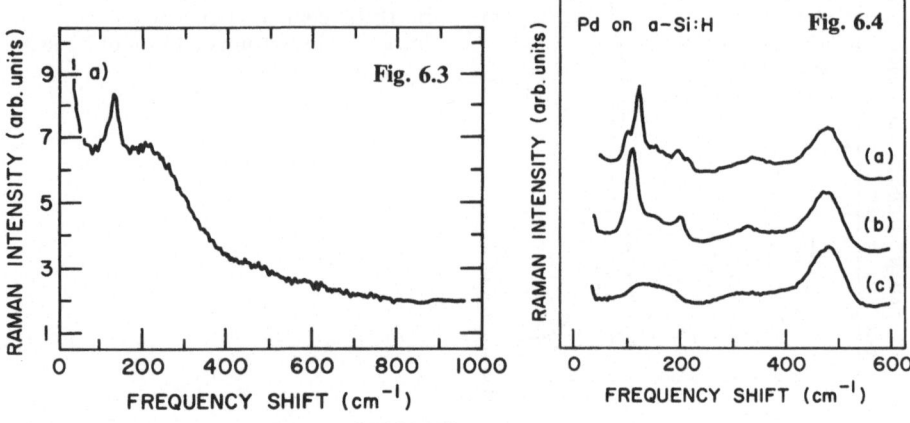

Fig. 6.3. IERS spectrum of 6 nm of Ti [6.37]

Fig. 6.4. IERS spectrum of 10 nm of hydrogenated amorphous Si. (*a*) Covered by 2 nm of Pd at room temperature, (*b*) after annealing at 300 °C and (*c*) when the amorphous Si is covered by an aged, native oxide before the room temperature deposition of Pd [6.38]

2 nm of Pd deposited on the amorphous Si reacted to form a crystalline Pd_2Si phase. Heating the sample above 300 °C resulted in the consumption of the unreacted Pd in forming additional silicide (Fig. 6.4b). There was a difference between the Pd_2Si phase created at high temperatures and the initially reacted phase which was detected through a small shift in the energies of the phonons, a shift that would have been difficult to detect with electron spectroscopies. They were able to show that the presence of an aged native oxide on the amorphous Si could suppress the initial reaction of the deposited Pd (Fig. 6.4c). The deposition of Pd on the aged native oxide of amorphous Si produced only the Raman spectrum of amorphous Si with no trace of a reacted silicide phase. Electrical measurements were made on similarly prepared structures, and the I–V characteristics were correlated with the Raman studies to show how the surface chemistry affected the behavior of the silicide-amorphous Si junctions [6.39]. It was demonstrated that the non-ideality of the I–V characteristics of diodes formed from Pd deposited on aged amorphous Si films that were fully oxidized could be correlated with the suppression of the initial formation of the silicide phase by the oxide barrier.

The 1983 review article of *Chang* and *Long* (CL) [6.2] clearly showed that vidicon and diode array detectors could produce at least tenfold enhancements in the experimental sensitivity of Raman scattering systems. This means that the experiments of Nemanich et al. could be performed without the enhancement of the signal levels due to the multilayer structures. A wide variety of experiments, therefore, become possible with such multichannel detectors. A new generation of such detectors has even greater sensitivity. We now consider how these new detectors perform, in order to understand what new experiments can be done.

6.3 Multichannel Detectors and Raman Spectroscopy

The article by *Chang* and *Long* [6.2] in Volume 2 of this series provides an introduction to a number of types of multichannel analyzers that were available for use at that time (early 1980s). It also gives an introduction to the vocabulary of spatially resolved photon detectors. The silicon intensified target (SIT) and photo-diode arrays discussed in CL had multichannel capabilities for the acquisition of optical spectra. The ability to accumulate information in 500 to 1000 channels simultaneously provided a substantial improvement in sensitivity over that of a single channel detector.

The quantitative improvement in performance between the SIT vidicons and the single channel photon counting photomultipliers, however, was less than expected based on just the increase in the number of active channels. While cooling a conventional photomultiplier results in a decrease in the dark count, it produces a degradation of the read-out behavior of the vidicons ("lag"). The noise generated by the process of reading out the charge stored in the silicon target, combined with their relatively high dark counts at room temperature, from the photocathode and the silicon target, meant that the total background due to both dark count and read-out noise per channel of the vidicon was significantly greater than the dark count for a cooled single channel photomultiplier. In addition, for experiments using optical sources in the visible range, the quantum efficiencies of the available multi-alkali photocathodes in SIT detectors were lower than the quantum efficiencies obtained from GaAs photocathode photomultipliers. As a result, the factor of 20–30 increase in signal-to-noise due to the parallel detection in the vidicon was balanced by a factor of 2 decrease in the signal level due to the lower quantum efficiency of the photocathode and a factor of 3 to 5 increase in the noise due to the higher dark count and read-out noise. The net result was a factor of only 3–5 increase in sensitivity under certain circumstances.

Similar considerations apply to the performance of the cooled intensified photodiode arrays described in CL. While the dark count per pixel can be lower than the dark count of a photomultiplier tube, the noise generated by the read-out of the diode array was comparable to or greater than the dark count of a good photomultiplier tube. Because these detectors also use thin-film photocathodes in transmission, the quantum efficiency of the detector is below that of a GaAs photocathode photomultiplier. While the diode arrays have 500 to 1000 channels in one direction, they have only a limited number of channels of finite height in the orthogonal direction. Therefore, they have difficulty in efficiently detecting light in situations where the source is not a point but a line. As indicated earlier, a major constraint in Raman scattering experiments is the intensity of the light incident on the sample. A standard solution to this problem is the use of cylindrical lenses to produce a line focus on the sample that matches the height of the entrance slits to the monochromator. If the active area of the detector has a *height that is only a fraction of the height of the entrance slits of the* monochromator, then a substantial fraction of the scattered light is not detected and the signal level is reduced accordingly. Since the vertical extent of the spectrum, when excited by a line focus, often exceeds 2 cm, if the active area of the detector is only 1 or 2 mm in height the useful efficiency of the detector will be reduced by almost an order of magnitude.

As indicated in the introduction, recent advances in optical detector technology have produced substantial improvements in the signal-to-noise ratios that can be obtained for the low light levels that occur in Raman scattering experiments involving surfaces and interfaces. This section describes the operation and performance of two of these new detectors, the mepsicron and the CCD. Both of these detectors are interesting because they provide two-dimensional multichannel detection capabilities with considerable enhancements in performance per channel over the peformance obtained from even the best conventional photomultiplier. This combination means that enhancements in signal-to-noise of 2 to 3 orders of magnitude over the performance of a conventional single channel photomultiplier are possible. This is more than an order of magnitude better than can be achieved using the older detectors. It is these improvements in performance that make practical the use of Raman scattering as a probe of surfaces and interfaces.

6.3.1 The Imaging Photomultiplier (Mepsicron)

The dark count from a cooled single channel photomultiplier can be as low as 1–5 counts per second (cps) with a quantum efficiency as high as 20%. Because the output of the photomultiplier is analyzed by pulse counting electronics to discriminate against electron pulses generated both within the electron multiplier chain and by external perturbations such as various types of radiation events, the dark count is dominated by the thermal emission from the photocathode and is proportional to its area. This area is usually much larger than the spatial extent of the signal which is defined in a scanning monochromator by the exit slits. The area of the photocathode is typically several cm^2, while the light incident on the photocathode can have a cross section of about 1 cm in height and 100 µm in width. Use of a photomultiplier with a 1 cm × 100 µm photocathode would result in a decrease in the dark count proportional to the decrease in area, about a factor of 500 or so. If the dark count from the full photocathode was 5 cps, then the count from the active region of the photocathode would be less than 0.01 cps. If, instead of having such a small photocathode, (which would be very hard to align) there was a full sized photocathode, and it were possible to determine where on the photocathode the optically excited photoelectrons emerged, the detector would have a multi-channel capability, with the number of channels equal to the area of the photocathode divided by the spatial resolution. The dark count per channel would be proportional to the dimensions of a pixel or resolution element of the detector. For a 2.5 cm diameter photocathode, the ability to spatially resolve the

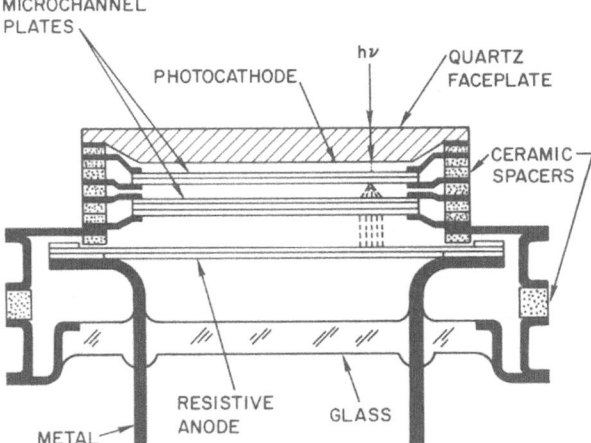

Fig. 6.5. Schematic of a microchannel plate photomultiplier with a position sensitive resistive anode often referred to as a mepsicron (Microchannel Electron Position Sensor with Time [chronos] Resolution)

photoelectrons excited by a 2 mm high beam of light with 100 microns resolution would result in a $500 \times$ improvement in the signal-to-noise over the use of the same photocathode with no spatial resolution of the incident light.

The mepsicron has the properties of the above idealized detector. A schematic diagram [6.40] of this detector is shown in Fig. 6.5. Both conceptually and operationally, this detector is relatively simple, including only a photo-cathode, a series of microchannel plate electron multipliers and a position sensitive resistive anode. A summery of the performance characteristics of such a tube is given in Table 6.3. A "naked" version of this device minus the photocathode and the glass envelope has been widely used in ESCA (Electron Spectroscopy for Chemical Analysis) where energetic photoelectrons excited by the x-rays are amplified by the microchannel plates and detected by the resistive anode. The behavior of this detector and the optimization of its performance to the levels given in Table 6.3 have been discussed in a number of papers [6.3, 40]. We summarize these and outline the principles of operation by following a signal through the detector.

The photocathode converts the incident photons into electrons with an efficiency v_d. The spectral range of interest determines the particular choice of photocathode material with different cathodes resulting in enhanced sensitivity over particular spectral regions. In the ultraviolet, both thin-film multi-alkali and bi-alkali photocathodes show quantum efficiencies well above 10%. This performance decreases in the visible with $v_d < 10\%$ being common. Use of thin-film multi-alkali photocathodes, which have enhanced sensitivity in the red, results in higher dark counts at a given photocathode temperature, so that either higher backgrounds have to be tolerated or the detector must be operated at lower temperatures to reduce the dark count. However, unlike the other

Table 6.3. Performance specifications for typical MEPSICRON and CCD detectors

Parameter	MEPSICRON	CCD
Working diameter	25 mm up to 50 mm	12.5 mm up to 50 mm
Spectral range	200–800 nm > 850 nm available	300–900 nm (backthinned, Back illuminated) 500–900 nm (front illuminated)
Quantum efficiency v_d	0–0.15	0–0.7
Spatial resolution	42–100 µm	27 µm
Dark current	10^{-4} cps/pixel at $-30\,°C$	0.01 cps/pixel at $-130\,°C$
Read out noise	0	20–50 electrons (specifications as low as $5e^-$
Max. count rate	50 000 cps	Very high
Max. pixel count rate	100 cps	Very high
Dynamic range	$> 10^6$	5×10^4
Max. signal	Unlimited	$4 \times 10^5\,e^-$ per pixel
Spatial variation	< 0.05	< 0.07 (excluding defective pixels)
Time resolution	< 500 ps	5 s
Failure modes	high light levels	Static voltages and fields
Lifetime	10^{13} c/pixel	Unlimited

multichannel detectors described in CL, where cooling does reduce the dark count but also significantly perturbs the operation of the detector [6.2], the operation of this detector is largely insensitive to its temperature.

To simplify the tube, the electron multipliers in Fig. 6.5 are located directly behind the photocathode. The distance between the front surface of the first microchannel plate and the photocathode is less than 1 mm. Since the photoelectrons emitted from the photocathode have little kinetic energy, for accelerating voltages between the photocathode and the first microchannel plate in excess of a few hundred volts any lateral motion of the photoelectrons with respect to their origin will be less than a few tens of microns. Early designs of this type were susceptible to a rapid loss of sensitivity due to the catastrophic degradation of the photocathode by ion bombardment from the microchannel plates. This was solved by overcoating the front surface of the microchannel plate stack with a thin oxide to reduce the sputtering of the plate surfaces by the accelerated photoelectrons. Photocathodes treated in this manner have remained stable for long periods of time [6.41]. The presence of the oxide film requires a bias voltage between the photocathode and the electron multipliers of 200 to 400 V so that the photoelectrons from the cathode have enough energy to initiate the electron cascade [6.42]. The use of still higher voltages results in a degradation of performance due to "halation" or the excitation of electrons out of the photocathode by secondary electrons created in the collision of a photoelectron with the surface of the microchannel plate.

The behavior and characteristics of microchannel plate electron multipliers have been described in CL. A general description can be found in [6.43]. The electron gain of the microchannel plates and the characteristics of the resistive anode and the electronics used to read out the charges deposited on the anode determine the spatial resolution of the detector. *Rees* et al. [6.42] and *Firmani* et al. [6.44] studied how the parameters of maximum spatial resolution, minimum pulse height width, and minimum noise can be optimized for a given linear resistive anode using the number and arrangement of microchannel plates and their bias voltages as free parameters. They determined that optimal performance was obtained by using 3 to 5 microchannel plates at relatively low gains rather than fewer plates at high gains. The use of fewer plates at high gains was hampered by ion feedback and space charge effects. Ion feedback, due to the acceleration of heavy ions emitted from the microchannel plate walls, rendered the gain of a single plate unstable for gains above 10^4. The problem of space charge at high electron gains arose from the small cross sectional areas ($150 \, \mu m^2$) of the channels in the microchannel plates and the large number of electrons in a channel at gains $> 10^5$. In addition, the high resistance microchannel plates were unable to supply enough carriers to a single channel at high gains. For these reasons, five microchannel plates are used in the device shown in Fig. 6.5. The plates are oriented in the so-called V and Z configurations so that the channel directions vary from one plate to the next, minimizing the ion feedback problem by limiting the straight paths in the channels. In addition, the two sets of plates (stacks) in Fig. 6.5 are spaced with a drift region between them. Electrons emerging from the V plates can spread laterally in space, so that as they enter the Z plates they use a larger number of channels, thereby reducing the magnitude of the charging effects within a single channel.

Given the above design, the operation of the detector shown in Fig. 6.5 requires the determination of bias voltages for the various stages of the tube. While values for a set of optimal voltages have been published [6.44], the variation from tube to tube of parameters such as the wall resistance of the microchannel plates has meant that at present, each tube has to be optimized individually. Fortunately, for a given resistive anode and the commercially available read-out unit, the optimization does not involve any large trade-offs in performance.

In the device shown in Fig. 6.5, the electron cloud is collected by a resistive anode which consists of a conducting layer on an insulating substrate. The *resistance* of the anode is highly uniform. The position on the anode of the centroid of the charge cloud generated by a single photoelectron can be obtained by measuring the fraction of the total charge deposited on the anode that accumulates at each of four appropriately placed electrodes along its edges. The behavior of the low distortion resistive anode as a two dimensional imager, including its noise properties, has been analyzed by *Lampton* and *Carlson* [6.45]. Because of the centroiding properties of the resistive anode, the spreading of the electron cloud does not severely degrade its spatial resolution. However, the magnitude of the gain can affect the spatial resolution since the uncertainty in the spatial position is related to thermal fluctuation charge noise in the anode. *Firmani* et al. [6.40] showed that spatial resolution on a single amplified pulse on the order of 50 microns could be achieved for a one inch diameter detector.

The finite response time of the system electronics required to analyze the pulses incident on the resistive anode and the RC time constant of this anode means that there is a limiting maximum count rate for the commercially available five microchannel plate detector. If two electron pulses are deposited on the resistive anode within several microseconds of each other, then they cannot be distinguished, resulting in the detection of a single pulse of unusually large amplitude, subject to rejection by the discriminator circuit, and a centroid position that is the average of the two pulses. Thus, the maximum count rate is of the order of 10^5 cps over the whole face of the detector. In addition, there is a maximum count rate per pixel or resolution element of the detector of about 100 cps, due to the finite rate at which the microchannel plates can supply charge.

As in the case of photon counting photomultipliers, careful control of the gain of the electron multiplier stages to produce a narrow pulse height distribution means that maximum advantage can be taken of the pulse height discrimination capabilities of the electronics that analyze the charge collected at the edges of the resistive anode. By summing the charges, it is possible to discriminate against almost all events not initiated by a photon incident on the photocathode, since the other events will generate either unusually large amplified pulses, as in the case of large scale ionization processes occuring in the tube, or very small pulses due to events initiated within the electron multipliers themselves. Such discrimination allows for a low dark count over the area of the photocathode, determined only by the thermal emission from the cooled photocathode. When divided by the number of resolution elements of the detector, the dark count per pixel achieves the very impressive figure given in Table 6.3. The product of the dark count per pixel and the number of pixels is about 50 cps, which is an unexceptional figure for a cooled 1 inch diameter photocathode.

The read-out for the microchannel plate photomultiplier with a position sensitive resistive anode is commercially available. As each electron pulse strikes the resistive anode, its position and amplitude are analyzed. For a pulse of appropriate magnitude which is accepted by the discriminator, the read-out unit generates two signals describing the (x, y) centroid position of the pulse on the resistive anode. When attached to a spectrograph in which the spectrum is dispersed along the horizontal axis and there is negligible curvature of the image of the entrance slits in the focal plane, the horizontal position information can be

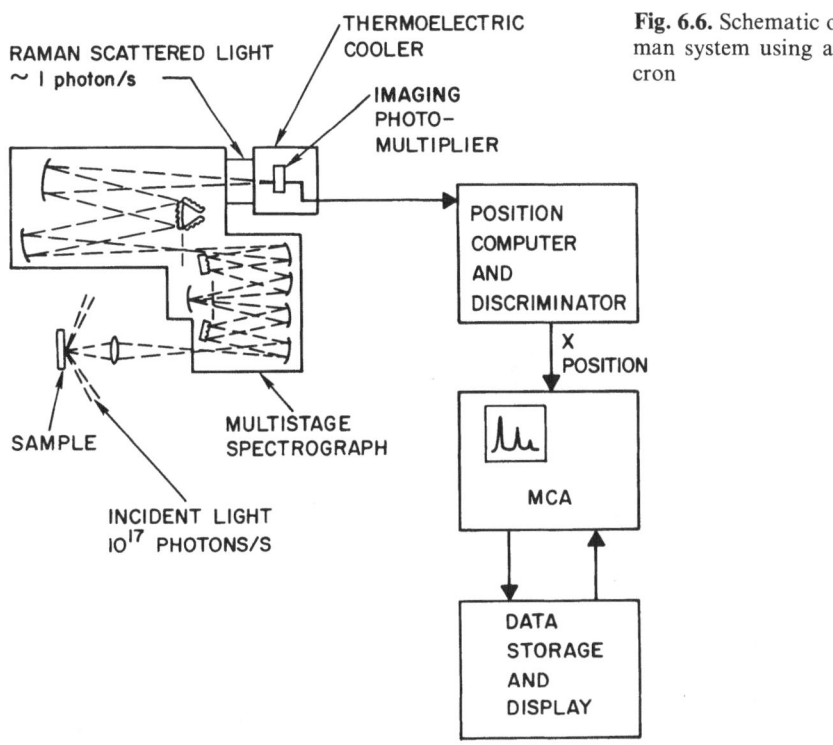

Fig. 6.6. Schematic of a Raman system using a mepsicron

accumulated in a multichannel analyzer or small computer and the spectrum presented as a histogram of the pulse positions. The (x, y) output can be displayed directly in two dimensions on an oscilloscope and stored as such if needed, i.e., if corrections for imaging distortions due to the monochromator, etc. are required.

A schematic for the use of such a detector in a Raman scattering experiment is shown in Fig. 6.6. (A pre-amplifier between the mepsicron and the position computer/discriminator is not shown). In this figure, the position information is temporarily stored in a multichannel analyzer and the spectrum accumulated in the analyzer is transferred to a computer for permanent storage. Figure 6.6 shows that the operation of this detector is extremely simple. The mepsicron emits a pulse every time a photon creates a photoelectron. The operation of the detector consists of exposing the detector to the signal and allowing the output from the position computer to accumulate in the multichannel analyzer until the spectrum meets some user criteria. The low dark count and the large dynamic range of the multichannel detector system in Fig. 6.6 allowed it to easily obtain the results shown in Fig. 6.1.

In Fig. 6.7, we show the dark count arising from such a cooled detector. The dark count was accumulated over a period of one hour using the system in Fig. 6.6. The detector has a multi-alkali photocathode with useful spectral

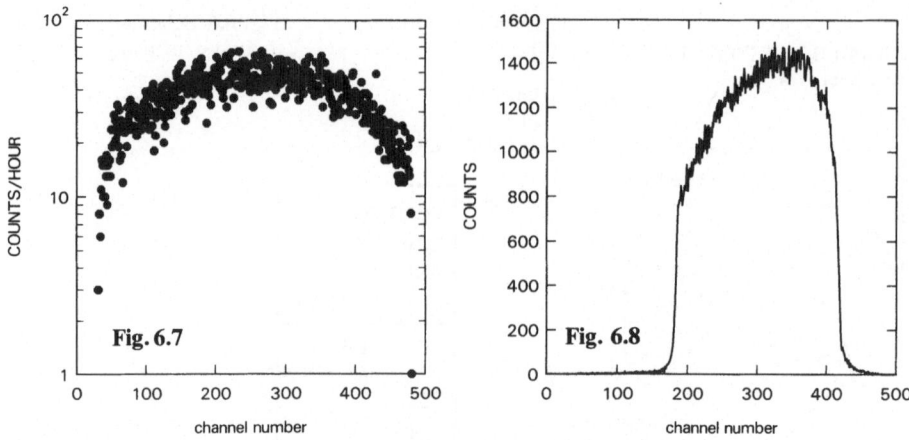

Fig. 6.7. Dark count of a mepsicron obtained for a 1 hour integration
Fig. 6.8. Response of a mepsicron to uniform illumination (from a monochromator) of its photocathode over 300 channels

sensitivity for photon energies below 6 eV (limited by the quartz envelope) and above 1.5 eV. The results in Fig. 6.7 were obtained with the detector cooled to $-20\,°C$ by a thermo-electric cooler. The channel height used in Fig. 6.1 was about 3/4″. The maximum dark count per channel was 70 counts, or less than 0.02 cps/channel. This is between 200 and 500 times better than the dark count of a good single channel photomultiplier.

Figure 6.8 shows the response of this detector to a tungsten light shining on a white sheet. The full system in Fig. 6.6 was used and the band pass of the triple spectrograph was set so that the turn on and turn off of the band pass could be seen in the spectrum. The spectral band was about 400 cm^{-1}, centered about 5000 Å. There is a significant variation in the response of the system towards the edges of the band pass which is due to both the spectrometer and the detector. However, the variation in the response of the tube across its face from channel to channel is close to being noise limited. This high degree of uniformity reflects the quality of the evaporated photocathode and the photon counting nature of the detector, where small variations in gain across the microchannel plates are not considered by the pulse counting read-out unit. This is in contrast to all of the charge storage detectors such as the vidicon, diode array, and CCD to be discussed, where variations in the individual pixels used to store the photoelectrons translate into substantial variations in sensitivity. In addition, the mepsicron has no "bad" pixels or rows or columns as are often found on solid state detectors due to the complex processing required for their fabrication.

The major strengths of the mepsicron are its operation as a true photon counting detector complete with discriminator, the resulting low dark count, and its ease of operation. With the exception of the setting of the bias voltages which determine the gain in the microchannel plates and the spatial resolution of the detector as well as the noise level, its operation is no more difficult than that of a regular photomultiplier. It should be noted that, like other photomultipliers, the use of high voltages to obtain electron multiplication means that the mepsicron is subject to catastrophic damage if exposed to high light levels and blind spots if these high levels are confined to a small area of the cathode.

6.3.2 The Charge Coupled Device Detector

Recently, astronomers have found wide use for charge-coupled-device photon detectors [6.46]. CCD detectors differ substantially from conventional photo-multipliers, the early multichannel detectors discussed in CL, and the mepsicron device described above. The CCD's most outstanding quality is its extremely high quantum efficiency, which far exceed the quantum efficiencies of photo-cathodes in the visible. Values for v_d of over 70% have been reported. Such high efficiencies are invaluable in detecting the small signals characteristic of Raman scattering from thin layers and surfaces. When combined with the multichannel imaging capabilities of the CCD, it is clear that such detectors will produce significant improvements in performance over the best single channel detectors and the earlier multichannel detectors described in CL.

Table 6.3 [6.47] also presents a compilation of some of the parameters that describe the performance of a low light level imaging CCD detector. There have recently been a number of reviews of CCD photon detectors [6.46, 47, 48]. While the operation of the CCD is ultimately based on the creation of an electron-hole pair in a semiconductor, the device is quite different from the previously discussed semiconductor optical detectors. The CCD is based on the ability of semiconductor device fabrication technology to create a two dimensional array of potential wells that can store electrons. If photons with energies greater than the band gap are incident on this two dimensional array, then the free electrons of the electron-hole pairs generated by the absorption of light can be trapped in the wells. The optical image is stored in the pattern of trapped carriers. The CCD's advantage over other imaging semiconductor devices such as vidicons, which also store photogenerated carriers in localized potential wells, comes from the manner in which the image pattern is read-out of the device. This process can involve read-out noise on the order of 1000 electrons or more for a vidicon or a diode array. Given such high levels of read-out noise, "photon-counting-like" behavior can only be obtained by using image intensifier stages which also contribute to the dark count and reduce the quantum efficiency of the detector to that of a photomultiplier. For the CCD, the read-out noise can be of the order of 5–100 electrons, or comparable to the dark count of a conventional photomul-tiplier. With such low read-out noise, the CCD does not require an intensifier stage and can directly use the high quantum efficiency of Si to photogenerate electrons and holes, resulting in a very powerful detector.

The read-out of the CCD is based on its ability to function as an efficient shift register. The variation in the density of stored charges across the CCD duplicates the spatial variation in the intensity of the light falling on the chip, normalized by

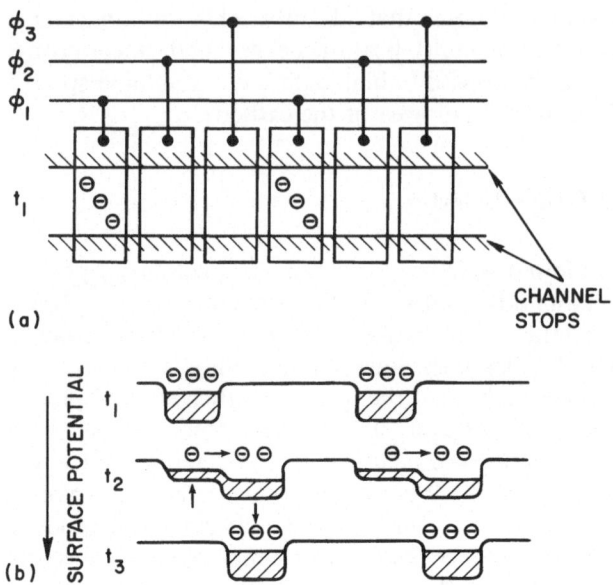

Fig. 6.9. Schematic of the process by which a 3 phase CCD gates information. The sequential modifications of the potentials ϕ_1, ϕ_2, and ϕ_3 applied by the three electrodes which define each well, result in the shift of the charge stored in a well to the adjacent well

factors related to the wavelength dependence of the optical constants of the chip and the collection efficiency of the individual wells. For a two-dimensional array of potential wells, a well at (x_0, y_0), where x_0 and y_0 are the spatial coordinates with respect to the repeat distances between the wells in the horizontal and vertical direction, can be read out by a low but finite noise output amplifier located at $(0, 0)$ after a series of $x_0 + 1$ horizontal shifts to the left and $y_0 + 1$ vertical shifts down. A few of the steps involved in such a process are shown in Fig. 6.9 [6.48]. This figure shows that by using a set of discrete electrodes, it is possible to vary the surface potentials ϕ_m for $1 < m < 3$ over and adjacent to the stored charge and localize charge in a single well and then shift the stored charge in a given direction. The remarkable aspect of CCD technology is that this process can now be done with near 100% efficiency.

The degrees of freedom involved in the choice and use of a CCD detector include (1) its spatial resolution, (2) the number of electrons that can be stored in a well, (3) the rate at which electrons are thermally generated, (4) the efficiency with which the charges are transferred from one well to another, (5) the read-out noise associated with the output amplifier on the chip, (6) the operating temperature of the device, (7) the way in which it is read out, including the read out pattern itself, (8) the manner in which it is exposed to light, including the integration time on the chip, (9) the operation of the electronics on the chip, and (10) the normalization of the optical response of the chip. A number of other free parameters exist but will not be discussed here.

The spatial resolution of the CCD is determined by the physical size of the potential wells. Given the need to be able to vary the potentials defining the well in order to move the photoexcited electrons, it is possible to work at lowered spatial resolution by combining adjacent wells (binning). Typically, well sizes of about 25 to 50 microns are available which determine the spatial resolution, and in conjunction with the dispersion of the spectrometer, the spectral resolution of the system. The recently published results which were obtained by using either the CCD or the mepsicron have been limited by the monochromator resolution [6.3, 47]. The spectral range of the CCD is currently limited by the fact that the commercially available imagers are based on intrinsic Si where the optical transition across the indirect gap is near 1.1 μm. Even then, the useful spectral range depends on a number of details regarding the preparation of the chip. If conventional means for defining the potential wells are used, then illumination of the detector from the front or the side results in a substantial loss of quantum efficiency, because of the finite transmission of the electrodes which reduces the intensity of the light reaching the body of the device. One solution to this problem is to thin the substrate to thicknesses of the order of 1 μm so that the detector can be efficiently illuminated from the substrate side for photon energies up to 4 eV. This, however, can introduce other problems including mechanical instabilities, fringing from interference effects inside the chip and due to its bonding to a transparent substrate for mechanical stability and charge trapping at the back surface. Some of these will be discussed later.

The wells in a CCD have both a finite size in real space and a finite depth in energy. As a result, they can only store a limited number of electrons. As a well is filled by carriers, the well potential becomes shallower due to the charges of the trapped electrons. The maximum number of carriers that can be stored in a CCD well varies from 10^5 to 10^6. The decrease in the well potential also modifies the efficiency with which the wells collect the free electrons, resulting in a non-linear response, which is obviously debilitating for any measurement with a background. These non-linearities are avoided by restricting the integration times so that the maximum charge storage is not approached. As will be seen, a variety of other conditions also restrict the length of time that a signal can be integrated on a CCD detector. This reduces the ultimate signal-to-noise ratio of the detector because it increases the frequency of read-outs, and each read-out adds noise.

The 1.1 eV band gap of Si means that at room temperature there are many thermally generated electron-hole pairs. These contribute to the dark count and fill the wells at a rate at which the integration time between read-outs is of the order of seconds. Since the CCD cannot discriminate between thermally generated and optically generated electron-hole pairs, it is necessary to cool the CCD to liquid-nitrogen temperatures in order to reduce the dark count sufficiently for low light level, long integration time studies. The number of thermally generated electron-hole pairs decreases exponentially as $-E_g/k_B T$. *Murray* et al. [6.47] showed that operation at temperatures near $-100\,°C$ reduced the thermally generated dark count to levels approaching those obtained from the microchannel plate detector described above.

The requirement of low temperature operation complicates the process by which the photogenerated electrons are read out of the CCD. As Fig. 6.9 showed, this is done by sequentially modifying the potentials that define the wells so that the trapped carriers can be transferred from one well to another and eventually read out by the output amplifier. Since the multichannel character of the CCD is valuable only when the number of channels is on the order of several hundred or greater, the charge transfer process can occur hundreds of times before the trapped charge representing a particular piece of the image reaches the amplifier. Room temperature charge transfer efficiencies in excess of 99.99% are commonly cited. Unfortunately, this figure decreases with decreasing temperature. If 99.9% of the charge in a well is transferred at each step, then after 512 steps only 60% of the charge in the inital well reaches the output. Aside from changing the signal stored in the particular well, it also results in an increase in the background charge level of all the wells transited. Figure 6.1 showed that the surface signals of interest in this article can often involve side-by-side lines whose intensities differ by more than two orders of magnitude. Inefficient charge transfer would make the weak lines hard to resolve.

Murray et al. [6.47] found that the charge transfer efficiency (CTE) was signal level-, direction of transfer (in the two-dimensional detector plane)-, and temperature-dependent. At −150 °C, they found that CTEs in one direction as high as 99.99% were obtainable from the device they used. However, they also obtained values as low as 99.8% in the orthogonal direction, which would have produced unacceptable losses and excessive smearing due to the charge trails. The mode of operation of the detector had to be adjusted to account for this variation to obtain usable results.

As mentioned, the electrons transferred out of the potential wells are read out of the CCD by an on-chip amplifier. Using a single read-out device removes the possibility of differences in signal levels due to differences in amplifier performance. Typical figures for the read-out noise associated with the on-chip amplifiers are between 5 and 100 electrons. Since each read-out event generates this much noise, there is a premium on minimizing the number of read-outs. This of course can conflict with the requirement for periodic read-out due to well-filling.

Because the CCD accumulates charge in the wells until they are read out, the device has no way of discriminating against carriers generated by sources other than the photons of interest. As discussed above, the contribution of thermally derived carriers can be reduced by cooling the detector. Another source of carriers is various types of radiation events. Ionizing radiation due to cosmic rays, radioactive species in the device itself, etc. can create enormous numbers of carriers in individual wells, in fact, often saturating them. These events are highly localized, usually affecting only a single well or pixel. Such ionization events are easily ignored in photon counting photomultipliers since the electron pulses they create usually exceed the threshold level of the discriminator. As long as the numbers of wells or pixels affected by these events are small, they can be easily identified in the read-out data from the CCD as very large individual spikes.

However, this presents another constraint on integration time and also the format of integration. The number of such events increases with integration time, thus, integration times of the order of an hour or less are favored. Since the effect is localized to single pixels, its identification is facilitated by the use of the smallest well size possible.

For any imaging photon detector, a characterization of the response across the face of the detector is necessary. In the case of the CCD detector, there can be significant structure in its spatial response [6.47]. Physical defects due' to problems in the fabrication of the CCD result in anomalous pixels. These include elements with low or no optical efficiency, or with high dark counts, or which are always fully saturated, etc. There can also be interference fringes in the response of the detector due the backthinning of the CCD and the bonding of the thinned chip onto a transparent substrate for mechanical support. Even in the case of a detector that is not backthinned, the multilayer structure of the chip can result in the introduction of interference fringes into its response. Murray et al. discovered that the on-chip amplifier was also a source of weak luminescence when operating. The emission from the amplifier was detected by the pixels of the CCD. Appropriate steps were taken to turn off the amplifer while the signal was integrated on the CCD and to avoid using the region of the CCD chip near the read-out device in obtaining a signal. Luminescence has also been observed due to the clocking of the gates and from bad pixels [6.48].

As in the case of the mepsicron, read-out electronics for the CCD detector are now available commercially. This allows the accumulation of data on the CCD chip, its read-out and storage in memory, and the implementation of various normalization procedures. All of the details of the operation of the CCD must be directly controlled using this electronics. A schematic for a Raman scattering system using a CCD detector is given in Fig. 6.10 [6.47]. The front half of the experiment, in particular, the excitation and the monochromator are common to both Figs. 6.6 and 6.10. Because the CCD must be operated at lower temperatures than the mepsicron, it requires the use of a liquid nitrogen dewar, in contrast to the thermoelectric cooler used by the mepsicron. Note also that in contrast to the microchannel plate photomultiplier where information goes only from the detector to the multichannel analyzer, for the CCD, information goes both to and from the device controller. A computer is required both for the storage of the data from the detector and for the operation of the detector. The current state of CCD fabrication is such that the actual use of a given chip requires individual optimization of the various parameters which contribute to the performance of the device. This includes a pixel by pixel evaluation of the response of the CCD. This optimization should be done in the context of the actual use of the detector since the real signal levels and the level of any background emission help to determine the chosen operating parameters.

The great strengths of the CCD detector are its ability to obtain spectroscopic information in a multichannel format, to do so with quantum efficiencies in the visible which can exceed 50%, and with a noise level that can be comparable to or better than that of single channel photomultiplier. Because it integrates charge, it

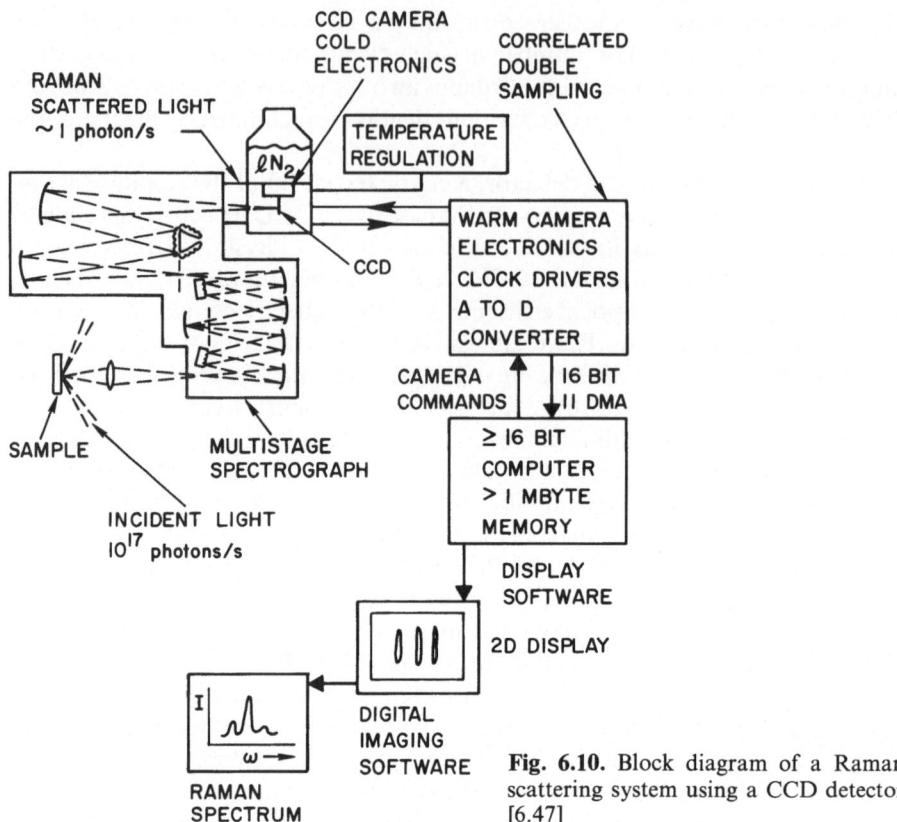

Fig. 6.10. Block diagram of a Raman scattering system using a CCD detector [6.47]

can be used with low repetition rate, high peak power sources. As a solid state detector with no high voltages, it is less sensitive to damage from exposure to strong light than photomultipliers (including mepsicrons). It is very sensitive to high static voltages and electric fields and can be destroyed by static discharges. Its optimization requires considerably more effort than a conventional single channel detector.

Recent applications of CCD detectors include multi-order Raman spectroscopy from silicon [6.49]. The weak multi-order phonon modes of Si are used as an optical source for the comparison of a high quantum efficiency CCD, a mepsicron, and an intensified linear photodiode array. First order through fourth order Raman shifts are assigned to known peaks in the density of states of the appropriate Si phonon branches.

6.3.3 Some Comparisons

Both of the above types of detectors afford substantial improvements in sensitivity over the old types of multichannel optical detectors and several orders of magnitude improvement in performance over the best single channel photomultipliers. Because of generic differences, there are particular applications for which each is singularly suited or unsuited. The fact that the mepsicron is a photon counting device means that it is of little use in an experiment involving a high peak power, low repetition rate source where more than one photon can be emitted at one instant of time on the microsecond scale. In addition, the advantages of this detector become less apparent in experiments where the background noise is not detector-limited, since one of the major advantages is the low dark count. Another disadvantage of this detector is the spatial resolution, which is 2 to 3 times poorer than in the CCD.

The CCD detector operates at considerably higher noise levels. In experiments where the background emission is less than 10^{-2} cps and where the signal levels are below 10^{-1} cps, its high quantum efficiency is balanced by the dark count and the problems associated with long integration times. The CCDs high quantum efficiency in the visible is also balanced by relatively low efficiency in the ultra-violet, due to the strong absorption of Si. Its optimal operation appears to require more attention than the microchannel plate device.

6.3.4 Raman Scattering Methodology

Table 6.2 showed that only a small fraction of the Raman scattered photons in an experiment are ever detected. The detectors described above improve this situation by removing the need for an exit slit on the monochromator. However, considerable losses remain in the collection optics and the filter and analyzer stages of the monochromators.

At the present time, the use of collection optics with optical speeds near $f1$ is fairly common. While faster collection systems can be built, the requirement that they be matched to the optical speed of the monochromators used as prefilters and analyzers limits the utility of such optical systems. This is because most high resolution commercial monochromators have optical speeds below $f5$ and there are limits to the minimum size that the light incident on the sample can be focused. Aside from their large f numbers, the inefficiency of existing monochromators is due to the large numbers of mirrors and multiple diffraction gratings in these instruments. The replacement of the normal aluminum coatings by dielectric coatings can result in a factor of 2–3 improvement in performance within the useful spectral range of the dielectric mirrors. A somewhat smaller improvement can be obtained for photon energies below 3.5 eV through the use of silver mirrors. Similarly, a careful matching of the spectral range in which one works to the optical configuration of the spectrometer being used and the spectral dependence of the grating response can result in grating efficiencies as high as 0.9 [6.25]. If these steps are taken for a standard triple monochromator, the throughput can reach 60%. It is clear that considerable (2–5X) improvements can be made on the low throughputs that characterize the operation of most multistage monochromators, although the current solutions would result in the ability to work efficiently only over limited spectral regions.

Efforts have been made to use other types of filters to discriminate against the elastic scattering in Raman experiments and dispense with some of the mirrors and gratings of the multiple monochromators. These have been shown to be promising for conditions of relatively low elastic scattering intensities and large frequency shifts. *Campion* et al. [6.10] were able to use standard, colored glass filters and a single monochromator in their experiments on molecular adsorbates on polished metal surfaces. There, rejection ratios for the laser line of 10^{10}, and transmission ratios of greater than 50% were achieved 1000 cm^{-1} away from the laser line. *Flaugh* et al. [6.50] have shown that colloidal crystals can be used as narrow band filters with extinction ratios in excess of 10^4 and transmission ratios greater than 70%.

The improvement of the sensitivity of the detectors and the throughput of the optical systems used in Raman scattering experiments discussed above result in the ability to detect light levels orders of magnitude lower than previously possible. At these low levels, it is necessary to re-evaluate all sources of stray light in a laboratory, since contributions that were previously undetectable can now be observed. An example of this is the non-lasing fluorescences from lasers which were traditionally blocked by narrow band interference filters and/or small prism and grating monochromators. The amount of filtering required to suppress these lines given a system with 2 to 3 orders of magnitude enhanced sensitivity will be considerably greater than previously needed. Emission from indicator lights on equipment and from display screens inappropriately situated can also result in substantial background signals.

6.4 Surface, Interface, and Thin-Film Raman Studies

In this section, we describe some recent applications of Raman spectroscopy to the study of surfaces, interfaces, and thin films. This is not a comprehensive review of the existing literature but a survey of the kinds of problems that Raman scattering can now be used to address. It is only concerned with studies where multichannel detectors have been used. In this section we show that Raman spectroscopy (1) has the ability to study adsorbates on solid surfaces in ultra-high vacuum situations, complementing the existing techniques of surface science, (2) can be used to characterize adsorbates on surfaces in air, thereby going beyond the capabilities of electron spectroscopies which have traditionally dominated surface vibrational spectroscopy, (3) can monitor chemical reactions both at and in surface layers, raising the possibility of its application to in-situ studies, (4) can characterize the effects of chemical processing on the surface of a solid, including the loss of long range order, (5) detect the interface between two distinct materials including interfacial reactions forming new chemical phases, and (6) characterize thin films at the monolayer level, thereby providing insight into the processes by which these films grow and the interaction of the bulk and surface properties of the films.

6.4.1 UHV Surface Studies

One of the chief advantages of Raman spectroscopy as a surface probe is its ability to work in environments which are forbidden to the traditional tools of surface vibrational spectroscopy. However, given the large data base that has been developed on the behavior of atoms and molecules at surfaces, any new surface vibrational spectroscopy has to prove itself by obtaining results which can be independently confirmed by the traditional techniques. Aside from the issue of general credibility, such studies on well characterized systems are needed to resolve questions concerning the effects of adsorption processes on the magnitudes of the Raman scattering cross sections of different materials and the Raman scattering selection rules which are used to characterize the symmetries of the adsorbates.

Campion et al. [6.10, 51, 52] combined both vidicons and diode array detectors with an ultra-high vacuum system to demonstrate that Raman spectra could be obtained from molecular monolayers adsorbed onto smooth metal surfaces. Angles of incidence and scattering were optimized, collection optics with a speed of $f0.95$ were used, and the inelastically scattered light was analyzed by an optically fast single monochromator with a throughput in excess of 50%. The elastically scattered light was minimized by careful polishing of the substrate. The Raman frequency shifts were so large ($> 1000 \text{ cm}^{-1}$) that it was possible to use colored glass cut-off filters to reject the elastically scattered light from the surface. Incident laser power levels of 150 mW focussed into a 300 micron by 100 micron spot producing an incident flux of about 500 W/cm^2 were used for these experiments.

Initial studies were performed with the molecule nitrobenzene. The scattering cross section for the NO_2 stretch mode of this molecule is 10.3 $\times 10^{-29} \text{ cm}^2 \text{ sr}^{-1}$ per molecule at 4880 Å [6.23]. The substrate used was Ni (111). Although like many transition metals, Ni is known to produce weak cathodoluminescence [6.53] due to Fermi level to d band transitions, this did not pose a significant problem for these measurements. Because Ni has a simple fcc crystal structure, it does not produce any first order Raman scattering. Its metallic properties mean that the penetration depth of light is so small that higher order scattering processes are unimportant. The well characterized surface preparation procedures for Ni (111) guarantee a reproducible surface. The large imaginary part of the dielectric response function rules out the possibility of any electromagnetic enhancement of the incident and scattered optical fields due to the excitation of a surface resonance [6.5].

The condensation of multilayers of nitrobenzene on Ni produced a Raman spectrum similar to the Raman spectrum of nitrobenzene in solution. The various lines in Fig. 6.11 [6.10] include the dominant NO_2 stretch mode at 1345 cm^{-1}, a strong C–C ring stretch mode at 1575 cm^{-1} and a trigonal ring breathing mode at 1000 cm^{-1}. A sketch of the physisorbed molecule is also provided in Fig. 6.11. The well defined frequencies of these modes clearly show the ability of Raman scattering to identify species adsorbed on surfaces.

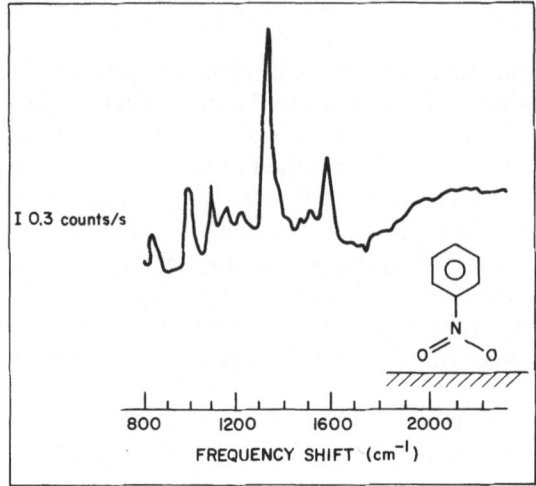

Fig. 6.11. Raman spectrum of 5 nm of nitrobenzene adsorbed on Ni (111) at 100 K [6.10]

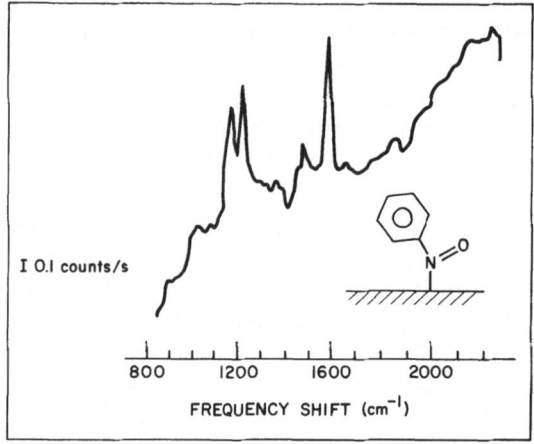

Fig. 6.12. Raman spectrum of about 1 monolayer of nitrobenzene adsorbed on Ni (111) at 100 K [6.10]

Campion et al. reported that the intensity of the spectrum in Fig. 6.11 was consistent with the calculated thickness of the condensed nitrobenzene layer given the exposure of the Ni (111) surface to nitrobenzene and the scattering cross section of the latter. This showed that the presence of the metal surface did not significantly change the magnitude of the non-resonant Raman scattering cross section of the adsorbed nitrobenzene. They then obtained a Raman spectrum from a monolayer of adsorbed nitrobenzene [6.10]. Unlike the previous thick layer case where most nitrobenzene molecules had only other nitrobenzene molecules as nearest neighbors, in this case, most of the nitrobenzene molecules interacted directly with the Ni surface. The Raman spectrum clearly showed this interaction (Fig. 6.12). The dominant NO_2 stretch mode at 1345 cm^{-1} and the C–C trigonal ring breathing mode at 1000 cm^{-1} were absent, while the weak lines above 1200 cm^{-1} became strong. Earlier x-ray

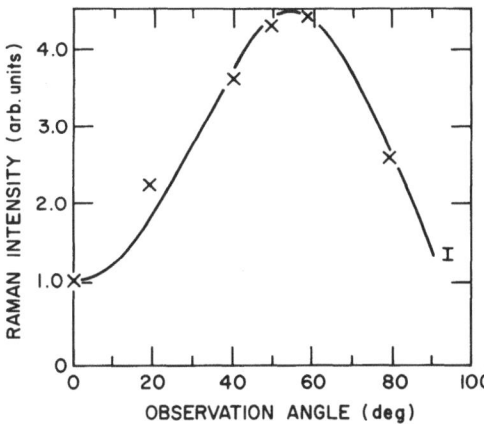

Fig. 6.13. Angular dependence of the scattered Raman intensity from a monolayer of benzene adsorbed onto Ag (111) for scattered light polarized normal to the surface [6.54]

photoemission studies of the absorption of nitrobenzene on Ni showed that the nitro group of nitrobenzene dissociated upon chemisorption to Ni, producing a free oxygen atom and nitrosobenzene. Campion et al. independently confirmed this result by directly absorbing nitrosobenzene onto the Ni surface and obtaining a spectrum reported to be similar to Fig. 6.12. The bonding scheme of the nitrobenzene to Ni is also shown in Fig. 6.12.

Given the ability to observe unenhanced Raman scattering from a thin layer condensed on a metal surface, *Mullins* and *Campion* [6.54] were able to quantitatively verify the predictions of *Greenler* and *Slager* [6.29] for the enhancement of the Raman scattering intensity of an adsorbed molecule by a reflecting substrate for light polarized normal to the surface. They measured the angle dependence of the intensity of the inelastically scattered light due to the 992 cm^{-1} ring breathing mode of benzene adsorbed on smooth Ag (111). The theoretically derived (solid line) and experimentally measured [6.54] results (crosses) are shown in Fig. 6.13 as a function of the angle from the normal to the surface. While the results in Fig. 6.13 show that the classical theory of Greenler and Slager explains the effect of the conducting surface on the Raman scattering, they also show that the presence of a metallic surface can produce changes in the selection rules for adsorbed molecules. Vibrational modes producing induced Raman dipoles in the plane of the conductor would scatter weakly, even if the dipole were large.

6.4.2 Surface Adsorption in Air

The work of Campion et al. demonstrated that Raman scattering could supplement the traditional tools of surface vibrational spectroscopy in studying the adsorption of molecules onto surfaces in vacuum. *Dierker* et al. [6.11, 47] demonstrated the capabilities of their CCD detector Raman scattering system as a tool for the study of surfaces and interfaces in a non-vacuum environment, a

situation where the traditional tools of surface science have difficulty function-
ing, by using it to characterize Langmuir-Blodgett films.

The preparation of molecular monolayers at the air-water interface and their
transfer onto a substrate by dipping into a trough with a moving barrier is a well
defined, non-vacuum method for the reproducible generation of monolayer
films and ultra-thin insulating multilayers. The magnetic, optical, and structural
properties of these Langmuir-Blodgett films have been widely studied [6.55]. The
symmetry properties of the Raman scattering from vibrational modes, in
particular, the depolarization ratio [6.56], have been used to quantitatively
describe the orientational order of the molecules in the layers. The intensity ratio
of different Raman active lines can reflect the degree of order present in a sample,
with the intensity ratio of the symmetric and asymmetric C–H stretch modes of
the long chain fatty acids which form the spines of these molecules, changing
significantly on the transition from a solid, ordered phase, to a liquid, disordered
one [6.57].

Raman scattering has been widely used to characterize these films. However,
their monolayer character, even though a monolayer can be a few nanometers
thick, has meant that previous measurements relied either on resonant Raman
scattering [6.58] or surface plasmon polariton-enhanced scattering [6.59] to
increase the intensity. In the former case, the resonant Raman technique limited
the range of materials that could be studied and also introduced uncertainties
associated with the resonant Raman process since deviations from the normal
selection rules are often observed in resonance. The surface-wave-enhanced field
technique is applicable to a wide range of film materials but severely constrains
the choice of substrates. It also introduces a number of coupling parameters
characterizing the interaction of the induced Raman dipole with the surface
modes which can complicate the analysis of the data.

Dierker [6.11, 14] showed that the enhanced signal-to-noise ratios obtained
from a state-of-the-art CCD detector meant that Raman scattering could be
used to directly characterize multilayer Langmuir-Blodgett films. They studied
Cd-stearate films deposited on clean fused silica and Si (111) using the standard
Langmuir-Blodgett technique. The Raman measurements were made with a
liquid nitrogen cooled, backthinned CCD bonded to a glass substrate. At the
photon energies used, the quantum efficiency of the CCD was over 70%. The
Raman scattering was excited by about 500 mW of 5145 Å light focussed on
the sample to a 5×10^{-4} cm^2 line. The scattered light was filtered and analyzed
by a conventional triple monochromator with an instrumental resolution of
10 cm^{-1}. The various procedures [6.47] described in Sect. 6.3 for the low noise,
high efficiency use of the CCD, including appropriate binning of pixels, system
response corrections etc., were all employed.

Figure 6.14 presents the Raman spectra of *Dierker* [6.11, 14] for frequency
shifts near 2850 cm^{-1} of multilayer films of cadmium stearate deposited on fused
silica substrates. The monolayer film whose spectrum is shown at the bottom is
about 2.5 nm thick. The spectral region shown in this figure embraces both the
symmetric (2843 cm^{-1}) and asymmetric (2882 cm^{-1}) C–H stretch vibrations of

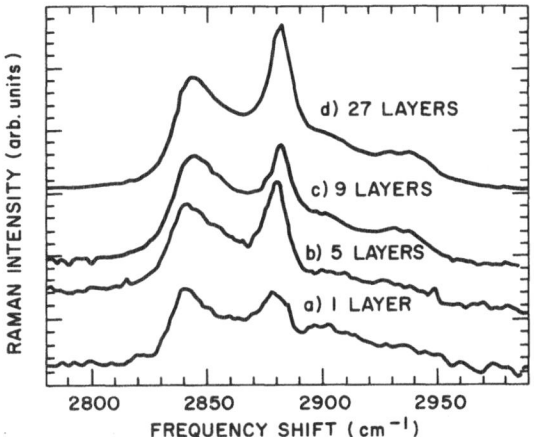

Fig. 6.14. Raman spectra of 4 Cd-stearate Langmuir-Blodgett films of different thicknesses adsorbed onto quartz. (*a*) 1 layer, (*b*) 5 layers, (*c*) 9 layers, and (*d*) 27 layers [6.11]

the CH$_2$ groups which form the spine of the Cd-stearate molecules. These are the vibrational modes whose relative intensities have been used to measure the conformational order within the film. Similar results were obtained for the Raman spectra of a relatively thick sample (15 layers) as a function of sample temperature. The results showed that the intensity of the asymmetric C–H vibrational mode increased and the line width narrowed with increasing film thicknesses and decreasing temperatures. Changing the thickness of the layer did not cause an observable shift in vibrational transition, energy but cooling the sample produced a frequency change on the order of 5 cm^{-1}, which would have been difficult to resolve using non-optical spectroscopies. Dierker et al. interpreted the intensity dependence as arising from decreased conformational order, since. The asymmetric (2882 cm^{-1}) C–H stretch mode is relatively strong when the molecules are conformationally ordered and weak when they are disordered. The simultaneous changes in linewidths and peak positions as a function of temperature were used as evidence that the thermal changes in these films were due to dynamic effects. The failure to observe a similar shift in vibrational energy as the film thickness changed, even though the mode intensity and linewidth changed, suggested to Dierker et al. that the initially deposited layers contained metastable, static defects which disrupted their lateral packing.

The results of *Dierker* et al. [6.11, 47] show that Raman spectroscopy, when performed using the CCD detectors described previously, has the ability to detect and characterize molecular monolayers in air. Their work especially underlines the utility of a surface spectroscopic probe whose capabilities are such that frequency shifts of the order of 2–5 cm^{-1} can be detected and changes in linewidths of the order of 5 cm^{-1} can be resolved.

6.4.3 In situ Oxidation of Metals

This section shows how Raman scattering has been used to study the high temperature oxidation of metals. Raman spectroscopy has been applied to study the in situ processing of materials in an ambient atmosphere and at temperatures as high as 700 °C. Of interest here is the identification of the oxide phases that are grown during a thermal reaction on a metal surface and how these phases change upon annealing.

Hamilton and *Anderson* [6.12] studied the low pressure oxidation of the (100) surface of the ternary metal alloy, Fe-18Cr-3Mo. The Raman spectrum of the oxidized metal consists of a single line near 660 cm^{-1} which was not observed on the clean metal surface prior to oxidation at 700 °C in 6×10^{-5} Pa of O_2 for 20 s. By combining their Raman studies with Auger depth profiling, those authors found that the oxide layer responsible for this mode was between 2 and 3 nm thick. The identification of this 660 cm^{-1} line as a spinel phase of the thermally grown oxide was based on a comparison with Raman spectra obtained from powder phase samples of TCr_2O_4, where T was Fe, or Mn or Ni, all of which showed similar structure near 660 cm^{-1}.

The spectra were obtained using a triple monochromator and a diode array detector. The incident laser power on the sample was about 1 W with an excited spot area of about 10^{-3} cm^2. The angle of incidence of the exciting light was near the Brewster angle while the scattered light was collected normal to the surface of the sample. Because of the magnitude of the dark count and the read-out noise from the diode array, it was necessary to subtract the diode array background signal to observe the oxide scattering. In addition, at temperatures above 800 °C, the Raman spectra were superimposed on a background due to black body emission from the sample.

Hamilton and *Anderson* [6.12] found that substantial changes in the composition of the oxide grown on their metal surface occurred with changes in growth conditions. Figure 6.15 shows the Raman spectra resulting from growth of an oxide layer on Fe-18Cr-3Mo (100) at 600 °C for prolonged periods of time. The initial reaction is the formation of the spinel phase identified in the Raman spectra by the 660 cm^{-1} peak. For reaction times greater than 30 min, a new weak structure is observed near 550 cm^{-1}. Annealing of this film in vacuum at 700 °C produced a dramatic change in the Raman spectra which is shown in Fig. 6.16. The weak structure near 550 cm^{-1} and lines at lower energies around 300 cm^{-1} grew in strength while the 660 cm^{-1} signature of the spinel phase decreased in intensity. The three new lines shown in Fig. 6.16 appear in the Raman spectrum of Cr_2O_3. Hamilton and Anderson found that the transformation of the spinel oxide to chromia is consistent with the phase diagram for these oxides. This showed that upon annealing, the spinel oxides underwent solid state reactions of the form

$$3FeCr_2O_4 + 2Cr = 4Cr_2O_3 + 3Fe \ . \tag{6.2}$$

While these results showed that Raman scattering could characterize oxide

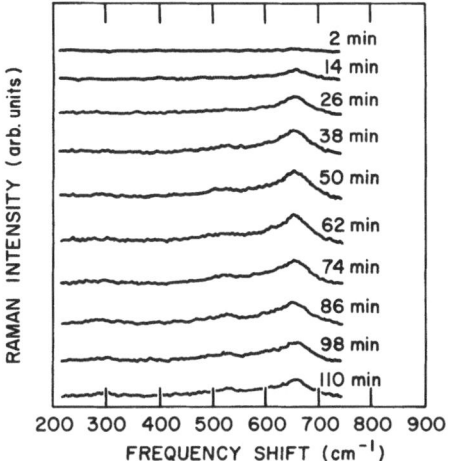

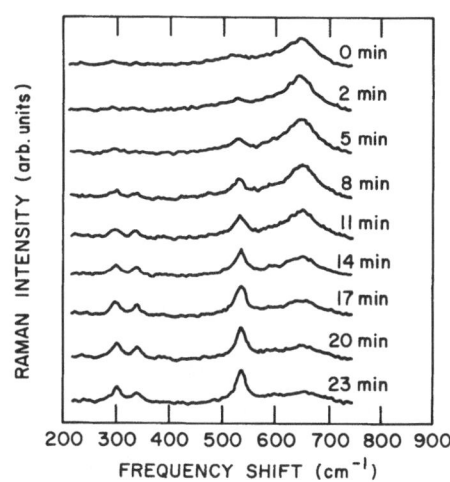

Fig. 6.15. The in situ Raman spectra of the oxidation of Fe-18Cr-3Mo (100) at 600 °C in 6.7×10^{-5} Pa oxygen [6.12]

Fig. 6.16. The in situ Raman spectra of the annealing of an oxidized Fe-18Cr-3Mo (100) surface at 700 °C [6.12]

formation on metal surfaces at the nanometer level, the use of metal substrates minimizes the problems associated with the Raman scattering from the substrate. The magnitude of this problem on other materials can be shown by considering the Raman spectrum of thermally grown SiO_2 on crystalline Si. Table 6.1 showed that Si is a relatively strong Raman scatterer compared to the transparent materials listed. Transparent, crystalline SiO_2 is a relatively weak scatterer and the amorphous or glassy phase of silicon dioxide is still weaker, as far as the peak scattering intensity is concerned. There is a scattering geometry for bulk Si, that of cross polarization on the (100) face with the incident electric field parallel to a (100) axis, where the first order scattering is symmetry forbidden and the second order contribution is rather weak. Unfortunately, the strongest component of the amorphous oxide spectrum is in the scattering geometry where the incident and scattered light are parallel polarized and the cross polarized spectrum of the oxide is at least an order of magnitude weaker than the parallel polarized spectrum. The consequence of these facts is given in Fig. 6.17 which shows the Raman spectra of thermally oxidized Si (100) obtained in a scattering geometry where the first oxide scattering is symmetry allowed and the first order substrate scattering symmetry forbidden [6.3]. However, the second order substrate scattering was still strong in this geometry. The spectra were excited using 5145 Å light. The Raman spectra of the surface oxide layers were degenerate with the much stronger second order scattering from the substrate. The oxide contribution, which is the weak shoulder near 500 cm^{-1} just below the symmetry forbidden first order scattering at 522 cm^{-1} (Fig. 6.17a) was almost impossible to detect against the substrate second order scattering. By carefully subtracting the Raman spectrum of a Si wafer covered by a thin oxide (also obtained in this scattering geometry) from the spectra shown in Fig. 6.17,

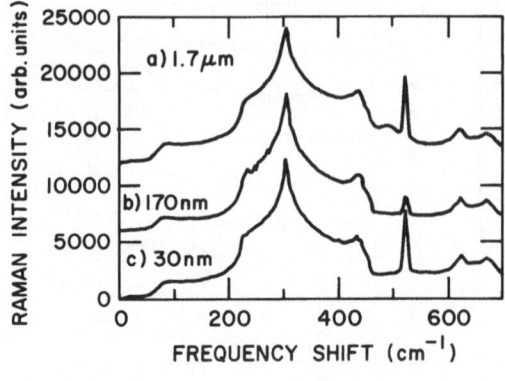

Fig. 6.17. Raman spectra of 3 different thicknesses of thermally oxidized Si (100), in the scattering geometry where the 2nd order scattering is symmetry-allowed and the 1st order scattering from the substrate is forbidden [6.3]

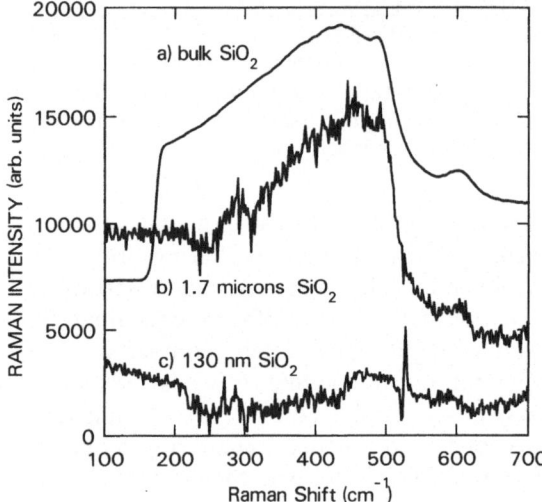

Fig. 6.18. Raman spectra of 1.7 μm and 130 nm of thermally grown SiO_2. The contribution to the spectra from the Si (100) substrate has been subtracted [6.3]. Also shown is the Raman spectrum of silicon dioxide glass

the Raman spectrum of 1.7 μm and 130 nm of SiO_2 are revealed in Fig. 6.18b, c [6.3]. It also shows the Raman spectrum of the bulk oxide (a). At 1.7 μm, the thermal oxide signal can be extracted from the Si substrate scattering. However, the apparent structure near $300\,cm^{-1}$ in Fig. 6.18b is due to errors in the subtraction of the second order scattering of the Si substrate. While there is some evidence for oxide induced scattering in Fig. 6.18c, it is clear that this will be difficult to identify for these thicknesses. Figure 6.18 shows that the background contribution due to Raman scattering from the sample substrate can greatly complicate the observation of Raman scattering from a surface layer.

6.4.4 Characterization of Substrates by Raman Scattering

The above sections demonstrated that Raman scattering can be used to characterize molecules laid down or reacted on a metal or semiconductor surface. It has always been assumed that the substrate is well characterized up to

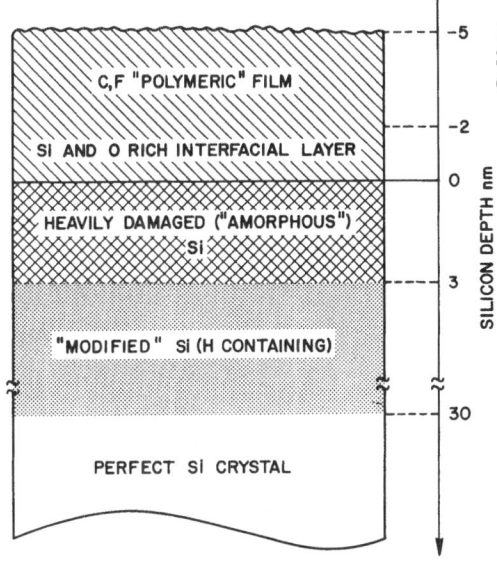

Fig. 6.19. Schematic of the profile of an Si wafer after reactive ion etching by a CF_4-H_2 plasma [6.18]

and including the surface and that the molecules interacting with the surface do not alter it. This need not be the case, however, and this section shows how Raman scattering can be used to help characterize structural changes that are introduced into the surface of a semiconductor by chemical processing such as reactive ion etching (RIE) of a thermal oxide and the growth of an oxide layer.

Exposure of a semiconductor surface to a flux of energetic particles generates atomic displacements near the surface. In recent years, semiconductor technology has made wide use of plasma processing for the removal of oxide layers on semiconductors. Unfortunately, the etching of an oxide layer by a plasma has been shown to be accompanied by the generation of lattice damage in the Si near the surface and the introduction of H atoms under the surface [6.18]. Figure 6.19 shows a schematic of the profile of a Si surface after an oxide layer has been etched off by a plasma. There are substantial deviations from the simple picture of an air-solid interface with a perfect lattice extending to the interface. At the surface, there is a C–F polymer layer. Just below the surface there is an amorphized region of pure Si and below that, a partially disordered region which contains hydrogen. The perfect Si lattice does not appear until 30 nm into the sample. In fact, devices fabricated on such plasma processed surfaces were shown to have many undesirable properties. Aside from the device problems it poses, the presence of a polymer layer at the surface complicates the application of many standard vacuum-solid surface probes since the polymer first has to be removed, a process which can further perturb the system. On the other hand, the polymer layer is transparent to Raman scattering. The presence of an amorphous layer and of hydrogen in the near surface layer is of interest to Raman spectroscopy since both can produce observable changes in the Raman spectrum of crystalline Si.

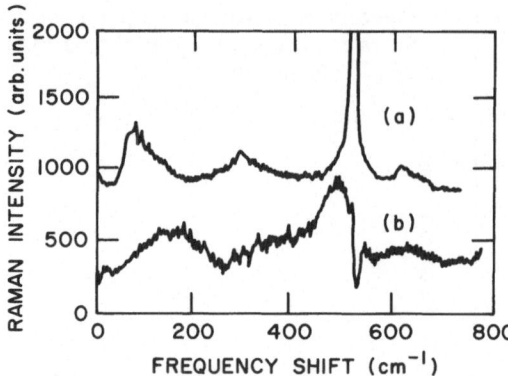

Fig. 6.20. Raman spectrum of a Si (100) substrate. (*b*) Raman spectrum of 6 nm of hydrogenated amorphous Si deposited on Si (100) after subtraction of the scattering from the substrate [6.14]

A comparison of the Raman spectra of hydrogenated amorphous Si and of bulk, crystalline Si shows that the two systems can be easily distinguished. The hydrogenated amorphous phase has structure near 2000 cm^{-1} which is due to Si–H vibrations. Both Si–H and Si–H$_2$ modes in hydrogenated amorphous Si have been identified [6.60]. The bulk crystal only shows very weak scattering below 2000 cm^{-1} due to fourth order Si scattering. The sharp, $\Gamma_{25'}$ symmetry k = 0 scattering at 522 cm^{-1} from the single crystal substrate, Fig. 6.20a, is distinct from the unpolarized broad band scattering obtained from the amorphous phase, which resembles the vibrational density of states of crystalline Si [6.61]. The Raman spectrum, (Fig. 6.20b) for frequency shifts between 50 and 800 cm^{-1}, of 6 nm of hydrogenated amorphous Si laid down on Si (100) and measured in the cross polarized scattering geometry where the first order phonon scattering from the substrate is weak, shows that in contrast to the problems associated with the detection of a thin SiO$_2$ layer on crystalline Si, the detection of a thin layer of amorphous Si on crystalline Si is quite simple. This is due to both the size of the Raman scattering cross section for amorphous Si and its polarization selection rules where the first order scattering is symmetry allowed and both one- and two-phonon scattering from the crystalline Si substrate are either symmetry forbidden or relatively weak for one geometry. Also shown in Fig. 6.20 is the Raman spectrum obtained from the Si (100) substrate in the forbidden scattering geometry. This shows that the structure at 522 cm^{-1} in Fig. 6.20b is due to errors in the subtraction of the forbidden first order scattering from the Si substrate. These results were obtained using a triple monochromator and a mepsicron. Incident laser power levels at 5145 Å on the sample were about 50 mW with the incident light focused to a 5 mm × 100 μm line and integration times were about 30 minutes.

Figure 6.21 [6.14] shows the Raman spectrum for frequency shifts below 600 cm^{-1} and between 1600 and 2300 cm^{-1} obtained from thermally oxidized Si (100) with 1000 Å of oxide exposed to a CF$_4$/H$_2$ plasma to etch away the oxide. The oxidized sample was exposed to the plasma 10 minutes longer than necessary to just remove the thermal oxide. The contribution to the Raman scattering from the crystalline substrate has been subtracted from the experi-

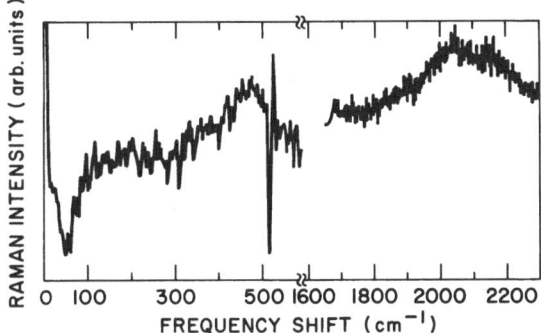

Fig. 6.21. Raman spectrum of Si (100) after reactive ion etching. Scattering due to the Si (100) substrate has been subtracted [6.14]

mentally measured spectrum. The low energy difference spectrum clearly shows broad band scattering centered about 480 cm^{-1}. This feature is similar to the disorder-induced Raman scattering shown in Fig. 6.20 due to the presence of amorphous Si on the sample. The feature around 2000 cm^{-1} is due to the vibrational modes of H-bonded Si atoms. These results demonstrate that the H atoms found below the Si surface in nuclear reaction studies [6.18] of these samples are bound to the Si atoms.

The previous section showed that a major limitation to the experimental sensitivity of Raman scattering as a surface probe was the strength of the scattering from the substrate. Considerable enhancements of the experimental sensitivity to a surface layer can be achieved by carefully choosing the excitation wavelength of the experiment to minimize the strength of the substrate scattering. The Raman scattering cross section of Si shows large changes with excitation energy due to the different critical points of the crystalline band structure [6.33]. A maximum near 3.4 eV occurs in the scattering strength when the incident and scattered light are in resonance with the E_1 gap of Si. At higher energies, there is a sharp drop in the scattering cross section with a minimum near 3.8 eV. The relatively small value of the scattering cross section at this energy, when combined with the high absorption coefficient ($\alpha_i = 1.2 \times 10^6$ cm^{-1}), maximizes the sensitivity of Raman scattering to any surface species with respect to the Si substrate. Although amorphous Si shares many similarities with crystalline Si, the loss of long range order in the amorphous phase has the effect of smoothing out the strong structures in the electronic and vibrational spectra of the crystal. The sharp minimum in the scattering cross section of the crystalline phase at 3.8 eV is absent so that the relative strength of the amorphous layer scattering with respect to the substrate scattering is greater at this energy than at lower photon energies. This is illustrated in Fig. 6.22 which shows the Raman spectrum of 6 nm of amorphous Si laid down on Si (100) for 3.8 eV excitation. The spectrum was obtained in the scattering geometry where the second order substrate scattering and the first order amorphous spectrum at the surface were allowed, but the first order substrate scattering forbidden. In Fig. 6.20, the Raman spectrum of the amorphous layer, even after careful

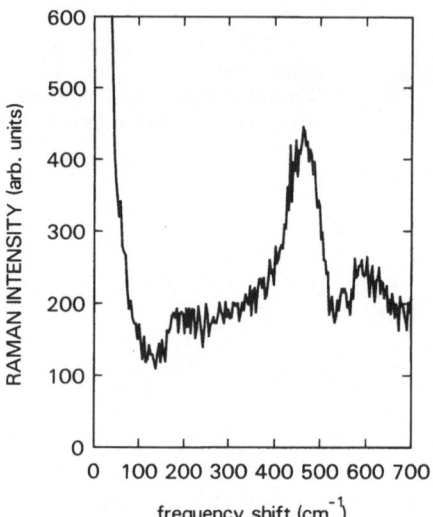

Fig. 6.22. Raman spectrum of 6 nm of hydrogenated amorphous Si laid down on Si (100), excited at 3.8 eV [6.62]

subtraction of the substrate scattering, still shows structure due to the very large contribution of the forbidden first order scattering from the substrate. However, at 3.8 eV, the Raman scattering from the amorphous layer can be clearly observed even without subtraction of the substrate scattering. It should be noted in this connection that the penetration depth of the light into Si between 2.4 eV and 3.8 eV changes significantly. The use of excitation wavelengths that are strongly absorbed provides another means for the enhancement of the surface contribution to the scattering over that of the bulk contribution. However, measurements of the Raman scattering from a 6 nm layer of amorphous Si on Si (100) excited by 3.5 eV light where the absorption coefficient is quite close to the absorption coefficient at 3.8 eV, verify that the minimization of the Si substrate scattering efficiency at 3.8 eV was largely responsible for the result in Fig. 6.22.

An enhancement of the Raman sensitivity to the presence of an oxide similar to that shown in Fig. 6.22 for an amorphous Si layer on Si (100) has been demonstrated in studies of thermal oxides grown on Si (100) and Ge (100) [6.62]. Figures 6.17, 18 showed that it was difficult to observe the vibrational spectrum of even a 1 μm thick oxide on Si (100) when the excitation energy was near 2.4 eV. Raman spectra of oxidized Si (100) excited using the 3259 Å light from a He–Cd laser in a scattering geometry which favored the observation of scattered light polarized parallel to the incident light and discriminated against the first order scattering from the Si substrate, showed a significant enhancement in sensitivity to the detection of the broad band oxide vibrational modes below 500 cm^{-1}. As the thickness of the oxide layer decreased below 200 nm, the scattering below 500 cm^{-1} changed in shape. Measurements made on these samples at lower excitation energies were unable to show any oxide-like Raman scattering for these thicknesses because of the background from the strong second order substrate scattering. The similarity of the additional scattering observed on the oxidized Si surfaces at an excitation energy of 3.8 eV for oxide thicknesses below 200 nm to the Raman spectrum obtained from amorphous Si led to the suggestion that it was due to the presence of a thin, disordered phase at the Si-oxide interface. Similar results have been obtained on oxidized Ge (100) [6.62].

6.4.5 Interfaces and New Phases in and on Crystalline Si

Recently, the use of multichannel detectors in Raman scattering has permitted the study of buried layers. These experiments involve the study of solid-solid interfaces buried under as much as 16 nm of material. Questions of interest include how two solids intermix, the nature of epitaxial growth at a solid-solid interface, the relationship between the microscopic structure of an interface and its electronic properties, etc.

Figure 6.23 shows the Raman spectrum of 3 nm of PtSi produced by the vacuum deposition and thermal reaction of 1.5 nm of Pt on UHV prepared Si (100) surfaces. The spectrum was identified by comparison with spectra previously published by *Nemanich* et al. [6.63] for thick samples of crystalline silicides of Pt. The characteristic vibrational modes of PtSi occur near 80 and 140 cm^{-1}. The Raman spectrum of the Si substrate shows no first or second order contributions for Raman shifts below 200 cm^{-1}. Therefore, there is no significant background scattering from the substrate to complicate the observation of the silicide vibrations. This is in contrast to amorphous Si where there is significant defect-induced acoustic phonon scattering below 200 cm^{-1} (Fig. 6.4). The symmetry allowed second order Raman scattering from the Si substrate provides a convenient scale for the normalization of the scattering strengths of the silicide vibrational modes. The silicide scattering intensities so normalized can be used to estimate the thicknesses of the silicide layers. The intensity of the Raman scattering from a 3 nm silicide layer is comparable to the strength of the second order acoustic phonon Raman scattering from the Si substrate. This means that the signals are accessible even to conventional

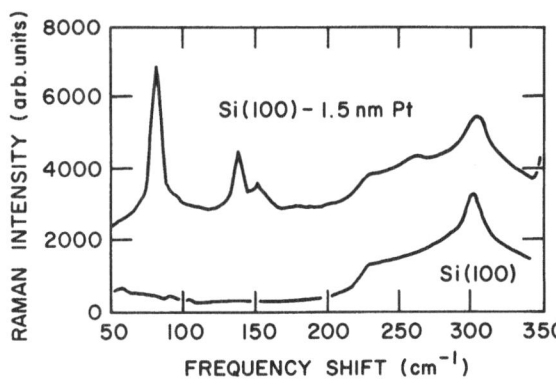

Fig. 6.23. Raman spectrum of 3 nm of PtSi reacted from 1.5 nm of Pt deposited on Si (100) in UHV. Substrate spectrum is also given for reference [6.13]

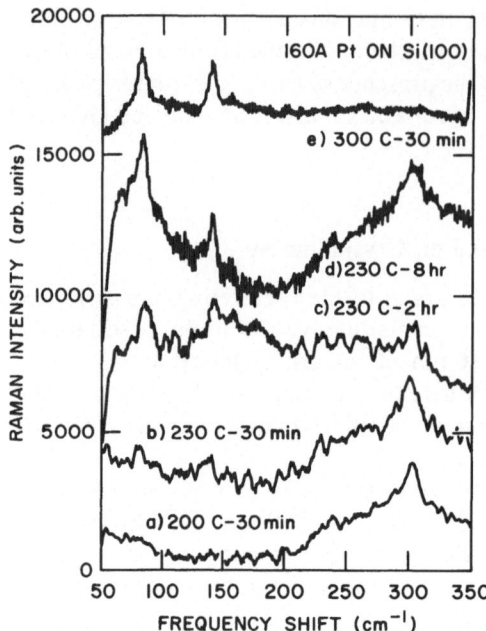

Fig. 6.24. Raman spectra of 16 nm of Pt deposited on Si (100) and annealed in He at (*a*) 200 °C for 30 min, (*b*) 230 °C for 30 min, (*c*) 230 °C for 120 min, (*d*) 230 °C for 480 min, and (*e*) 300 °C for 30 min [5.64]

detection schemes, although the decrease in the intensity of the signal from a buried layer enhances the usefulness of the multichannel detector systems.

Tsang et al. [6.64, 65] studied the formation of both PtSi and PdSi on crystalline Si. In both cases, the growth of the silicide at the boundary between the Si crystal and the unreacted metal was directly observed by Raman scattering for silicide thicknesses at the nanometer level. Figure 6.24 shows the Raman spectra obtained from a 16 nm film of Pt deposited by sublimation from a Pt filament onto polished, cleaned and UHV annealed Si (100). The spectra were obtained in air with the surface of the sample exposed to flowing He to reduce the strength of the Raman scattering from the atmospheric gases. The Pt-Si sample was annealed for various lengths of time at 200 °C and 230 °C in clean He to react the Pt and Si to form the silicide. Because Pt has the simple fcc structure, there are no Raman active vibrational modes which might obscure the silicide signal, even when the silicide layer is very thin compared to the Pt. The only contributions to the Raman scattering for frequency shifts below 200 cm^{-1} must be either defect-induced scattering from the substrate or from a silicide phase growing on Si (100). The Raman spectra show clear evidence of new lines near 80 and 140 cm^{-1} due to a reacted PtSi layer for annealing times at 230 °C as short as 30 minutes. The thermally reacted samples were analyzed by transmission electron microscopy to independently obtain the thicknesses of the reacted silicide layers. This technique was able to identify the reacted silicide for annealing times at 230 °C as short as one hour and yielded a value for the silicide thickness of about 2 nm. For diffusion limited growth of the silicide, the observation of the silicide phase by

Raman scattering after only 30 minutes of annealing at 230 °C meant that PtSi thicknesses below 1.5 nm could be detected. This value was consistent with the relative intensity of the silicide scattering with respect to the intensity of the second order acoustic phonon scattering from the substrate.

The evolution of the PtSi layer with annealing time was followed until all of the Pt was consumed. No evidence was found for the presence of a Pt_2Si phase (based on the calibration of [6.63]) or broadened PtSi lines indicative of a disordered PtSi phase. Raman studies of epitaxial PtSi grown on different Si surfaces showed that changes in the orientation of the PtSi produce changes in the relative intensities of the different Raman active modes, reflecting the symmetry properties of the Raman tensor. No significant changes in the relative intensities of the PtSi-derived modes were observed during the reaction of the thick Pt layer. The results of Tsang et al. showed that the formation of PtSi at 230 °C at their Pt-Si (100) interfaces proceeded directly through the epitaxial growth of the PtSi phase and not through an initial mixing reaction or a Pt_2Si phase.

Tsang et al. [6.65] were able to obtain similar results for the growth of PdSi on vacuum cleaved Si (111). As in the case of PtSi, the silicide lines are at energies below 250 cm^{-1} (Fig. 6.25e). Figure 6.25 shows the Raman spectra obtained from about 3.5 nm of Pd sublimated from a Pd filament in UHV onto freshly cleaved Si (111) and annealed in vacuum for 15 min at different temperatures. Once again, it was possible to monitor the formation of the silicide layer as the

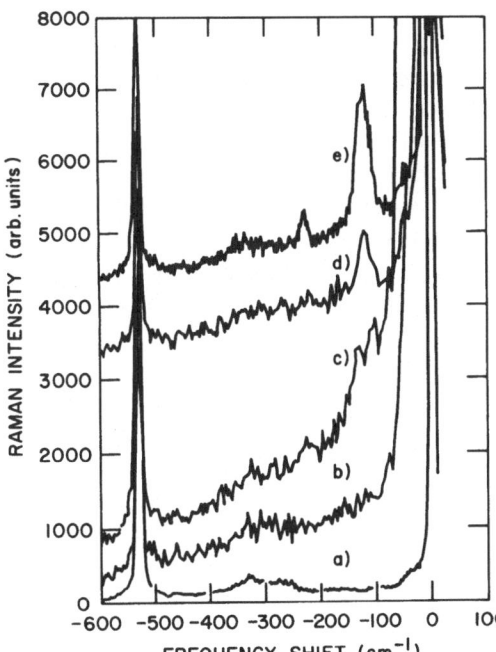

Fig. 6.25. Raman spectrum of 3.5 nm of Pd deposited on cleaved Si (111) and annealed at 3 different temperatures. (*a*) Uncoated substrate, (*b*) after evaporation of Pd at 300 K, (*c*) annealed at 110 °C, (*d*) annealed at 180 °C, (*e*) annealed at 390 °C. Each anneal took 15 minutes [6.65]

deposited Pd layer was reacted. The Raman spectra showed that the reaction proceeded directly via the epitaxial formation of the silicide without the formation of any intermediate phases. However, there was an initial reaction involving the first nanometer of sublimed Pd which formed a disordered silicide phase. Figure 6.25b shows an increase in the intensity of the scattering at energies below 150 cm^{-1} apparently due to the presence of a disordered phase. Upon annealing, the Raman spectrum of the disordered phase sharpened into the Raman spectrum of PdSi which grew monotonically until the Pd layer was consumed (Fig. 6.25c–e). By comparison with Raman spectra obtained from known thicknesses of PdSi on Si (111), it was shown that reacted silicide layers could be detected at the 1 nm level. *Nemanich* et al [6.66] carried out similar experiments on annealed Si (111) surfaces in UHV. They found a significant dependence of the initially reacted phases on the thickness of the deposited Pd layer. This has been attributed to critical size effects in the nucleation of the silicide phase. These authors have developed quantitative models of the reaction kinetics for the formation of these silicides on the basis of their experimental results [6.67].

The results of Tsang et al. for PdSi grown on cleaved Si (111) were obtained using the microchannel plate detection system with a triple monochromator. The cleaved Si surfaces employed in these experiments were considerably rougher than the polished and annealed surfaces used in the PtSi studies. The intensity of the elastic scattering from the surfaces was considerably stronger and it was necessary to use an iodine filter [6.68] to enhance the rejection by the optical system of the elastically scattered light so that measurements could be made for frequency shifts as small as 80 cm^{-1}.

Recently *Tsang* et al. [6.15] showed that multi-channel Raman scattering can be used to study the bonding between two different materials, providing a direct view of the interface between the materials. Using Si molecular beam epitaxy, samples were fabricated which consisted of an epitaxial layer of Si grown on silicon (100), a Ge layer consisting of from 1 to 10 atomic layers of Ge and a 10 nm cap layer of Si. For Ge layer thicknesses below about 5 atomic layers, the structure was epitaxial with no dislocations and the Ge under considerable strain, due to the 4% mismatch between the Si and Ge lattice constants ([6.17] and Sect. 3.3.4c). For thicker Ge layers, the strain was relieved by the generation of dislocations in the Ge film.

The values of the Raman scattering efficiencies in Table 6.1 suggest that the Raman signal from a monolayer of Ge should be observable. For the 2 monolayer sample described above, the Ge Raman signal is comparable in strength to the second order Raman scattering from the Si substrate (Fig. 6.1). For Ge thicknesses greater than a monolayer, the first order Ge layer scattering should therefore be stronger than the second order Si scattering in the scattering geometry where the second order scattering is weak. The results of *Tsang* et al. [6.15] obtained under 2.18 eV excitation for Si–Ge samples containing 1, 2, and 6 monolayers of Ge are shown in Fig. 6.26. These spectra were obtained in the scattering geometry where only $\Gamma_{25'}$ symmetry scattering was allowed. As

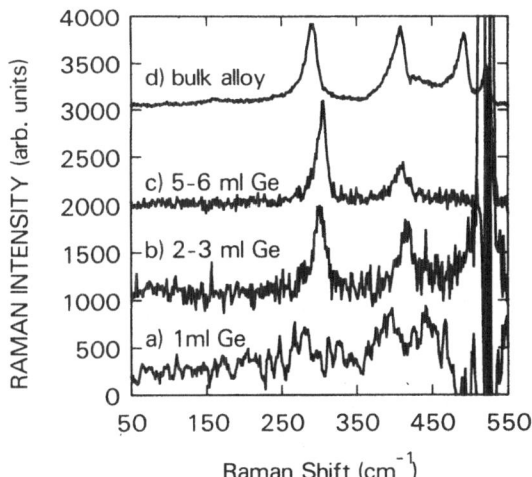

Fig. 6.26. Raman spectra of (*a*) 1 monolayer, (*b*) 2 monolayers, and (*c*) 6 monolayers of Ge epitaxially grown on Si (100) and covered by 10 nm of Si (100). The spectra were excited at 2.18 eV and the substrate scattering subtracted. The Raman spectrum of the Si-Ge alloy is given for reference (*d*) [6.15]

indicated in the discussion of Fig. 6.1, the vibrational modes near 300 and 520 cm^{-1} can be identified with the first order Raman scattering from bulk Ge and Si. Theoretical studies [6.69] of Ge impurities in Si have shown that a well defined Ge resonance vibration occurs near 400 cm^{-1}. Experimental studies of Ge-Si random alloys show a well defined mode between 380 and 420 cm^{-1} which has been attributed to Ge-Si bond vibrations [6.70]. In fact, the complete spectrum of Ge-Si alloys, which also shows strong contributions near 300 and 520 cm^{-1}, has been interpreted in terms of modes arising from nearest neighbor Ge-Ge, Si-Si, and the Ge-Si bond vibrations [6.34, 71]. Thus, the structure near 400 cm^{-1} in Fig. 6.26 is attributed to Ge-Si vibrations at the interfaces.

The relative intensity of the Ge-Ge vibrations with respect to the Ge-Si vibrations rises monotonically with increasing Ge layer thickness. This is to be expected since for a monolayer of Ge grown on Si (100), all of the nearest neighbor bonds are to Si atoms. For the approximately 2 layer sample, half of the bonds associated with each Ge atom are to other Ge atoms and half to Si atoms. As the Ge films grow thicker, the bonding becomes more and more Ge-like. This describes the evolution of the Ge-Ge and Ge-Si modes with increasing Ge thickness as seen in Fig. 6.26 and provides spectroscopic confirmation that the samples studied were thin continuous layers of Ge and not alloys of Ge and Si. The distinct energies for the Ge-Ge, Ge-Si, and Si-Si bonds mean that the Raman scattering near 400 cm^{-1} provides a direct probe of the Ge-Si interface in these samples. The variation of the positions of the Ge-Ge, Ge-Si, and Si-Si vibrations with composition has been used to provide information on the strains present in these samples [6.15].

6.5 Conclusions

A wide gamut of problems associated with the physics and chemistry of surfaces and interfacial layers are now accessible for study using Raman spectroscopy. These problems include bonding of molecules to a surface, the order in an adsorbed layer, chemical reactions at gas-solid and solid-solid interfaces, the effect of surface interactions on the structure of a substrate, and the evolution of the bulk properties from the surface properties as a film grows in thickness. These problems can be studied for surfaces under a wide variety of ambient conditions. While these problems are associated with relatively low signal levels, recent advances in optical detector technololgy, appropriate care in the actual measurements, and sensitivity to the physics behind Raman scattering can combine to produce usable results.

We have described the improvements in sensitivity that are now possible for Raman experiments using the current generation of position sensitive photo-multipliers and low noise charge coupled device detectors. Depending on the details of the particular experiments, both types of detectors can produce improvements in signal-to-noise of between 2 and 3 orders of magnitude over previous detectors. For experiments where the noise is detector dark count limited, the mepsicron can easily obtain, in less than one hour, a 400 channel spectrum at the 0.1 cps signal level. The high quantum efficiency of the CCD detector means that small signal spectra can be acquired with exceptional speed.

Although this article has concentrated on the use of these recent advances in optical detector technology in the study of Raman scattering from surfaces and interfaces, these advances also have major implications for a variety of other applications of Raman spectroscopy. These include the use of Raman spectros-copy in microprobe studies [6.72] and the study of time resolved phenomena [6.73]. In addition, these detectors have been shown to be well suited for hyper-Raman scattering measurements [6.74].

The ability of Raman spectroscopy to label chemical phases, detect physical changes in systems due to factors such as stress, and the ease of focusing laser light to the diffraction limit has resulted in the use of Raman spectroscopy as a tool for microprobe studies. A major problem in these studies is the amount of power required to produce an observable Raman spectrum which, as seen in Sect. 2.6.2, is often on the order of 50 milliwatts or more, and the small spot size of the laser in microprobe studies [6.72] which can approach 1 μm. At these sizes and power levels, the laser power incident on a sample can exceed $50 \, \mathrm{MW/cm^2}$. This can destroy many samples. The new detectors, with their increased sensitivity, allow spectra to be obtained at laser excitation levels of the order of tens of microwatts, greatly reducing the likelihood of laser induced sample damage.

Kash et al. [6.73] have shown that these detectors are ideally suited for time resolved Raman scattering experiments involving the pump-probe technique and sub-picosecond lasers. Because most of the power incident on the sample is in the pump beam, power levels in the probe beam can be too small for convenient measurements by traditional single channel detection systems. The enhanced sensitivity of the multichannel detectors means that the spectra excited by the low probe beam powers can be readily detected. Recently, *McMullan* et al. [6.75] have shown how these detectors can be simply gated. Temporal resolution of the order of less than 500 ps has been reported, and further improvements are likely. The ability to operate the microchannel plate detector with standard time correlated photon counting circuitry improves the sensitivity of this technique by at least two orders of magnitude.

The hyper-Raman effect [6.76] provides a tool for looking at very small Raman shifts without interference from elastic scattering and for observing vibrational modes that are Raman inactive. These advantages are balanced by the fact that unlike Raman scattering, which is described by third order perturbation theory, hyper-Raman scattering occurs in fourth order perturbation theory involving two incident photons, a scattered photon, and a scattered phonon. The result of this is extremely low signal levels for nonresonant excitations. Experiments using the hyper-Raman effect and single channel detectors have often involved measurement times on the scale of a full day for a single spectrum. *Vogt* [6.74] used a mepsicron to study hyper-Raman scattering from $SrTiO_3$ and showed that the times required to acquire a spectrum were reduced by a factor of 20 from 1 day to 1 h.

A long standing subject of contention on the part of observers of science is whether ideas and theories or experimental tools drive science. It is clear that the experiments that are done reflect the capabilities of the equipment that is available at the time. The Raman scattering experiments described in these volumes, most of which have been done in the last two decades, reflect the existence of reliable, reasonably powerful, monochromatic lasers, high rejection multiple monochromators, and photon counting photomultipliers. The advent of position sensitive photon detectors whose sensitivity per pixel is comparable or superior to the best single channel photomultiplier greatly extends the range of inelastic light scattering experiments that can be performed. While the improvements in performance described here are impressive, even as this article is being written, significant further advances in performance are being reported. These include the extension of the performance of the microchannel plate photomultipliers to higher count rates and of the spectral response of the CCD's into the infrared. [6.77]. We should also mention that mepsicron detectors are playing an important role in Raman investigations of the new high-T_c superconductors [6.78].

Acknowledgements. I thank P. Avouris, J.R. Kirtley, J.A. Kash, G. Rubloff, G. Oehrlein, and S.S. Iyer, IBM Research Center, Yorktown Heights, N.Y. for the loan of equipment, the preparation of samples, and discussions concerning the use of multichannel detectors in Raman scattering from surfaces and interfaces. I also thank C.A. Murray, AT & T Bell Laboratories and W. Acker, Yale University for information on CCD imagers.

References

6.1   C.E. Hathaway: "Raman Instrumentation and Techniques", in *Raman Scattering*, ed. by A. Anderson (Marcel Dekker, New York 1971) p. 182
6.2   R.K. Chang, M.B. Long: "Optical Multichannel Detection", in Light Scattering in Solids II, ed. by M. Cardona, G. Güntherodt, Topics Appl. Phys. Vol. 50 (Springer, Berlin, Heidelberg 1983) Chap. 3
6.3   J.C. Tsang: "Raman Spectroscopy of Thin Films on Semiconductors", in *Dynamics at Surfaces* ed. by B. Pullman, J. Jortner (Reidel, Dordrecht 1984) p. 379
6.4   G. Abstreiter, U. Claessen, G. Tränkle: Solid State Commun. **44**, 673 (1982)
6.5   A. Otto: "Surface Enhanced Raman Scattering, Only an Electromagnetic Resonance Effect", in *Light Scattering in Solids IV*, ed. by M. Cardona, G. Güntherodt, Topics Appl. Phys., Vol. 54 (Springer, Berlin, Heidelberg 1984) Chap. 6
6.6   H.J. Stolz, G. Abstreiter: J. Vac. Sci. Technol. **19**, 380 (1980)
6.7   J.A. Cape, W.E. Tennant, L.G. Hale: J. Vac. Sci. Techol. **14**, 921 (1977); R.L. Farrow, R.K. Chang, S. Mroczkowski, F.H. Pollak: Appl. Phys. Lett. **31**, 768 (1977)
6.8   G.P. Schwartz, A.Y. Cho: J. Vac. Sci. Technol. **19**, 1607 (1981)
6.9   H. Shen, F. Pollak: Appl. Phys. Lett. **45**, 692 (1984)
6.10   A. Campion, J.K. Brown, V.M. Grizzle: Surf. Sci. **115**, L153 (1982)
6.11   S.B. Dierker: Proc. 10th, Int. Conf. Raman Spectroscopy, ed. by W.L. Peticolas, B. Hudson (U. of Oregon, Eugene, OR. 1986) p. 6–8
6.12   J.C. Hamilton, R.J. Anderson: High Temp. Sci. **19**, 307 (1985)
6.13   J.C. Tsang, Y. Yakota, R. Matz, G. Rubloff: Appl. Phys. Lett. **44**, 430 (1984)
6.14   J.C. Tsang, G. Oehrlein, I. Haller, J.S. Custer: Appl. Phys. Lett. **46**, 589 (1985)
6.15   J.C. Tsang, S. Iyer, S.L. Delage: Appl. Phys. Lett. **51** 1732 (1987)
6.16   A. Pinczuk, E. Burstein: "Fundamental of Inelastic Light Scattering in Semiconductors and Insulators", in *Light Scattering in Solids I*, ed. by M. Cardona, Topics Appl. Phys., Vol. 8 (Springer, Berlin, Heidelberg 1983) Chap. 4
6.17   F. Cerdeira, A. Pinczuk, J.C. Bean, B. Batlogg, B.A. Wilson: Appl. Phys. Lett. **45**, 1138 (1984)
6.18   G.S. Oehrlein, R.M. Tromp, J.C. Tsang, Y.H. Lee, E.J. Petrillo: J. El. Chem. Soc. **132**, 1441 (1985)
6.19   M.V. Klein: "Raman Studies of Phonon Anomalies in Transition Metal Compounds", in *Light Scattering in Solids III*, ed. by M. Cardona, G. Güntherodt, Topics Appl. Phys., Vol. 51 (Springer, Berlin, Heidelberg 1982) Chap. 5
6.20   J.M. Gibson, H.J. Gossmann, J.C. Bean, R.T. Tung, L.C. Feldman: Phys. Rev. Lett. **56**, 355 (1986)
6.21   M.A. Renucci, J.B. Renucci, R. Zeyher, M. Cardona: Phys. Rev. B**10**, 4309 (1974)
6.22   S.K. Chan, V. Heine: J. Phys. F**3**, 795 (1973)
6.23   Y. Kato, H. Takuma: J. Opt. Soc. Am. **61**, 347 (1971)
6.24   N. Wada, S.A. Solin: Physica **105B**, 353 (1981)
6.25   Milton Roy Company, Rochester, New York
6.26   D.E. Aspnes, A. Studna: Phys. Rev. B**27**, 985 (1983)
6.27   M. Born, E. Wolf: *Principles of Optics* (Pergamon, Oxford 1980) p. 620
6.28   E. Anastassakis, Y.S. Raptis: J. Appl. Phys. **57**, 920 (1985); A. Anastassiadou, Y. S. Raptis, E. Anastassakis: J. Appl. Phys. **59**, 627 (1986)
6.29   R.G. Greenler, T.L. Slager: Spectrochim. Acta A**29**, 193 (1973)
6.30   J.R. Kirtley, T.N. Theis, J.C. Tsang, D.J. DiMaria: Phys. Rev. B**27**, 4601 (1983)
6.31   A. Mooradian: Phys. Rev. Lett. **22**, 185 (1969)
6.32   P.A. Temple, C.E. Hathaway: Phys. Rev. B**7**, 3685 (1973)
6.33   J.B. Renucci, R.N. Tyte, M. Cardona: Phys. Rev. B**11**, 3885 (1975)
6.34   J.B. Renucci, M.A. Renucci, M. Cardona: In Proc. 2nd Int. Conf. Light Scattering in Solids, ed. by M. Balkanski (Flammarion, Paris 1971) p. 326

6.35 G.A.N. Connell, R.J. Nemanich, C.C. Tsai: Appl. Phys. Lett. **36**, 31 (1980)
6.36 A.E. Bell, F.W. Spong: IEEE J. QE **14**, 486 (1978)
6.37 R.J. Nemanich, C.C. Tsai, G.A.N. Connell: Phys. Rev. Lett. **44**, 273 (1980)
6.38 R.J. Nemanich, C.C. Tsai, M.J. Thompson, T.W. Sigmon: J. Vac. Sci. Technol. **19**, 685 (1981)
6.39 C.C. Tsai, R.J. Nemanich, M.J. Thompson: J. Phys. Colloq. **42**, 1077 (1981)
6.40 C. Firmani, L. Gutierrez, E. Ruiz, G.F. Bisiacchi, L. Salas, F. Paresce, C.W. Carlson, M. Lampton: Astron. Astrophy. **134**, 251 (1984)
6.41 I.P. Csorba: Appl. Opt. **18**, 2440 (1979)
6.42 D. Rees, I. McWhirter, P.A. Rounce, E.F. Barlow: J. Phys. E: Sci. Instrum. **14**, 229 (1981)
6.43 M. Lampton: Sci. Amer. **245**, 62 (1981)
6.44 C. Firmani, E. Ruiz, C.W. Carlson, M. Lampton, F. Paresce: Rev. Sci. Instrum. **53**, 570 (1982)
6.45 M. Lampton, C.W. Carlson: Rev. Sci. Instrum. **50**, 1093 (1979)
6.46 J. Kristian, M. Blouke: Sci. Amer. **247**, 66 (1982)
6.47 C.A. Murray, S.B. Dieker: J. Opt. Soc. A**3**, 2151 (1986)
6.48 J.R. Janesick, T. Elliot, S. Collins, M.M. Blouke, J. Freeman: Opt. Eng. **26**, 692 (1987)
6.49 W.P. Acker, B. Yip, D.H. Leach, R.K. Chang: J. Appl. Phys. **64**, 2263 (1988)
6.50 P.L. Flaugh, S.E. O'Donnell, S.A. Asher: Appl. Spectroscopy **38**, 847 (1984)
6.51 A. Campion: J. Vac. Sci. Technol. B**3** (5), 1404 (1985)
6.52 V.M. Hallmark, A. Campion: Chem. Phys. Lett. **110**, 561 (1984)
6.53 B.G. Papanicolaou, J.M. Chen, C.A. Papageorgopoulous: J. Phys. Chem. Sol. **37**, 403 (1976)
6.54 D.R. Mullins, A. Campion: J. Phys. Chem. **88**, 8 (1984)
6.55 G.L. Gaines: Thin Sol. Fi. **99**, ix (1983)
6.56 S. Jen, N.A. Clark, P.S. Pershan, E.B. Priestley: Phys. Rev. Lett. **31**, 1552 (1973)
6.57 W. Knoll, M.R. Philpott, W.G. Golden: J. Chem. Phys. **77**, 219 (1982)
6.58 J.F. Rabolt, R. Santo, J.D. Swalen: Appl. Spectrosc. **34**, 517 (1980)
6.59 J.F. Rabolt, R. Santo, N.E. Schlotter, J.D. Swalen: IBM J. Res. Dev. **26**, 209 (1982)
6.60 M.H. Brodsky, M. Cardona, J.J. Cuomo: Phys. Rev. B**16**, 3556 (1977)
6.61 J.E. Smith, M.H. Brodsky, B.L. Crowder, M.I. Nathan, A. Pinczuk: Phys. Rev. Lett. **26**, 642 (1971)
6.62 J.C. Tsang: Appl. Phys. Lett. **49**, 1796 (1986)
6.63 R.J. Nemanich, T.W. Sigmon, N.M. Johnson, M.D. Moyer, S.S. Lau: in *Laser and Electron Beam Solid Interactions and Materials Processing*, ed. by J. Gibbons, K. Hess, T.W. Sigmon, (North-Holland, Amsterdam 1981) p. 541
6.64 J.C. Tsang, R. Matz, Y. Yakota, G.W. Rubloff: J. Vac. Sci. Technol. A**2**, 556 (1984)
6.65 J.C. Tsang, W.A. Thompson, G.S. Oehrlein: J. Vac. Sci. Technol. B**3**, 1129 (1985)
6.66 R.J. Nemanich, C.M. Doland: J. Vac. Sci. Technol. B**3** (4), 1142 (1985)
6.67 R.J. Nemanich, R.T. Fulks, B.L. Stafford, H.A. Vander Plas: Appl. Phys. Lett. **46**, 670 (1985)
6.68 G. Devlin, J.L. Davis, L.L. Chase, S. Geschwind: Appl. Phys. Lett. **19**, 138 (1971)
6.69 A.A. Maradudin: "Theoretical and Experimental Aspects of the Effects of Point Defects and Disorder on the Vibrations of Crystals-1", in *Solid State Physics*, Vol. 18, ed. by F. Seitz, D. Turnbull (Academic, New York 1966) p. 273
6.70 D.W. Feldman, M. Ashkin, J.H. Parker: Phys. Rev. Lett. **17**, 1209 (1966)
6.71 W. Byra: Solid State Commun. **12**, 253 (1973)
6.72 M. Delhaye: In *Raman Spectroscopy, Linear and Nonlinear*, ed. by J. Lascombe, P.V. Huong (Wiley, Chichester and New York 1982) p. 223
6.73 J.A. Kash, J.C. Tsang, J.M. Hvam: Phys. Rev. Lett. **54**, 2151 (1985)
6.74 H. Vogt: Proc. 10th Int. Conf. *Raman Spectroscopy*, ed. by W.L. Peticolas, B. Hudson (U. of Oregon, Eugene, OR. 1986) p. 11-5
6.75 W.G. McMullan, S. Charbonneau, M.L.W. Thewalt: Rev. Sci. Instrum. **58**, 1626 (1987)

6.76 H. Vogt, G. Rossbroich: Phys. Rev. **B24**, 3086 (1981)
6.77 "Imagery Comes to Infrared Astronomy", Science, **236**, 1525 (1987)
6.78 L. Rau, R. Merlin, M. Cardona, Hj. Mattausch, W. Bauhofer, A. Simon, F. Garcia-Alvarado, E. Morán, M. Vallet, J.M. González-Calbet, M.A. Alario: Solid State Commun. **63**, 839 (1987)

7. Brillouin Scattering from Metallic Superlattices

Marcos H. Grimsditch

With 13 Figures

In this chapter, the use of Brillouin scattering as a tool for the study of metallic superlattices is discussed, and results of scattering from phonons and magnons are reviewed. For phonons, experimental results show unexpected behavior in the elastic properties; possible origins for the anomalies are briefly discussed. The experimental magnon data are found to be in good agreement with theoretical predictions.

7.1 Background

The current interest in the properties of layered materials, although certainly fueled by possible technological applications, is also attributable to a number of novel basic properties which arise due to the additional periodicity introduced into the samples [7.1]. In this article we review the role that light scattering experiments have played in the investigation of elastic and magnetic properties of such layered metallic systems. The word superlattice will be used in a very general sense i.e., any system in which an additional periodicity has been incorporated by successive deposition of two materials. This definition includes both lattice-matched and non-lattice-matched systems, and systems which are composed of materials that form solid solutions (and hence usually form structures with a composition modulation), as well as those that do not mix (and hence have "sharp", well-defined interfaces between the layers).

It is somewhat surprising that in examining metallic systems light scattering should have turned out to be such a valuable tool. The main reason for its widespread application is that since superlattices are usually grown in the form of thin films, many other techniques (e.g., neutron scattering, conventional ultrasonic experiments, etc.) are not easily applicable. Consequently, even though the information obtained from light scattering experiments is incomplete, often it has been the only information available. An advantage of light scattering is that it probes excitations whose wavelengths are comparable to typical modulation wavelengths of superlattices. From this standpoint, it is ideally suited to the study of a region where interesting phenomena might be expected.

The advances in the Brillouin scattering technique which have made it a powerful tool in applications to solid state physics are now well known [7.2], and will not be discussed here. Although the application of the technique to the study of elastic and magnetic properties of opaque materials and thin metallic films has been covered in recent review articles [7.2, 3], a few of the aspects discussed there will be repeated here, to the extent that they contribute to the understanding of results on superlattices.

7.1.1 Coupling Mechanisms of Light to Excitations: Phonons and Magnons

The coupling of light of a given frequency to excitations in a medium to produce scattered radiation at a different frequency is due to the changes in the polarizability of the medium produced by the excitation. Conversely, any excitation which does not change the polarizability of the medium will not couple to light and will not be observed in a Raman or Brillouin experiment. Without going into microscopic details, the changes in polarizability (α) can be written phenomenologically as:

$$\alpha = \alpha_0 + \sum_i \frac{\partial \alpha}{\partial q_i} q_i ,\tag{7.1}$$

where $\partial \alpha / \partial q_i$ is the change in the polarizability induced by the normal mode i of the system. In most cases, a microscopic calculation of the $\partial \alpha / \partial q_i$ is extremely complex. For long wavelength acoustic excitations, however, they can be related to macroscopic properties which can be determined experimentally. For example, the change in bulk polarizability produced by acoustic phonons is related to the elasto-optic constants which govern the changes in index of refraction as a function of strain. For magnetic excitations the situation is similar, and the intensity of the scattered light is related to various magneto-optic coefficients [7.4].

A somewhat different form of coupling between light and excitations can also take place at the surface of a material. If an excitation produces a time-dependent corrugation of the sample surface, then there are parts of space which will be alternately occupied by the material and by vacuum. This effectively produces large polarizability changes which scatter the incident radiation (ripple mechanism).

In transparent materials, the dominant scattering mechanism is through the change in polarizability produced in the bulk of the materials. However, in going to less and less transparent materials the contribution from the surface ripple mechanism becomes more and more important. Since magnetic excitations do not produce a corrugation of the surface (magneto-striction effects are usually negligible), the coupling occurs only through polarizability changes in the bulk material. In metals where the penetration depth of light is typically only a few hundred angstroms, the fact that any scattering from magnetic excitations is observed at all is due to the "very large" magneto-optic constants. For phonons in metals, the elasto-optic contribution to the scattering is usually considerably smaller than that due to the ripple mechanism. It must be emphasized that even in cases where the *coupling* takes place at the surface or in the first few hundred angstroms, the excitations which produce the corrugation or the polarizability changes usually represent excitations characteristic of the bulk material.

7.1.2 What Excitations Couple to Light?

For the purpose of this chapter, the discusssion will be limited to the case of metals where the penetration depth of light is considerably smaller than its wavelength. As was mentioned in the previous section, the coupling to magnons and phonons in these materials occurs via different mechanisms, so that different conditions must be satisfied for each type of excitation. These are discussed in detail in [7.2] and are briefly summarized below.

a) Magnons

As occurs for all scattering processes in the vicinity of a surface, it can be shown that conservation of wavevector parallel to the surface must hold. This allows the magnon wavevector parallel to the surface ($q_\parallel$) to be calculated for a specific scattering geometry, viz.

$$q_\parallel = k_0(\sin\theta_i + \sin\theta_s) \tag{7.2}$$

where k_0 is the wavevector of the incident radiation and θ_i and θ_s are the angles between the incident and scattered light and the surface normal (n), respectively [(7.2) is valid if k_i, k_s, and n are coplanar]. Perpendicular to the surface, the radiation is strongly attenuated so that the conservation of wavevector requires only that the perpendicular component of the wavevector of the excitation ($q_\perp$) satisfies

$$0 \leq q_\perp \lesssim \frac{2\pi}{l_0}, \tag{7.3}$$

where l_0 is the penetration depth of the light.

For illustrative purposes we consider the case of an isotropic ferromagnet of saturation magnetization M, gyromagnetic ratio γ, spin wave stiffness D, in an applied field H. The frequency of bulk magnons (ω_B) of wavevector q propagating perpendicular to H is given by [7.4]:

$$\omega_B = \gamma[(H + Dq^2)(H + 4\pi M + Dq^2)]^{1/2}. \tag{7.4}$$

Apart from bulk modes, at the surface of a ferromagnet an additional excitation is present which is known as the Damon-Eshbach mode [7.5] or a surface magnon. The frequency (ω_s) of a surface magnon of wavevector $q_\parallel$, propagating perpendicular to H is given by [7.4]:

$$\omega_s = \gamma[H + 2\pi M + D'q_\parallel^2] \tag{7.5}$$

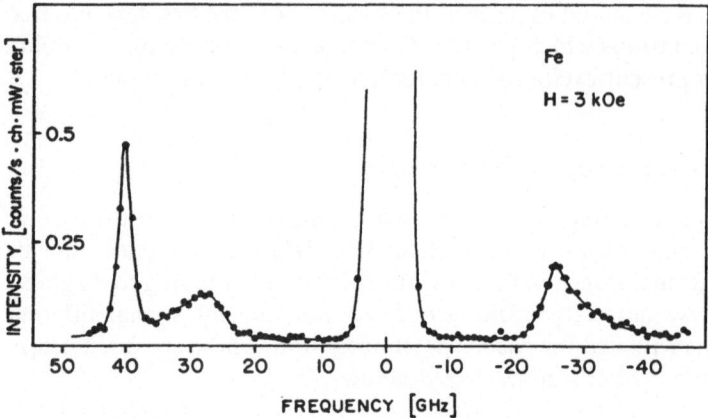

Fig. 7.1. Brillouin spectrum of magnons in bulk Fe. The sharp feature is the Damon and Eshbach mode (surface magnon), the broad features are bulk magnons. From [7.6]

where D' is the surface spin wave stiffness which could be different from that in the bulk. From (7.4–5) and the fact that $q_{\parallel}$ is a well-defined quantity (7.2) while $q^2 = q_{\perp}^2 + q_{\parallel}^2$ is not, it is easy to understand the observed experimental spectra such as that shown in Fig. 7.1 [7.6]. The sharp feature is due to the surface magnon (7.5) while the broad asymmetric lines come from the bulk magnons (7.4) broadened by the range of q values given in (7.3).

b) Phonons

As for the magnons, the component of wavevector parallel to the surface is given by (7.1). However, since the interaction takes place *at* the surface there is no

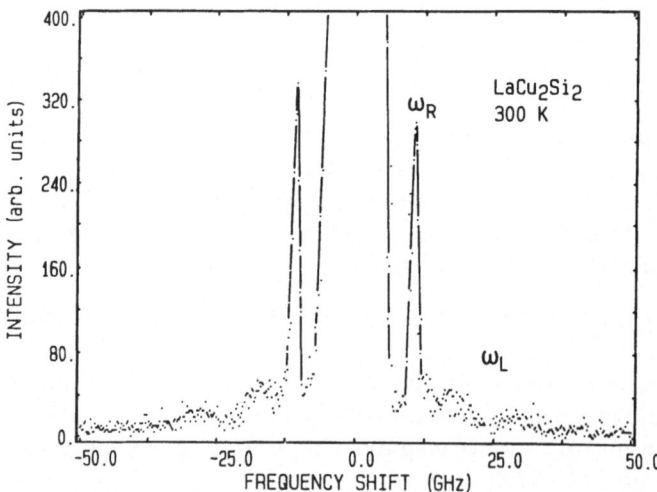

Fig. 7.2. Brillouin spectrum of polycrystalline $LaCu_2Si_2$. The sharp features are due to a surface phonon, the broad features to bulk phonons. From [7.7]

constraint placed on $q_\perp$. Consequently, any phonon which corrugates the surface and satisfies (7.1) will be observed in a light scattering experiment.

A typical Brillouin spectrum obtained from an elastically isotropic metal is shown in Fig. 7.2 [7.7]. The broad features arise from phonons in the bulk reflected at the free surface. During the reflection process the surface is corrugated and, since phonons of all $q_\perp$ can contribute, the features are broad. The sharp feature is due to the presence of a surface wave (also known as the Rayleigh wave). Surface waves are solutions of the elastic wave equations in a semi-infinite solid subjected to the boundary conditions of a free surface. They are characterized by an amplitude that decreases exponentially away from the surface. Their velocity (v_s) is usually closely related to that of a bulk transverse wave polarized perpendicular to the free surface (v_T). For a complete discussion of surface waves see [7.8].

7.2 Elastic Properties of Superlattices

7.2.1 Theory

The calculation of the elastic response of a stratified medium in terms of the elastic properties of its constituents is a conceptually simple problem (see Chaps. 2 and 3). The wave equations in each layer can be written in terms of the elastic constants (C_{ij}) of that layer, and then solved with the appropriate boundary conditions at each interface. An analytical solution can only be obtained in certain cases of long wavelengths or layers composed of high symmetry materials, but for the general case, numerical methods are required. Various stages of generalization can be found [7.9–15]. A concept that results from these studies is that of "effective elastic constants"; that is, in the long wavelength limit (where the excitation wavelength or the length scale over which a deformation is produced is larger than the modulation wavelength of the superlattice), the stratified medium can be described as a homogeneous medium with a well-defined elastic constant tensor. Expressions for the components of this tensor in terms of the C_{ij} and the thicknesses of the individual layers d_i are given in [7.9–15]. One particular expression which will be useful in later discussion is

$$\Lambda/C_{44} = d_1/C_{44}(1) + d_2/C_{44}(2) ,\qquad (7.6)$$

where the modulation wavelength (Λ) satisfies

$$\Lambda = d_1 + d_2 .\qquad (7.7)$$

[Equation (7.6) is valid, provided the individual layers have at least orthorhombic symmetric with a principal axis along the superlattice normal.] A striking feature of (7.6), which also holds for all other components of the effective elastic constant tensor, is that the C_{ij} of the superlattice do not depend on the modulation wavelength but only on the relative fraction (f_i) of each component ($f_i = d_i / \Lambda$). It is this simple feature, obtained in all theoretical approaches [7.9–15], that is found to be in disagreement with many experimental results described below.

7.2.2 Measurement of Elastic Properties of Thin Films

The determination of the elastic properties of films thinner than a few microns poses severe experimental difficulties because their dimensions are smaller than the wavelengths generated with ultrasonic techniques. Because of this, it has been necessary to develop special techniques to investigate the elastic response of these systems. Most techniques (e.g. bulge tester, vibrating reed, torsional measurements, Young's modulus, etc.) involve the removal of the film from the substrate on which it was deposited; after removal, the film is subjected to some macroscopic deformation from which the elastic properties can be derived. The difficulty of these types of experiments, coupled with the fact that since they require removal from the substrate they are destructive in nature, led to an initial reticence to accept the first measurements on superlattices [7.16] that showed a strong modulation-wavelength dependence of the elastic properties. Since then, a large number of studies has been performed using the methods mentioned above [7.17–22], and most of them indicate some anomaly in an elastic property.

The use of Brillouin scattering has turned out to be a powerful technique in the study of elastic properties. In metals, where the strongest coupling is to surface waves with wavelengths of ~ 3000 Å, the situation is such that if the superlattice film is thicker than ~ 7000 Å, there is a negligible contribution from the substrate and measurements can be made with the sample "as prepared". Furthermore, provided the modulation wavelength is $\lesssim 500$ Å, the superlattice

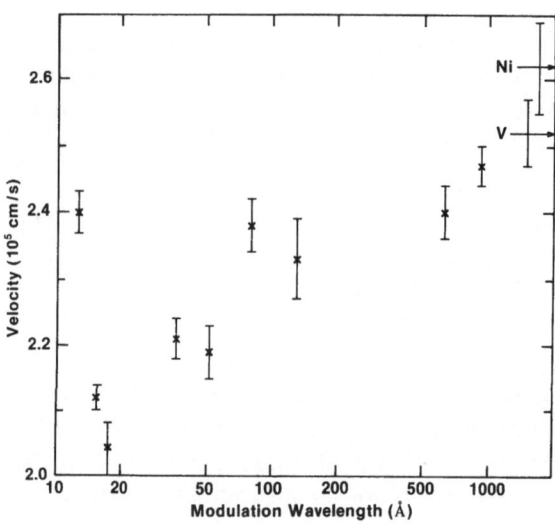

Fig. 7.3. Surface wave velocity versus modulation wavelength for V/Ni superlattices. From [7.27]

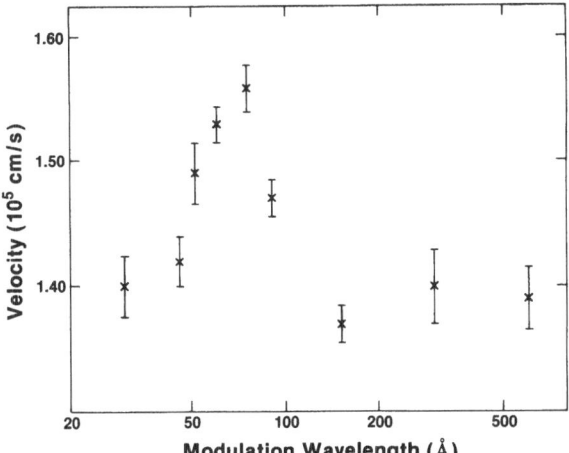

Fig. 7.4. Surface wave velocity versus modulation wavelength for Au/Cr superlattices. From [7.30]

can be treated as a homogeneous medium and an effective elastic constant determined. That the concept of a effective constant also holds for surface waves has been verified numerically in [7.23].

To date, the superlattices that have been studied using Brillouin scattering are Nb/Cu [7.24, 25], Mo/Ni [7.26], V/Ni [7.27], Fe/Pd [7.28], Mo/Ta [7.29], and Au/Cr [7.30]. Except for Fe/Pd, all these systems show an anomalous behavior in the surface wave velocity. Contrary to the expectation that the velocity should be independent of Λ, behavior like that shown for V/Ni in Fig. 7.3 [7.27] and for Au/Cr in Fig. 7.4 [7.30] is found.

At first it was thought that the origin of the anomalies could be the fact that at each interface a layer with different elastic properties would be formed. Since this hypothesis predicts that the magnitude of the anomaly should only depend on the number of interfaces per unit length, measurements were carried out on three series of Mo/Ni samples with $d_{Ni} = 3d_{Mo}$, $d_{Ni} = d_{Mo}$, and $3d_{Ni} = d_{Mo}$ [7.26]. The prediction was not borne out by experiment. However, the results, interpreted on the basis of an extension of (7.6), showed that the softening was probably due to the Ni layers that for some reason exhibited softer elastic properties than when in bulk form.

7.2.3 Possible Explanation for the Anomalous Behavior

In [7.26] it was also found in x-ray studies of the Mo/Ni superlattices that the average interplanar spacing increased as the modulation wavelength decreased. *Schuller* and *Rahman* [7.31] performed a molecular dynamics calculation which assumed that the expansion takes place in the Ni layers. They showed that the elastic anomaly could be understood as a consequence of the lattice expansion. This conceptually simple idea, i.e., that changes in lattice constant produce changes in the elastic properties, was then cast in a simple, albeit phenomenological, form in [7.30, 32]. Based on the Murnaghan equation of state, they arrive

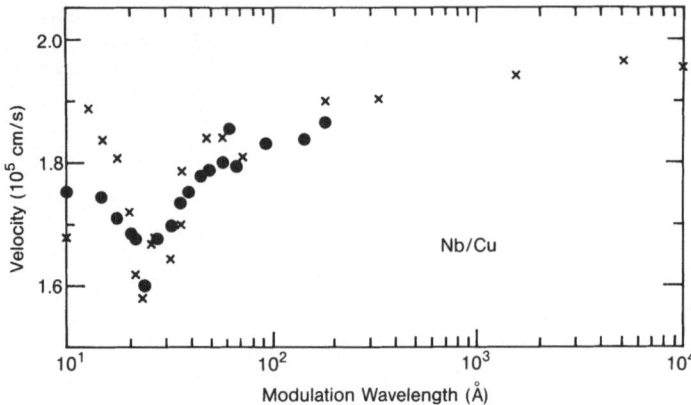

Fig. 7.5. Surface wave velocity versus modulation wavelength for Nb/Cu superlattices. The crosses are experimental points and the dots are obtained from (7.8). From [7.25]

at an expression of the form

$$v = v_0 \left(\frac{a_0}{a} \right)^9 , \qquad (7.8)$$

where v is a velocity and a the lattice constant. The subscript 0 indicates the reference state. This very simple expression *with no adjustable parameters* (v_0 can be calculated from bulk values of the C_{ij}) explains the anomalous behavior in all systems for which x-ray data exist, viz. Nb/Cu [7.24, 25], Mo/Ni [7.26], Mo/Ta [7.29] and Au/Cr [7.30]. The results of (7.8) applied to Nb/Cu are shown in Fig. 7.5; the crosses are the experimental results obtained with Brillouin scattering and the dots are obtained from solutions of (7.8) using the lattice constant given in [7.25].The astonishingly good agreement between the calculations of [7.31], the results of (7.8), and experimental results leaves little doubt as to the origin of the elastic anomalies, namely changes in lattice constants.

The origin of the changes in lattice constant, however, is still not well understood. The most obvious solution is that strains at each interface cause the lattice changes. This hypothesis can be discarded on the following grounds: In Mo/Ni it was found that changes perpendicular to the layering are greater than those in the plane, contrary to what would be expected if strains in the plane were the driving mechanism; in Au/Cr the lattice matching in the plane is better than a few parts per thousand so that the observed $\sim 8\%$ expansion of Cr perpendicular to the layers is, again, not explainable; finally, in a calculation for Cu/Ni, *Banerjea* et al. [7.33] have shown, using third order elastic constants, that interfacial strains cannot explain the observed anomalies.

Arguments based on electronic effects, such as the folding of the Brillouin zone due to the layering, have also been proposed as the driving mechanism [7.34–37a]. These are discussed in [7.37a] but as they stand now, these mechanisms do not seem capable of providing a general solution to the problem.

Table 7.1. Measured elastic properties of superlattices and differences in the work functions of the constituent materials. (Uncertainties of up to ± 0.2 eV are possible in the values of $\Delta\phi$)

Superlattice	Ref.	$\Delta C/C$[%]	$\Delta\phi$[eV][a]
V/Ni	[7.27]	45	1.2
Au/Cr	[7.30]	30	0.8
Mo/Ni	[7.26]	50	0.6
Nb/Cu	[7.24, 25]	35	0.5
Mo/Ta	[7.29]	10	0.3
Fe/Pd	[7.28]	0	0.0

[a] From [7.38]

A different mechanism, which has recently been proposed [7.37b] is based on the charge transfer which occurs when two metals with different work functions are placed in contact. This mechanism, in spite of its very simplistic nature, has several attractive features: it predicts an anomaly even for a free electron model, the anomaly is predicted to occur in the right modulation wavelength regime (~ 30 Å), the predicted changes in lattice constants are of the same order of magnitude as those observed experimentally, and finally, the magnitude of the anomaly is predicted to be proportional to the difference in work functions – a fact also roughly borne out by experiment, as can be seen in Table 7.1. A complete calculation, where the specific electronic structures of the constituent materials are considered, has not yet been performed.

7.2.4 Continuing Work and Future Directions

In the Brillouin scattering work discussed in the preceding section, the reported anomalies were in the velocity of the surface wave. This is because in these metals, the only excitation that couples to light and gives a well-defined spectral feature is the surface mode. Since other measurement techniques [7.16–22] have found anomalies in other elastic moduli, it would be of interest to obtain all the elements of the elastic constant tensor. Potentially, Brillouin scattering can also be used to acquire such information. It is known that if a thin film of a material is deposited on a substrate, then the broad continuum shown in Fig. 7.2 splits into well-defined peaks. These peaks correspond to guided modes in the film and are similar in nature to the Lamb modes of a free standing plate – hence their name: generalized Lamb modes or Sezawa modes. The position of these Sezawa modes depends on the density and the elastic constants of the film as well as on those of the substrate. For systems for which all these constants are known, the peak positions can be accurately predicted [7.2, 39–42]. Solving the reverse problem, however, (i.e., obtaining the C_{ij} from the experimentally observed peak positions), is not straightforward. Even in cases where superb experimental data exist, such as that shown in Fig. 7.6 [7.29], it has been found that it is not possible to perform a fit in which all the C_{ij} are free parameters [7.29, 43–47]. In spite of

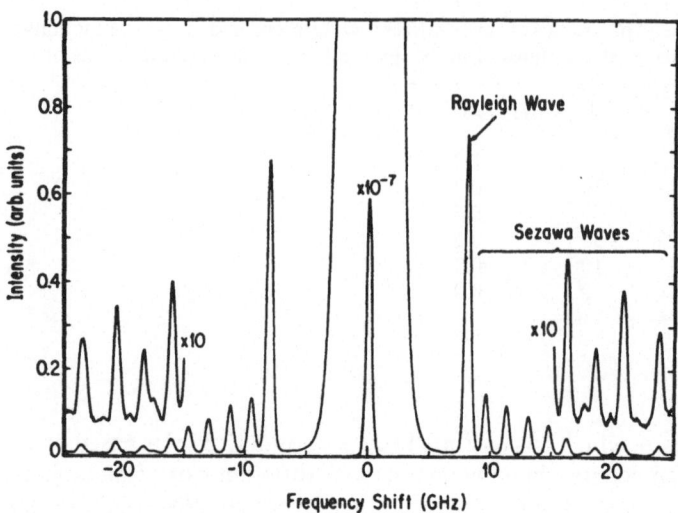

Fig. 7.6. Brillouin spectrum of Sezawa modes in a Mo/Ta superlattice. From [7.29]

this difficulty, for the two superlattice systems investigated with this technique [7.29, 46], there are indications that other elastic constants also show anomalous behavior.

A variation of the above technique, which may turn out to be useful in the study of superlattices, is Brillouin scattering from unsupported films [7.48]. In a recent article on unsupported Au films [7.49] it was found that since the peak positions depend only on the properties of the film itself (and not on those of a substrate) it was possible to perform fits to many of the elastic constants. Preliminary work on unsupported Nb/Cu superlattices is hampered by the low laser power required so as not to damage the samples.

7.3 Magnetic Excitations in Superlattices

Since many aspects of this topic are covered in great detail in Chaps. 2 and 8 of this book, this section will be restricted to a very specific case; namely, experimental results on superlattices composed of ferromagnetic and non-magnetic layers in which the magnetic layers are ferromagnetically aligned. The applied magnetic field (H) is taken to be parallel to the magnetization (M) and to lie in the plane of the sample.

7.3.1 Theory

Complete theoretical calculations of the magnetic excitations in a ferromagnetic superlattice are given in [7.50–52] and some details can also be found in Chaps. 2

and 8. Here, for the sake of easy reading, plausibility arguments will be given so that the formulas and expressions to be quoted will be easier to follow.

Consider an isolated ferromagnetic slab of thickness d: the modes that exist in such a slab are closely related to those in a semi-infinite medium (7.4, 5). The position of the surface mode is slightly modified with respect to (7.5) and is given by

$$\omega_s' = \gamma \left[(H + 2\pi M)^2 - 4\pi^2 M^2 \exp(-2q_{\parallel} d) \right]^{1/2} . \tag{7.9}$$

[In (7.9) the term $D'q_{\parallel}^2$ has been dropped because for typical values of $q_{\parallel}$ it is negligible.] The "bulk"-like modes in the film are still given by (7.4) but the allowed values of q are now restricted to

$$q = n\pi/d , \tag{7.10}$$

where n is an integer (note the similarity to the confined optical phonons described in Chap. 3). Spectra from such an isolated ferromagnetic film [7.53] are shown in Fig. 7.7, clearly displaying how the broad bulk feature of a semi-infinite medium, Fig. 7.1, splits into sharp features due to the standing spin waves. The problem of a superlattice then is, to investigate the behavior of these modes as many films, separated by a distance d_0, are brought closer and closer together. The complexity of the problem has necessitated certain approximations in the theoretical treatment [7.50–52], the most important one being the neglect of exchange (viz $D = 0$). This approximation leads to the fact that all spin wave modes in a thin film, given by (7.4) and (7.10), are assumed to be at the frequency

$$\omega_B' = \gamma \left[H(H + 4\pi M) \right]^{1/2} . \tag{7.11}$$

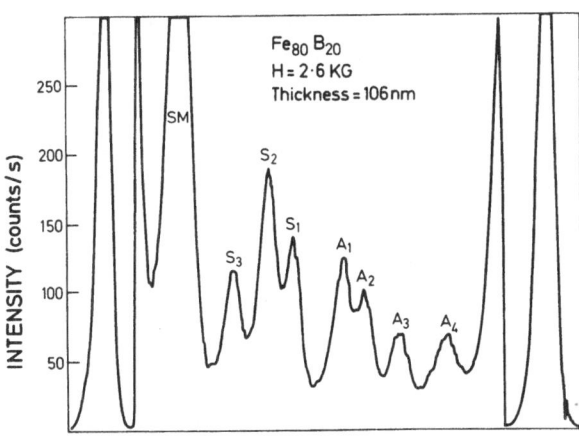

Fig. 7.7. Brillouin spectrum from magnons in a thin film of a ferromagnet. The peak labeled SM is the surface magnon, the peaks labeled S_i and A_i are the Stokes and anti-Stokes standing spin waves. From [7.53]

A comparison of Fig. 7.7 [7.53] with the "prediction" of (7.11) that all spin wave modes are at the same frequency, shows how drastic this approximation is. For the surface-like modes in each film, however, the omission of exchange has a negligible effect and hence, any mode resulting from their interaction should be well-described by the theory.

Consider a system of ferromagnetic films of thickness d separated by nonmagnetic layers of large thicknesses d_0. In each film there is a mode of frequency ω'_s (7.9) and many modes at ω'_B (7.11). All the latter modes have at least one node within the layer and, as a consequence, the magnetization (m) of a given mode averages to zero across the film. This is not true for the ω'_s mode. Consequently, in a theory in which exchange is neglected so that the only coupling is through dipolar fields, it is not unreasonable that spin wave modes ω'_B do not couple for any value of the thickness of the separating layer. The surface-like modes at ω'_s, however, will be coupled by their dipolar fields and consequently, must split into a band as d_0 is reduced. The limit as $d_0 \to 0$ is also predictable without calculation since it must correspond to a semi-infinite ferromagnet with one surface mode at ω_s (7.5) and all other modes at ω'_B. This behavior is illustrated in Fig. 7.8. For a quantitative analysis one must, of course, perform the full calculation of [7.50, 51]. The results of these calculations can be expressed mathematically as follows: the standing spin waves are indeed found not to couple and remain at the frequency given by (7.11). The surface-like modes (ω'_s) do couple and form a band of modes (Fig. 7.8) characterized by (see Chap. 2)

$$\omega_b = \gamma \left[H(H + 4\pi M) + 4\pi^2 M^2 w \right]^{1/2} , \tag{7.12}$$

$$w = \frac{2 \sinh(dq_\parallel) \sinh(d_0 q_\parallel)}{\cosh\left[(d + d_0)q_\parallel\right] - \cos\left[(d + d_0)q_\perp\right]} , \tag{7.13}$$

$$0 \leq q_\perp \leq \pi/(d + d_0) . \tag{7.14}$$

The most striking feature arising from the calculations is that, for $d > d_0$ (i.e. the magnetic layers thicker than the nonmagnetic layers), a mode splits off from the band at a frequency

$$\omega_s = \gamma \left[H + 2\pi M \right] . \tag{7.15}$$

This mode has a surface-like characteristic of the superlattice itself, meaning that its amplitude decreases exponentially away from the free surface. It also has the peculiar property that its frequency is *equal* to that of a surface wave in a bulk material of magnetization M (note: it is *not* that expected for a material of average magnetization $\bar{M} = dM/(d + d_0)$).

What is expected to happen if exchange is included? It seems likely that since exchange is a very short range interaction it will not affect the coupling between layers until d_0 becomes smaller than ~ 2 lattice spacings. (This case is discussed in detail in Chap. 8). Furthermore, since it has a negligible effect on ω'_s, all

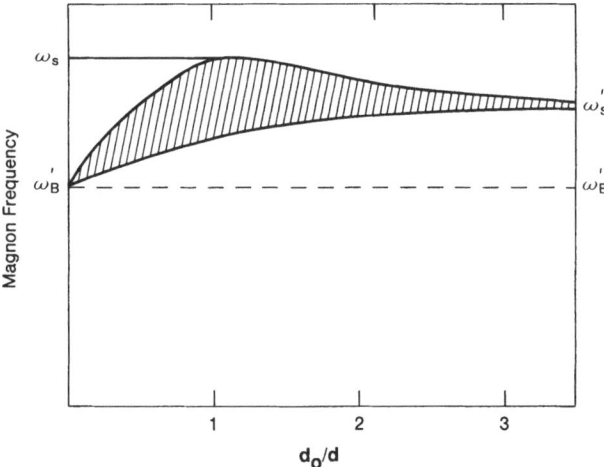

Fig. 7.8. Schematic of magnon modes in superlattices: ω_B' are the standing spin wave modes (exchange neglected) in each layer which do not couple, the shaded area is the band arising from coupling of surface magnons in each layer, ω_s is a surface-like magnon of the superlattice as a whole

modes derived from it, i.e. (7.12–15), will remain unchanged. The standing spin wave modes (ω_B') will change frequency dramatically [see (7.4, 11)]. However, since their spatial characteristics will not radically change, the dipolar coupling between modes in adjacent layers will remain small and, as a consequence, one might expect the standing spin wave modes in each layer of a superlattice to have the same frequency as those in an isolated film [see (7.4, 10)]. If this is the case, in superlattices containing Ni, Fe, or Co with layer thicknesses ~ 100 Å, the frequency of the lowest standing spin-wave mode will be well above those of ω_s' and ω_b, and will probably not be observed in a Brillouin experiment unless particular care is taken.

7.3.2 Experimental Results

The first experimental evidence on the magnetic excitations in a system of 10 layers of Cu/Fe was presented in [7.50], where it was shown that the observed peaks in such a system fall into the region predicted by theory. A more complete study of Mo/Ni superlattices followed [7.54–56], in which many of the detailed properties of the excitations were investigated. The most striking feature is the appearance of a new mode for $d > d_0$. Figure 7.9 shows the spectra obtained from three different samples with $d = 3d_0$, $d = d_0$, and $3d = d_0$. With reference to Fig. 7.8, this behavior is easy to understand. When $3d = d_0$, only one peak, corresponding to the band (ω_b' [7.12] is observed (recall that all spin-wave modes at ω_B' have been moved up in frequency due to exchange and are not observed), for $d = d_0$ only one peak at ω_b is expected, although perhaps somewhat broadened, and for $d = 3d_0$, one observes two distinct peaks. (The asymmetry in

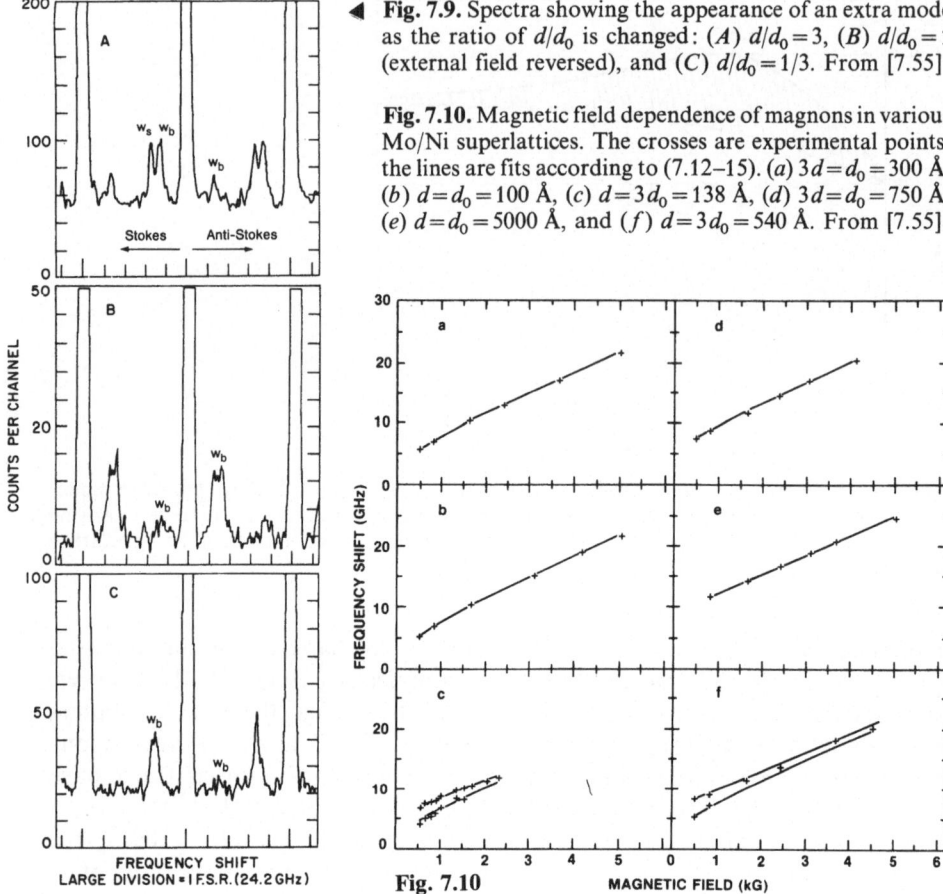

Fig. 7.9. Spectra showing the appearance of an extra mode as the ratio of d/d_0 is changed: (*A*) $d/d_0 = 3$, (*B*) $d/d_0 = 1$ (external field reversed), and (*C*) $d/d_0 = 1/3$. From [7.55]

Fig. 7.10. Magnetic field dependence of magnons in various Mo/Ni superlattices. The crosses are experimental points, the lines are fits according to (7.12–15). (*a*) $3d = d_0 = 300$ Å, (*b*) $d = d_0 = 100$ Å, (*c*) $d = 3d_0 = 138$ Å, (*d*) $3d = d_0 = 750$ Å, (*e*) $d = d_0 = 5000$ Å, and (*f*) $d = 3d_0 = 540$ Å. From [7.55]

Fig. 7.10

Fig. 7.9 between Stokes and anti-Stokes lines, and the switch in intensity from one to the other as the applied field direction is reversed, is a well-known feature of surface magnons and will not be discussed here.)

In order to ascertain that the observed peaks are well described by (7.12–15), their dependence on the applied magnetic field was investigated. This behavior is shown in Fig. 7.10 for a number of samples. The lines are fits to the experimental points with (7.12, 15) using M and w as fitting parameters. These fits produced results [7.55] in agreement with independent measurements of M and values of w consistent with (7.13).

The dependence of the mode frequencies on $q_{\parallel}$ was also studied and results for various samples are shown in Fig. 7.11. The lines are calculated from (7.12, 13) while the dots represent the experimental results. In spite of the small frequency changes expected for changes in $q_{\parallel}$, and the relatively large experimental errors, there is reasonably good agreement between theory and experiment.

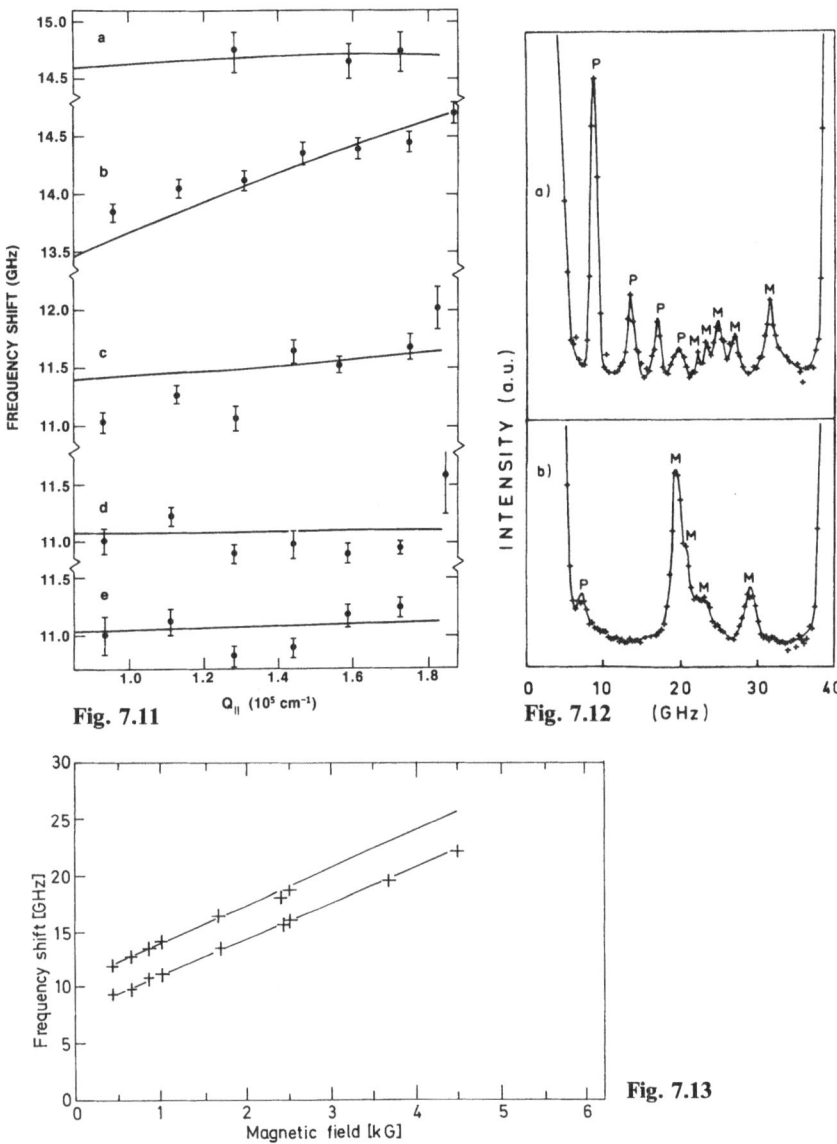

Fig. 7.11

Fig. 7.12   (GHz)

Fig. 7.13

Fig. 7.11. Wavevector dependence of magnons in Mo/Ni superlattices. Solid lines are fits according to (7.12–14). (a) $d = d_0 = 2500$ Å, (b) $d = d_0 = 550$ Å, (c) $d = d_0 = 250$ Å, (d) $d = d_0 = 100$ Å, and (e) $d = 3d_0 = 249$ Å. From [7.55]

Fig. 7.12. Brillouin spectra of (a) Co/NbC and (b) Co/NbD. M and P refer to magnons and phonons respectively. From [7.57]

Fig. 7.13. Magnetic field dependence of magnons in a $d = d_0 = 550$ Å Mo/Ni superlattice. The upper line is identified as being due to a standing spin wave mode. From [7.55]

More recent measurements on Co/Nb [7.57] and Fe/Pd [7.58] have shown that in systems with a finite number of layers, the band of modes actually splits into separate peaks as originally reported in [7.50]. Spectra from [7.57] illustrating this effect, are shown in Fig. 7.12 for two different samples. The peaks labeled P and M correspond to phonons and magnons, respectively. The positions of these peaks as a function of applied field, are well accounted for by theory.

The only reported observation of standing spin waves in a superlattice is in [7.55]. In a $d_0 = d = 550 \pm 30$ Å Mo/Ni superlattice, apart from the peak due to the band, an additional peak at a higher frequency was also observed. From a fit to the magnetic field dependence of this extra line (shown in Fig. 7.13) and the aid of (7.4, 10) the thickness of the layer was determined to be 490 ± 30 Å. This reasonably good agreement is evidence that the hand-waving arguments about standing spin waves in superlattices, given at the end of the previous section, which argue that they should be in the same position as in a single film, may be correct (cf. the results for confined phonons in Chap. 3). Thus, we conclude that within experimental error, experiments to date confirm the predictions of theory regarding magnetic excitation in superlattices.

7.3.3 Discussion

In spite of the fact that all experimental evidence to date supports the theoretical calculations, the question of whether the present theories will be valid for all layer thicknesses still remains. In the limit that the layers become a few atomic layers thick, it seems reasonable to expect the material to behave like a crystal with a (perhaps large) unit cell spanning one layer of each material. It is known that for most crystals (e. g. Fe, Ni, EuO, Co) the long wavelength excitation (in a theory in which exchange is also ignored) is given by (7.11) and (7.15) but with the magnetization M replaced by the average magnetic moment per unit volume. In the case of a superlattice this is $\bar{M} = dM/(d+d_0)$. Hence, the intuitive assumption that for very thin layers the material should behave like an "ordinary" crystal is in disagreement with the results of the complete super-lattice theory. This discrepancy is not serious in itself, since the theories presented in [7.50–52] treat each layer as a *homogeneous medium* and conse-quently cannot be expected to be valid when layers of atomic dimensions are reached. If that is so, at what thicknesses does the theory break down? The measurements reported in the previous section indicate agreement between theory and experiment down to layer thicknesses of ~ 10 Å [7.58]. Since this distance is just a few atomic planes, there does not seem to be much room for a transition from one type of description to the other. The solution to this dichotomy must await further theoretical and experimental investigation.

References

7.1 For a review of the field see various chapters in *Synthetic Modulated Structure Materials*, ed. by L.L. Chang, B.C. Giessen (Academic, New York 1985)
7.2 J.R. Sandercock: In *Light Scattering in Solids III*, ed. by M. Cardona, G. Güntherodt, Topics Appl. Phys., Vol. 51 (Springer, New York 1982)
7.3 P. Grünberg: Prog. Surf. Sci. **18**, 1 (1985)
7.4 W. Wettling, M.G. Cottam, J.R. Sandercock: J. Phys. C: Sol. State Phys. **8**, 211 (1975); see also Chaps. 2 and 8 of this volume
7.5 K.W. Damon, J.R. Eshbach: J. Phys. Chem. Sol. **19**, 308 (1961)
7.6 J.R. Sandercock, W. Wettling: J. Appl. Phys. **50**, 7784 (1979)
7.7 R. Mock, G. Güntherodt: J. Phys. C: Sol. State Phys. **17**, 5635 (1984)
7.8 G.W. Farnell: In *Physical Acoustics*, ed. by W.P. Mason (Academic, New York 1969)
7.9 D.A.G. Bruggeman: Ann. Phys. (Leipzig) **29**, 160 (1937)
7.10 S.M. Rytov: Akust. Zh. **2**, 71 (1956) [English transl.: Sov. Phys-Acoust. **2**, 68 (1956)]
7.11 T.J. Delph, G. Herrmann, R.K. Kaul: J. Appl. Mech. **45**, 343 (1978)
7.12 J. Sapriel, B. Djafari-Rouhani, L. Dobrzynski: Surf. Sci. **126**, 197 (1983)
7.13 F. Nizzoli: In *Proc. 17th Int. Conf. Physics of Semiconductors*, ed. by J.D. Chadi, W.A. Harrison (Springer, New York 1985) p. 1145
7.14 M. Grimsditch: Phys. Rev. **B31**, 6818 (1985)
7.15 M. Grimsditch, F. Nizzoli: Phys. Rev. **B33**, 5891 (1986)
7.16 W.M.C. Yang, T. Tsakalakos, J.E. Hilliard: J. Appl. Phys. **48**, 876 (1977)
7.17 H. Itozaki: Ph.D. Thesis, Northwestern University (1982)
7.18 B.S. Berry, W.C. Pritchet: Thin Sol. Fi. **33**, 19 (1976)
7.19 D. Baral, J.B. Ketterson, J.E. Hilliard: J. Appl. Phys. **57**, 1076 (1985)
7.20 L.R. Testardi, R.H. Willens, J.T. Krause, D.B. McWhan, S. Nakahara: J. Appl. Phys. **52**, 510 (1981)
7.21 T. Tsakalakos, J.E. Hilliard: J. Appl. Phys. **54**, 734 (1983)
7.22 G.E. Henein, J.E. Hilliard: J. Appl. Phys. **54**, 728 (1983)
7.23 A. Kueny, M. Grimsditch: Phys. Rev. **B26**, 4699 (1982)
7.24 A. Kueny, M. Grimsditch, K. Miyano, I. Banerjee, C.M. Falco, I.K. Schuller: Phys. Rev. Lett. **48**, 166 (1982)
7.25 I.K. Schuller, M. Grimsditch: J. Vac. Sci. Technol. **B4**, 1444 (1986)
7.26 M.R. Khan, C.S.L. Chun, G.P. Felcher, M. Grimsditch, A. Kueny, C.M. Falco, I.K. Schuller: Phys. Rev. **B27**, 7186 (1983)
7.27 R. Danner, R.P. Huebener, C.S.L. Chun, M. Grimsditch, I.K. Schuller: Phys. Rev. **B33**, 3696 (1986)
7.28 P. Baumgart, B. Hillebrands, R. Mock, G. Güntherodt, A. Boufelfel, C.M. Falco: Phys. Rev. **B34**, 9004 (1986)
7.29 J.A. Bell, W.R. Bennett, R. Zanoni, G.I. Stegeman, C.M. Falco, F. Nizzoli: Phys. Rev. **B35**, 4127 (1987)
7.30 P. Bisanti, M.B. Brodsky, G.P. Felcher, M. Grimsditch, L.R. Sill: Phys. Rev. **B35**, 7813 (1987)
7.31 I.K. Schuller, A. Rahman: Phys. Rev. Lett. **50**, 1377 (1983)
7.32 M. Grimsditch: Mat. Res. Soc. Symp. Proc. **77**, 23 (1987)
7.33 A. Banerjea, J.R. Smith: Phys. Rev. **B35**, 5413 (1987)
7.34 T.B. Wu: J. Appl. Phys. **53**, 5265 (1982)
7.35 P.C. Clapp: In *Modulated Structure Materials*, ed. by T. Tsakalakos, M. Nijhoff, NATO ASI Series, Applied Sciences (D. Reidel, Dordrecht, 1984) p. 455
7.36 W.E. Pickett: J. Phys. F: Met. Phys. **12**, 2195 (1982); to be published
7.37 R.C. Cammarata: Scripta Metal. **20**, 479 (1986); M. Grimsditch: Superlattices and Microstructures **4**, 677 (1988)
7.38 AIP Handbook, ed. by O.E. Gray (McGraw-Hill, New York 1972) pp. 9–173
7.39 V. Bortolani, F. Nizzoli, G. Santoro, A. Marvin, J.R. Sandercock: Phys. Rev. Lett. **43**, 224 (1979)

302   *M. H. Grimsditch:* Brillouin Scattering from Metallic Superlattices

7.40 J.R. Sandercock, F. Nizzoli, V. Bortolani, G. Santoro, A.M. Marvin, Le Vide, Les Couches Minces, **201**, 754 (1980)
7.41 J.R. Sandercock, F. Nizzoli, V. Bortolani, G. Santoro, A.M. Marvin: In *Recent Developments in Condensed Matter Physics*, ed. by J.T. Devreese, L.F. Lemmens, V.E. van Doren, J. van Royen, Vol. 2 (Plenum, New York 1982) p. 419
7.42 V. Bortolani, F. Nizzoli, G. Santoro, J.R. Sandercock: Phys. Rev. B**25**, 3442 (1982)
7.43 B. Hillebrands, P. Baumgart, R. Mock, G. Güntherodt, P.S. Bechthold: J. Appl. Phys. **58**, 3166 (1985)
7.44 L. Bassoli, F. Nizzoli, J.R. Sandercock: Phys. Rev. B**34**, 1296 (1986)
7.45 B. Hillebrands, R. Mock, G. Güntherodt, P.S. Bechthold, N. Herres: Sol. St. Commun. **60**, 649 (1986)
7.46 P. Baumgart, B. Hillebrands, R. Mock, G. Güntherodt, A. Boufelfel, C.M. Falco: Phys. Rev. B**34**, 9004 (1986)
7.47 F. Nizzoli, R. Bhadra, O.F. de Lima, M.B. Brodsky, M. Grimsditch: Phys. Rev. B**37**, 1007 (1988)
7.48 E.L. Albuquerque, M.C. Oliveros, D.R. Tilley: J. Phys. C: Sol. State Phys. **17**, 1451 (1984)
7.49 M. Grimsditch, R. Bhadra, I.K. Schuller: Phys. Rev. Lett. **58**, 1216 (1987)
7.50 P. Grünberg, K. Mika: Phys. Rev. B**27**, 2955 (1983)
7.51 R.E. Camley, T.S. Rahman, D.L. Mills: Phys. Rev. B**27**, 261 (1983)
7.52 G. Rupp, W. Wettling, W. Jantz: Appl. Phys. A**41**, 3348 (1986)
7.53 M. Grimsditch, A. Malozemoff, A. Brunsch: Phys. Rev. Lett. **43**, 711 (1979)
7.54 M. Grimsditch, M.R. Khan, A. Kueny, I.K. Schuller: Phys. Rev. Lett. **51**, 498 (1983)
7.55 A. Kueny, M.R. Khan, I.K. Schuller, M. Grimsditch: Phys. Rev. B**29**, 2879 (1984)
7.56 I.K. Schuller, M. Grimsditch: J. Appl. Phys. **55**, 2491 (1984)
7.57 G. Rupp, W. Wettling, W. Jantz, R. Krishnan: Appl. Phys. A**37**, 73 (1985)
7.58 B. Hillebrands, P. Baumgart, R. Mock, G. Güntherodt: Phys. Rev. B**34**, 9000 (1986)

8. Light Scattering from Spin Waves in Thin Films and Layered Magnetic Structures

Peter Grünberg

With 17 Figures

Essentially there are three methods to detect spin waves, namely, inelastic neutron scattering, microwave absorption (MA), and inelastic light scattering (LS). Neutron scattering has its main advantage in the large accessible wavevector range, but generally, its sensitivity is too low for application to thin films. This is where MA and LS come in. MA is a well-established method in this area. The value of LS for the exploration of spin waves in thin films and layered magnetic structures has become increasingly apparent within the past five years. This will be the topic of this article.

The type of LS with small frequency shifts we are dealing with here is sometimes also called Brillouin scattering (BS) (for a note concerning this nomenclature see the end of Sect. 8.2.1). In this series the last articles covering among other topics the status of BS from spin waves was the one in Volume III by J. R. Sandercock [8.1] and the one by D. L. Mills (Chap. 2 of this volume). Further reviews are cited in [8.2–6]. The basic experimental setup for the application of this method to spin waves is shown in Fig. 8.1. We see here the backscattering geometry, which is most frequently employed. The sample is between the pole pieces of a magnet. The spectrometer is a Fabry-Perot interferometer. Figure 8.1 shows a three-pass arrangement with one interferometer, but one can also use other arrangements or even two interferometers in tandem. More details on this are contained in [8.1].

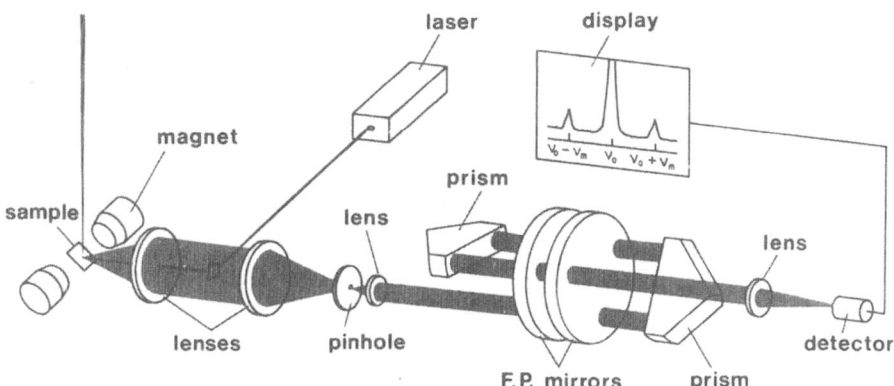

Fig. 8.1. Experimental arrangement for the measurement of light scattering from spin waves using a three-pass single interferometer

The method turns out to be particularly useful in the case of metallic films because of their strong optical absorption. We will elaborate on this in Sect. 8.2 which summarizes the important aspects of the scattering mechanism and intensities, including both theory and experiments. The other advantage of LS, at least compared with MA, is the available wavevector range. Apart from situations where the wavevectors, q, of the spin waves are determined by the sample geometry (for example standing modes in thin films) MA probes only waves with $q \approx 0$. For LS the available range of q is still small but finite $(q \approx 10^5 - 10^6 \text{ cm}^{-1})$. This has an important effect on the types of modes seen by the two methods. Also, the modes seen by MA do not display dipolar coupling in layered structures. These differences will be explained in more detail in Sect. 8.1. We will briefly outline the mathematical procedure, but more emphasis will be given to a physical interpretation of the most important results. In Sect. 8.3 we will discuss what can be learned from the observed mode frequencies in the case of single- and multilayers. Of particular importance will be the magnetic coupling effects which can best be studied on double layers. Finally, in Sect. 8.4 we will discuss the interesting modifications of magnetic properties that can be obtained by layering and the role of LS in their exploration.

8.1 Continuum Theories of Spin Waves in Single Films and Multilayered Structures

In this section the starting equations and approximations for a continuum theory of spin waves in layered structures relevant for LS experiments will be outlined. We use this also for a definition of the parameters which will be employed in the description of the experimental results (see in particular Table 8.1). All equations and parameters are given within the SI system. The standard treatment starts from the equation of motion (see e.g. [8.5, 7–9] and Chap. 2 of this volume)

$$dM/dt = \gamma M \times B_{\text{eff}} \tag{8.1}$$

where M [A/m] is the sample magnetization and B_{eff} [Vs/m^2 = T] is an effective magnetic induction. Instead of M we will also frequently use the quantity $J[\text{T}] = \mu_0 M$, called the magnetic polarization. The gyromagnetic ratio γ is related to the spectroscopic splitting factor g via $\gamma = -g\mu_B/\hbar = -g \times 8.79 \times 10^{10}$ [T^{-1} s^{-1}]. In contrast to microwave absorption, LS is not very useful for the determination of linewidths due to its poorer resolution. Therefore in the following we neglect damping. In this case B_{eff} can be written in the form [8.8]

$$B_{\text{eff}} = B - \nabla_u E_{\text{ani}}/M + (2A/M^2)\nabla^2 M , \tag{8.2}$$

where $B = \mu_0(H + M)$ is the sum of all external, internal and demagnetizing fields. Volume anisotropy and exchange are included via the second and third term on the right-hand side of (8.2), where E_{ani} [J/m³] is the volume anisotropy energy density, A [J/m] is the exchange stiffness parameter and ∇_u denotes the gradient operator in which the differentiation variables are the components of the unit vector $u = M/M$. Instead of A, another parameter, D [Vs], called exchange constant, is sometimes used. It is related to A via $D = 2A/M$, in both the SI and cgs systems. For the anisotropy we obtain for example in a cubic system to first order $E_{ani} = K_1 (u_x^2 u_y^2 + u_y^2 u_z^2 + u_z^2 u_x^2)$ where the coordinates of u are taken along the cubic axis and K_1 [J/m³] is the first-order cubic-anisotropy constant. Likewise, a uniaxial anisotropy in direction n is described by $E_{ani} = -K_u (u \cdot n)^2$. In the general case we use K_b to describe a bulk anisotropy.

H and M are each split into a static part and a dynamic component, $h(r, t)$ or $m(r, t)$, and the time dependence is assumed to be of the form $m(r, t) = m(r)\exp(-i\omega t)$, $h(r, t) = h(r)\exp(-i\omega t)$. This is inserted into (8.1, 2) and only terms linear in $h(r)$ or $m(r)$ are retained. It turns out that for the fluctuations $h(r, t)$ and $m(r, t)$ seen here, Eddy-currents may be neglected, even in the case of metals. Inclusion of these effects would yield corrections on the order of one percent [8.9] which is practically unobservable in LS experiments. From this it follows that

$$\text{curl}\,[h(r)] = 0, \quad \text{div}\,[h(r) + m(r)] = 0 \tag{8.3}$$

is an excellent approximation and $h(r)$ can be expressed as $h(r) = -\text{grad}\,\psi(r)$, $\psi(r)$ is a scalar potential. Furthermore $h(r)$, $m(r)$, and $\psi(r)$ are assumed to be of plane wave form $\sim \exp(iq \cdot r)$.

At the surfaces, the boundary conditions which follow from Maxwell's equations have to be taken into account. From (8.3) one can obtain, in the usual way, the classical boundary conditions:

tangential component of $h(r)$ $\Big\}$ continuous at boundaries. $\qquad$ (8.4)
normal component of $h(r) + m(r)$

Furthermore, due to the exchange, there is a boundary condition of the form

$$M \times (\nabla_u E_{surf} - (2A/M)\partial M/\partial n) = 0 \tag{8.5}$$

which was found by *Rado* and *Weertman* (RW) [8.10] and will henceforth be called the RW condition. Here E_{surf} [J/m²] denotes the surface-anisotropy energy density and $\partial/\partial n$ is the partial derivative in the direction of a unit vector n which is normal to the surface of the ferromagnetic film and at each surface points towards the film interior. Note that surface anisotropy enters via the boundary condition and not via the equation of motion. An important case is an anisotropy perpendicular to the surface, which can be written as $E_{surf}^\perp = -K_s^\perp (u \cdot n)^2$. Note that the easy axis is along n when $K_s > 0$ and in the plane of the film otherwise. Detailed investigations have been performed on $\{110\}$ type surfaces of Fe. In addition to the perpendicular anisotropy just described, one finds a uniaxial anisotropy in the plane. It can be written in the form $E_{surf}^{\parallel} = -K_s^{\parallel}(u \cdot x)^2$ where x is a unit vector along the [001] direction of the {110} plane. Positive values of $K_s^{\parallel}$ mean, that the easy axis is along the in-plane [001] direction and for negative values it is along the in-plane [110]. In the general case we use K_s [J/m²] to describe a surface anisotropy. Within the "homogeneous magnetization approximation" surface anisotropy can be treated as an effective bulk anisotropy, K_b^s, via $K_b^s = 2K_s/d$, where the factor of 2 takes into account the presence of the two surfaces. The validity of this is discussed in [8.8, 11, 32].

The generalization of (8.5) to the boundary between two different ferro-magnets was discussed by *Hoffman* [8.12, 13]. If surface anisotropy can be neglected we obtain:

$$M_1 \times [(A_1/M_1)\partial M_1/\partial n + (A_{12}/M_2)M_2] = 0 , \qquad (8.6a)$$

$$M_2 \times [(A_2/M_2)\partial M_2/\partial n - (A_{12}/M_1)M_1] = 0 , \qquad (8.6b)$$

where the various quantities are defined as before and the two indices 1, 2 refer to the two magnetic materials which couple across the boundary with strength A_{12} [J/m²]. Here n is a unit vector normal to the interface directed towards the interior of the first film. For the limiting case of layers of the same material, which are in direct contact, we denote A_{12} by A_{12}^c and obtain for the relation between A_{12}^c and A [8.14]:

$$A_{12}^c = n_{latt}A/a , \qquad (8.7)$$

where a is the lattice constant. For interfaces parallel to {100} planes of cubic lattices we obtain $n_{latt} = 1$ for sc- and $n_{latt} = 2$ for fcc and bcc type lattices [8.14]. Note, that conceptually we are leaving here the field of continuum theories.

8.1.1 Single Films

a) Dipolar Approximation

The first calculation of spin wave modes with small but finite wavevector q in a single ferromagnetic slab was performed by *Damon* and *Eshbach* (DE) [8.15]. They neglected anisotropy and exchange. We discuss here only the result which they obtained for modes whose wavevector parallel to the sample plane, $q_{\parallel}$, is perpendicular to the polarization J which is also in the sample plane. These modes are observed in LS experiments if the plane of incidence is perpendicular to J. This situation is illustrated in Fig. 8.2a. The result of the DE theory is displayed in Fig. 8.2b. The upper branch corresponds to a surface mode, the so-called DE mode, the lower horizontal line describes the bulk modes. Both coincide at $q_{\parallel} = 0$, which corresponds to the uniform mode in the conventional MA experiment (in MA one can introduce a finite $q_{\parallel}$, and observe the DE mode by using special couplers [8.16]). The DE mode is a surface mode because of its

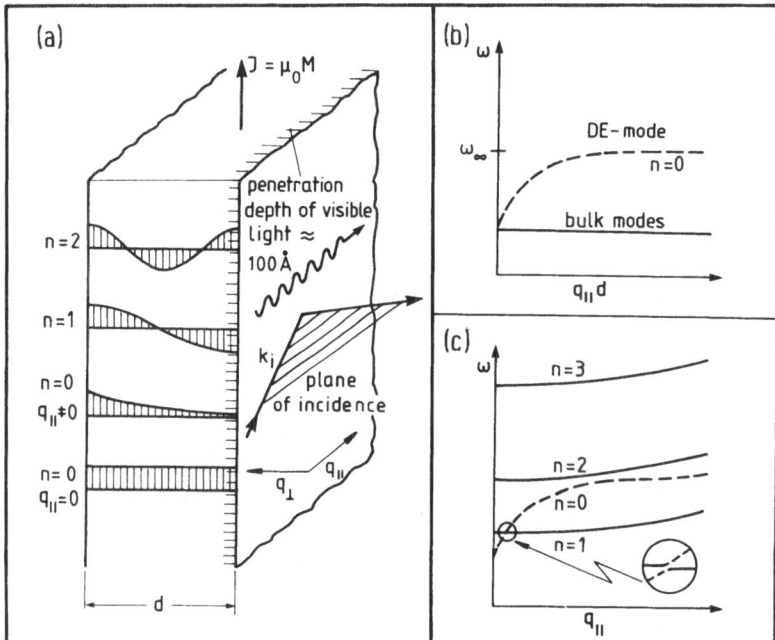

Fig. 8.2. (a) Amplitude profiles of spin waves in thin films and the scattering geometry; (b) mode frequencies in the dipolar approximation; (c) same as in (b) with inclusion of exchange

exponentially decaying amplitude profile as shown in Fig. 8.2a. Modes with opposite $q_{\parallel}$ have their maximum amplitudes at opposite surfaces. (Figure 8.2a shows only the mode profile for $+q_{\parallel}$.) In sufficiently thick samples only one propagation direction is present at each surface (see also Chap. 2 of this volume). The DE mode is therefore called nonreciprocal or unidirectional. For $q_{\parallel} \perp J$ and neglecting anisotropy, its frequency is given by

$$\omega = \gamma \left\{ [1 - \exp(-2q_{\parallel}d)] (J/2)^2 + B_0^2 + B_0 J \right\}^{1/2} . \tag{8.8}$$

In the limit of $q_{\parallel}d \to 0$ it coincides with the bulk modes given by

$$\omega = \gamma [B_0(B_0 + J)]^{1/2} . \tag{8.9}$$

Bulk anisotropy can be included in (8.8, 9) when B_0 is along an easy axis. In this case, only a term of the form $B_{an} = 2K/M$ needs to be added to B_0, where K is the anisotropy constant. In the other case, the formalism has been worked out in [8.17–19].

b) Inclusion of Exchange

The result of exchange on the modes seen in Fig. 8.2b is qualitatively displayed in Fig. 8.2c. The most important effect is that the bulk modes split and shift to higher frequencies. The different branches correspond to different standing-mode character, as displayed in Fig. 8.2a, and can also be characterized by their mode number n. This comes from a quantization of the wavevector $q_\perp$, perpendicular to the film plane. The DE mode here is labeled with $n=0$ because for $q_\parallel \to 0$ it becomes the uniform mode. For $q_\parallel \neq 0$ it has an exponentially decaying amplitude, as shown. Due to the exchange there is now also dispersion of all modes with respect to $q_\parallel$. As indicated, crossing points between different branches are avoided. This is called mode repulsion or anticrossing and is a result of the RW condition (8.5). It turns out that away from branch crossovers its effect on the mode frequencies is only weak and the frequencies are, to a good approximation, given by

$$\omega = \gamma\,[(B_0 + Dq^2)\,(B_0 + Dq^2 + J)]^{1/2} \, , \tag{8.10}$$

where $q^2 = q_\parallel^2 + q_\perp^2$ and $q_\perp = n\pi/d$ [8.7]. If B_0 is along an easy axis, then inclusion of bulk anisotropy is achieved by adding a term of the form $B_{an} = 2K/M$ to B_0. In the general case and for the inclusion of surface anisotropy the formalism described in [8.8, 9] has to be applied.

8.1.2 Multilayers

a) Dipolar Approximation

There has been a considerable amount of work on the collective excitations in multilayers in the dipolar approximation [8.20–26]. To demonstrate the essential features we concentrate on double layers consisting of single films with equal thickness d and polarization J and the case of infinitely many such layers. They are separated by nonmagnetic spacer layers of thickness d_0. Both inside the magnetic films and across the spacer layers only the dipole–dipole interaction is assumed to act as a restoring force on the precessing moments. Anisotropy is first neglected. As in Sect. 8.1.1 we consider only modes whose in-plane wavevector, $q_\parallel$, is perpendicular to J, which also lies in the film plane. Hence we start from the result for the single layer as displayed in Fig. 8.2b. It turns out that only the DE modes located on the single films produce dipolar fields outside of the films and hence interact. For the bulk modes we get flux closure inside the films and they are not affected by dipolar coupling. Hence the bulk modes are still at the same frequency as in Fig. 8.2b and if we choose $B_0 = 0$ as is done in Fig. 8.3 the horizontal bulk mode branch in Fig. 8.2b coincides with the abscissa. In Fig. 8.3 all frequencies have been normalized to the DE mode frequency of a halfspace, ω_∞, as indicated also in Fig. 8.2b.

Figures 8.3a, b show results for double layers in parallel, and in antiparallel alignment of the magnetizations in the two films. For the parallel alignment the gross features of the plot can be understood as follows. Since the two films are assumed to be identical, we have at large values of d_0, where they are decoupled, a double degeneracy at the DE frequency of the single film with thickness d. At

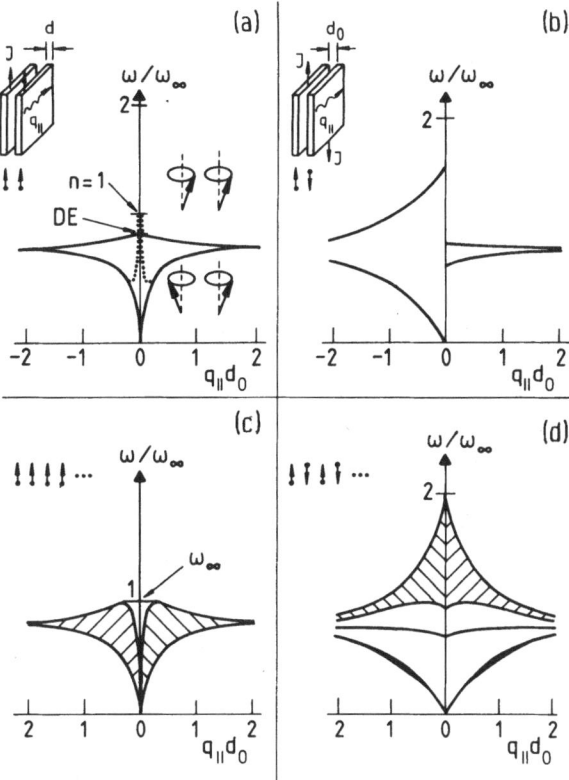

Fig. 8.3a–d. Mode frequencies of double and infinite multilayers in parallel and antiparallel alignment for $B_0 = 0$ and $q_\parallel = 1.82 \times 10^5\,\mathrm{cm}^{-1}$. Bulk branch coincides with the abscissa [8.27, 28, 29]

smaller values of d_0 to two DE modes start to interact and there is a splitting into two branches. In the limit of $d_0 \to 0$ we again obtain a single film, with thickness $2d$, whose DE mode is given by the upper branch. The other branch has to end up at the bulk modes because for a single film there is just one DE mode and the bulk modes with their high degeneracy. The meaning of the dotted line will be explained in Sect. 8.1.2b. For parallel alignment of the magnetization, the plot is symmetric about the ω-axis. This is a result of the twofold rotational symmetry which links $+q_\parallel$ and $-q_\parallel$. This symmetry is broken when the magnetizations on the two films are aligned antiparallel (Fig. 8.3b). As expected, we then get different branches for $+q_\parallel$ and $-q_\parallel$ because the two directions of the wavevector are no longer linked by symmetry. This asymmetry can be used to identify the antiparallel alignment (see Sect. 8.3.2c). Also remarkable is the high frequency of the highest branch for $q \to 0$. In the limit of large d it is at $2\omega_\infty$, twice as high as the corresponding value for parallel alignment.

The nature of these modes has also been studied in more detail [8.27, 28]. One of the results is indicated on the branches for parallel alignment in Fig. 8.3a. For the upper branch we are dealing with an in-phase precession of the magnetization on the two films whereas for the lower branch the precessions are 180° out of phase.

Figures 8.3c,d show the results for infinite multilayers with parallel and antiparallel alignment of the polarizations respectively [8.21, 22, 29]. We see here a close connection to Fig. 8.3a, b. For the parallel alignment the lower branch of Fig. 8.3a is now replaced by a band, which, for $d_0 \to 0$, merges with the bulk branch. This is to be expected because for $d_0 = 0$ we obtain a magnetic halfspace which has only the bulk modes (here at $\omega = 0$) and one surface mode at ω_∞. Somewhat surprising is the behaviour of the surface mode. It exists only for $d_0 \le d$, but there its frequency remains at ω_∞ independent of d_0. At $d_0 = d$ it merges with the bulk modes. For the antiparallel alignment a comparison of Fig. 8.3d with Fig. 8.3b reveals that bands are formed from the highest and lowest branch of the double layers. The lower band of Fig. 8.3d is remarkably narrow and hence the density of states is very high. The two intermediate branches stay non-degenerate and are associated with surface modes. For the infinite multilayer they are symmetric with respect to the ordinate, which has to do with the fact that in this limit, as for any odd number of single layers, such a multilayer again has twofold rotational symmetry around a vertical axis.

For the configurations discussed here, the effect of bulk anisotropy on the multilayer modes has been taken into account in [8.24, 26]. Surface anisotropy in the form of a thickness-dependent effective bulk anisotropy was considered in [8.26]. Due to this, at small values of d the mode frequencies are strongly modified.

b) Inclusion of Exchange

Intralayer exchange is included via the exchange term in the equation of motion (8.1, 2) and the RW boundary condition (8.5). If we consider in addition *interlayer* exchange than at the interface, the RW condition has to be replaced by the Hoffmann boundary condition (8.6). The result of such calculations for the low frequency part of the mode spectra of double layers is seen in Fig. 8.4. Surface and bulk anisotropy have been neglected. In Fig. 8.4a we see mode frequencies vs. A_{12} appropriate for Fe double layers (except for the Fe anisotropy which is neglected). The individual Fe layer thickness is assumed to be $d = 400$ Å and the Fe layers are coupled with strength A_{12} across an interlayer of a nonmagnetic material. The modes can be characterized by amplitude profiles as shown. The arrows indicate to which branch a certain amplitude pattern belongs, but it has to be emphasized, that the patterns on the left-hand side are strictly valid only for $A_{12} \to 0$ and those on the right hand side only for $A_{12} = A_{12}^c$ (8.7). For more details see [8.14].

An important distinction has to be made with respect to $q_\parallel$: For $q_\parallel = 0$, as is the case in MA experiments, the two lowest modes of the single films are the uniform mode and the first standing mode, as illustrated for $n = 0$ and $n = 1$ in Fig. 8.2a. From symmetry considerations it follows [8.29] that in order to obtain symmetry adapted double layer modes the profiles have to be combined in a symmetric and an antisymmetric fashion as shown on the left-hand side of Fig. 8.4a. (We adopt here the assignment of the terms "symmetric" and

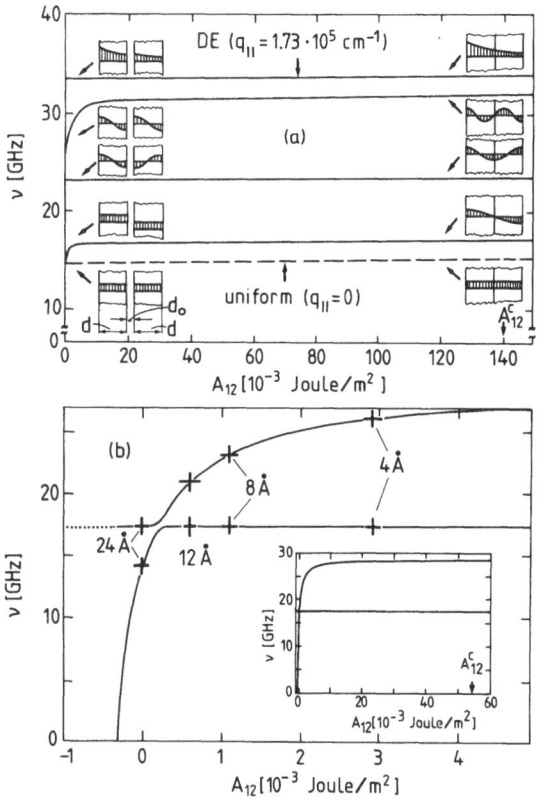

Fig. 8.4. (a) Frequencies of low-lying modes in Fe double layers with individual thickness $d=400$ Å, vs. interlayer exchange A_{12}. Mode profiles as shown are for $A_{12}=0$ and $A_{12}=A_{12}^c$ [8.14, 38]. (b) Frequencies of exchange mode and DE mode vs. A_{12} for permalloy double layers with individual thickness $d=100$ Å. Experimental points are for permalloy double layers interspaced with Pd of thickness d_0 as marked [8.14]

"antisymmetric" to the patterns seen in Fig. 8.4a used by Puszkarski [8.30]). In the fully coupled limit ($A_{12} = A_{12}^c$), where the two films are combined into a single one we are dealing once again with a uniform mode and the standing modes of a single film. Along the branches of Fig. 8.4a the symmetric (antisymmetric) mode character is conserved. Interlayer exchange as obtained from MA experiments has been treated both theoretically and experimentally in [8.31–35].

For $q_{\parallel} \neq 0$, as is the case in LS experiments, the uniform mode (dashed line) becomes the DE surface mode (branch marked "DE") with an upwards frequency shift whose size depends on $q_{\parallel}$, d, d_0. The mode profile changes from uniform to exponentially attenuated. In the same way as in Fig. 8.3a, we now also have to consider the dipole–dipole interaction and we have to realize that a change in A_{12} will be connected with a change of the interlayer thickness d_0. As will be seen in the experimental section, we obtain $A_{12} \neq 0$ only for $d_0 \lesssim 50$ Å. In this range of d_0 the dipolar coupling remains practically in saturation and is given by its value for $d_0 = 0$. Hence the branch marked DE is independent of d_0. Apart from branch crossings (see below) it is also independent of A_{12} in the same way as the dashed line from which it originates. Standing modes, even for $q_{\parallel} \neq 0$ are not strongly affected by the dipolar interlayer coupling [8.35, 36] which is due to the fact that their net precessing moment cancels. Therefore, in going from $q_\parallel = 0$ to $q_\parallel \neq 0$, to a good approximation, the only effect on the branches in Fig. 8.4a is to replace the uniform mode branch by the DE branch. A quantitative analysis of the exchange- and dipolar-coupled double layer has recently been performed in [8.14, 37, 38].

In order to obtain values A_{12} from LS experiments, it is appropriate to reduce the thickness of the ferromagnetic films as compared to those used for Fig. 8.4a. This has been done in Fig. 8.4b. Parameters appropriate for $Ni_{0.8}Fe_{0.2}$ (permalloy) have been chosen and the individual thickness, d, of the permalloy films is 100 Å. Due to the smaller d, the DE mode branch is shifted to smaller, and all standing modes to higher frequencies. In Fig. 8.4b therefore the DE mode is in the range of the lowest antisymmetric branch. For increasing A_{12} the upwards shift of the antisymmetric mode, which we also call the exchange mode is much stronger than in Fig. 8.4a which is important for experiments. The insert displays the result for a wider range of A_{12} values. As in Fig. 8.4a in the limiting case $A_{12} = A_{12}^c$ the exchange mode is rather insensitive to A_{12}. Real shifts occur only for much smaller values of A_{12}. There is a physical reason for this. We have to consider the two moments which interact across the interface as part of a long chain of moments. Making the interaction between them very (or even infinitely) strong causes them to precess always in unison but for the whole chain this is relatively unimportant because the phase change from one spin to the next is only small anyway. Weakening this interaction has a much greater effect because the chain finally falls into two weakly coupled parts.

Another interesting feature is the mode repulsion at branch crossings. As will be discussed in Sect. 8.3.1c, such a behaviour is related to the symmetry of a system and is more likely to occur if the symmetry is low.

The calculation has also been extended to negative values of A_{12}, which shows that the exchange mode goes to zero at some value. This means that, at this point, the system becomes unstable and one obtains canting of the static moments, as was shown by means of microscopic theory [8.39]. To the left of this point therefore the calculation becomes meaningless because one important assumption for the calculation shown in Fig. 8.4 is parallel alignment of the static magnetizations of the two films. (In this range the DE branch has been drawn as a dotted line). Anisotropy in this calculation was neglected. Apart from the mode-crossing point, the DE mode is thus, to a good approximation, given by (8.8) where we have to insert for d the total thickness of 200 Å. For $A_{12} = 0$ the exchange mode is given by (8.9) and for $A_{12} = A_{12}^c$ by (8.10). This follows from the qualitative discussion given above.

Multilayers consisting of more than two single films have also been considered [8.33–35, 40–42]. Since there is not much correlation with experimental results presented here we do not want to elaborate on this point. Note, however, that including interlayer exchange, the limiting case of Fig. 8.3c is a bulk ferromagnet and of Fig. 8.3d a bulk antiferromagnet. Recently, antiferromagnets have also been studied theoretically both for the semi-infinite case and for thin films, and the existence of surface modes was established [8.43, 44].

8.2 Scattering Intensities

8.2.1 The Nature of the Scattering

Light scattering from an excitation in a solid can always be observed if the excitation causes a fluctuation of the optical constants of the material. In the case of spin waves this is achieved via the magneto-optic interaction [8.5]. Classically, it can be understood as the fluctuation of the transverse polarizability of a medium due to the Lorentz forces caused by the fluctuation m $= m(r, t)$, in a way similar to the elasto-optic effect discussed for acoustic phonons in Chap. 3 of this volume. In the magnetic case, however, the polarizability is both perpendicular to the E-vector of the incoming light and to m. This is why the scattering tensor here is always off-diagonal. The scattered light is observed with its polarization perpendicular to that of the incident light. Furthermore one has to consider the transverse nature of electromagnetic waves. For geometries where m can only generate a polarizability parellel to the light propagation it remains in this simple picture ineffective and (weak) LS can only occur due to higher order magneto-optic effects. This is for example the case if the wavevector of the incoming light, k_i, and the static magnetization, M, are collinear. For the interaction of the light with the static component M, this geometry is very favourable and yields the Faraday effect in transmission and polar Kerr effect in reflection. The difference comes from the fact that $m \perp M$. For LS experiments it is therefore more favourable to use geometries where $k_i \perp M$ as shown in Figs. 8.1, 2a.

Hence, in LS from spin waves it is dominantly the electric component of the electromagnetic radiation, E, which interacts with m. In contrast to this, in MA by spin waves it is the magnetic component of the radiation which interacts. This difference is of practical importance for the observation of standing modes in thin metallic films. We have to consider that the penetration depth of visible light into metals is on the order of 10 nm, whereas for the magnetic component of a microwave it is on the order of µm. Both the microwaves and the visible light interact with the net fluctuating moment that they 'see". Let us assume that the film shown in Fig. 8.2a has a thickness of approximately 100 nm and is penetrated entirely by a microwave. From inspection of the standing mode patterns and the small penetration depth of visible light, as indicated in Fig. 8.2a, it is clear that the moment seen by a microwave cancels out, but not so for the moment seen by the light. Hence, thanks to the strong optical absorption, LS from spin waves in thin metallic films is an "allowed" effect, whereas for MA we need disturbances like inhomogeneities or pinning [8.30] in order to see an effect. These considerations suggest that, for the detection of standing modes in very thin films, LS might be superior to any other method. This was indeed shown by the observation of standing modes in films as thin as 4 nm [8.45], roughly an order of magnitude below the thickness where standing modes can still be detected by means of MA. Another reason for this lies in the high frequencies of such modes in very thin films. In order to observe them in an absorption experiment one would have to employ far infrared spectroscopy, with the associated small incident intensities. Because of the large frequency shifts, the experiments in [8.45] were performed with a Raman spectrometer. Hence we see that a distinction between Brillouin and Raman scattering in the present case is not justified by the underlying physics (as for example in the case of acoustic vs. optical phonons) and it is more appropriate to use the more general term light scattering.

8.2.2 Light Scattering Spectra from Single Films

a) Experiments on Thermally Excited Waves

The theory of light scattering from modes in single films including the effect of exchange has been worked out by applying the fluctuation-dissipation theorem [8.46, 47] and based upon concepts from statistical mechanics [8.9, 48]. A typical feature of LS from spin waves are Stokes (S)–anti-Stokes (aS) anomalies which are not due to thermal population. They can either be due to magneto-optic effects [8.1, 5] or they may reflect genuine spin wave properties. A typical example of the latter kind is observed in LS from the DE mode in a thick nontransparent sample where only scattering at one side of the elastic line, either S or aS, is observed [8.49]. The explanation for this behaviour comes from the fact that classically S- and aS-scattering are associated with opposite propagation directions of a collective excitation. Hence the result just described demonstrates the unidirectional propagation or "nonreciprocal behaviour" of the DE mode on the surface of a sufficiently thick sample, as described in Sect. 8.1.1a. The other DE mode, travelling in the opposite direction is localized on the backside and is not seen in this experiment.

Surprisingly in LS from thin Fe-films this anomaly is only partly lifted [8.50]. Here the two DE modes travelling in opposite directions should penetrate the sample entirely and, from this point of view, they should show up with equal intensity on the S- and aS-side, which however is not observed experimentally. It was shown in [8.50] that this effect has to do with a different interaction of the light with the two DE modes travelling in opposite directions.

Another cause of S-aS anomalies can be surface anisotropies, which can also effect the overall scattering intensities. This was theoretically investigated [8.47] for various interesting cases. One example, together with corresponding experiments, is shown in Fig. 8.5. A typical feature of these spectra is the intensity ratio of the first and second standing modes on each side of the elastically scattered laser line. We see that in Fig. 8.5a, b A_1 is stronger than A_2 but S_2 is stronger then S_1. The surface mode (SM) is observed on the Stokes side. This situation reverses upon reversal of the external field, as shown in Fig. 8.5c. Generally, the first standing mode is always weaker than the second on the side of the elastic line where the surface mode appears. This effect has been observed in all LS experiments from metallic films in this thickness range which have been

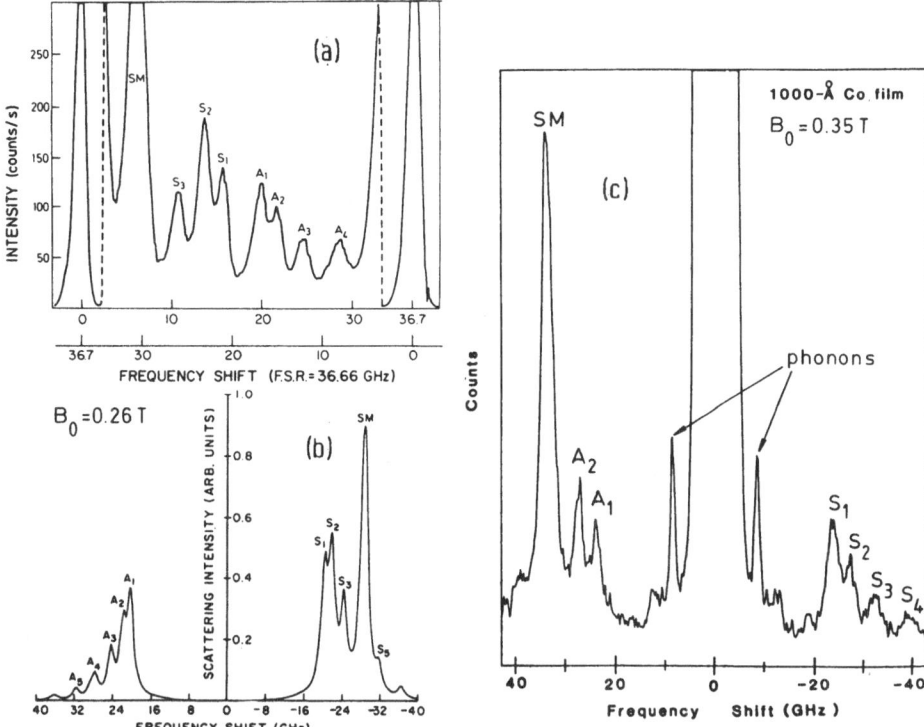

Fig. 8.5a–c. LS spectra from thin films, showing DE mode and various standing modes. Theory (**b**) [8.47] is compared with spectra from $Fe_{0.8}B_{0.2}$ (**a**) [8.51] and from Co (**c**) [8.52]. In (**a**) the peaks marked with S and A are shifted from the elastic order on the right-hand and left-hand sides respectively

published so far. (For references see Table 8.1.) A more detailed analysis [8.47] revealed that this is a consequence of the RW condition when $K_s = 0$. Hence, whenever this effect appears, it is a good indication that surface anisotropy is small. As will be seen at the end of Sect. 8.3.1b, the fact that these films were oxidized is of importance. From the results of [8.47], the general conclusion can be drawn that the scattering intensities react more sensitively to surface anisotropy than do the mode frequencies. Strong intensity changes also occur at points, where the DE and the $n = 1$ mode are close to degeneracy. This case has been investigated in [8.48].

Scattering intensities can also be changed due to multiple interference effects or for thin enough films on certain substrates [8.53]. These cases are characterized by a suppression of the metallic reflectivity of a sample. It is well known that antireflection coatings can be used to enhance magneto-optic effects and hence there can also be a considerable enhancement of LS intensities. The reflectivity is also suppressed when the film is thin enough and is on a substrate with a small reflectivity. This is because, in the limit of $d \to 0$, we must obtain the small reflectivity of the substrate. In such a case there is also an enhanced interaction of the light with the magnetic film which can lead to a remarkable increase of the LS intensity.

The fact that we are dealing here with spin waves and not for example with phonons is also of importance: Spin waves in thin magnetic films on nonmagnetic substrates are confined to the film. In contrast, for acoustic phonons we always get a coupling of film and substrate modes and hence the amplitudes are spatially distributed over the whole film–substrate system. The total amplitude of a mode is connected with the number n_m of excited bosons (here magnons), which is given by

$$n_m = \frac{1}{\exp(\hbar\omega/k_B T) - 1} .$$ (8.11)

Remarkably, this expression is independent of the sample volume. As long as ω and T are constant, the mode amplitude of a thick and a thin film are the same. For decreasing film thickness, the fluctuation $m(r, t)$ particularly that at the surface, thus has to increase. For films thick enough not to be penetrated completely by the light, this leads to an increase of the LS intensity when the film thickness d decreases. When d is below the penetration depth of the light the intensity will again decrease. For more detailed discussions of these effects see [Ref. 8.4, Sect. 5D], [8.53] and [Ref. 8.48, p. 308].

b) Experiments on Pumped Waves

LS from pumped spin waves allows direct investigation of spin wave decay. The concept is a straight forward and effective one. We pump particular spin waves by means of microwaves, and observe their amplitude increase or the waves into which they decay by means of LS. In the LS observation a diaphragm can be used to select waves with a particular wave vector. Earlier experiments on bulk samples are described in the review article of *Borovik-Romanov* and *Kreines* [8.2]. More recent work on thin films is due to *Wettling* and *Jantz* [8.54] and *Patton* and coworkers [8.55, 56]. The method is also suitable for the investigation of amplitudes and decay of spin waves in microwave device structures. The good focussing properties of the probing laser beam, allowing the study of spatial amplitude variations, is an interesting aspect. So far, most work has been on films or thin plates of yttrium iron garnet and similar compounds, which are characterized by very low spin wave damping. (This is why they are of interest for devices.) Usually strong excitation of surface modes with frequency ν_{DE} and decay into modes with $\nu_{DE}/2$ is observed. An example is shown in Fig. 8.6. In part (a) we see a LS spectrum obtained at low microwave pump power. The peaks labelled "S" come from the DE mode in a close to forward scattering geometry. In part (b), at high pump power, there is strong excitation of modes at $\nu_{DE}/2$, labelled "P" whose intensity is even stronger than that from the surface mode itself. Thin films of Fe [8.54] and of permalloy [8.56] have also been investigated by this method and an increase of the surface mode intensity was observed, but

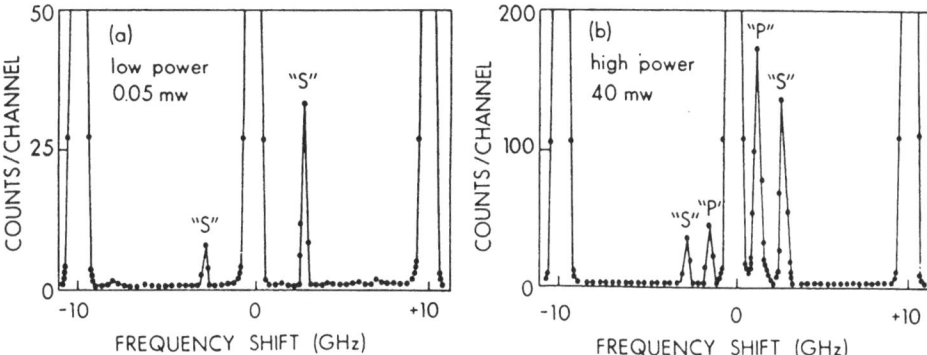

Fig. 8.6a, b. LS from spin waves in a YIG film and the effect of microwave excitation at 3 GHz with low, (**a**), and high, (**b**), power. The peaks labelled "*S*" and "*P*" correspond to the DE mode at the pump frequency and half frequency spin waves, respectively [8.55]

not the decay into $\nu_{DE}/2$ modes, which is due to the larger spin wave damping in the metals. This might change in the future, when higher driving powers will be applied.

8.2.3 Light Scattering Spectra from Multilayers

Scattering intensities from modes in multilayers have been worked out theoretically in [8.20] in the dipolar limit and can be compared with experimental results from various groups [8.21, 26, 57–59]. The gross features expected from Fig. 8.3c indeed show up in the spectra. In Fig. 8.7 we show for comparison LS

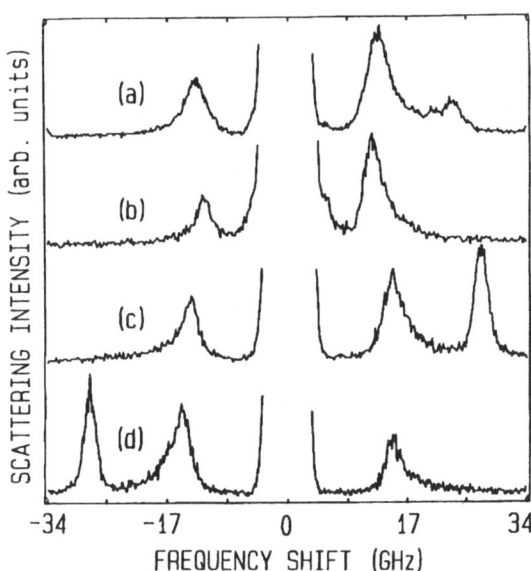

Fig. 8.7. LS from Fe/Pd superlattices in an applied magnetic field of 0.1 T: (*a*) $d = 21.9$ Å and $d_0 = 24.3$ Å, (*b*) $d = 41.7$ Å and $d_0 = 138.7$ Å, and (*c*), (*d*) $d = 41.0$ Å and $d_0 = 9.1$ Å. In (*d*) the direction of the applied field has been reversed as compared to (*c*) [8.26]

spectra from Fe/Pd multilayers. For $d_0 \approx d$ [trace (a)] a broad asymmetric band of collective spin waves is observed. The density of states is maximum at the lower band edge and decreases towards the upper band edge with some superimposed peaks due to the finite number of layers. For $d_0 \gg d$ [trace (b)] the band width is much reduced due to the reduced dipolar coupling across the nonmagnetic spacer layers. For $d_0 \ll d$, one also observes the surface mode, visible in traces (c) and (d) as an additional peak at higher frequencies. It is identified as nonreciprocal surface mode by its behaviour when B_0 is reversed, at which it switches from S to aS.

Recently an analysis of LS intensities has been performed for double layers including both intra- and interlayer exchange [8.37, 60]. Of particular interest are mode crossings as seen in Fig. 8.4c. The calculation presented in [8.60] yielded the result that far away from the cross points the DE mode is expected to be appreciably stronger than the exchange mode, as observed in the experiments. Close to the crossing region there is a point where they should be equally strong, a fact that results from the mode crossing. A more detailed analysis of this effect, however, will also require a distinction between S and aS scattering because on single layers it was found [8.61] that close to a mode crossing, one mode can be strong on the S- and the other on the aS-side.

8.3 Evaluation of Mode Frequencies

8.3.1 Single Films

a) Determination of Bulk Magnetic Parameters

For reasons described in Sect. 8.2.1 in LS experiments it is favourable to have M in the sample plane. Therefore B_0 is mostly applied in the sample plane and the light with wavelength λ_L is incident under an angle α with respect to the surface normal. The in-plane wave vector, $q_{\parallel}$, of the excited wave then is given by $q_{\parallel} = 2 \, (2\pi/\lambda_L) \sin \alpha$. If $\alpha = 45°$, (Fig. 8.1) we obtain $q_{\parallel} = 1.73 \times 10^5$ cm^{-1} for $\lambda_L = 514.5$ nm and $q_{\parallel} = 1.82 \times 10^5$ cm^{-1} for $\lambda_L = 488$ nm. Further details are given, e. g., in [8.62, 63]. From spectra such as those shown in Fig. 8.5a, c, mode frequencies can be determined and can be used together with the formalism outlined in Sect. 8.1.1 for a determination of magnetic parameters. In Table 8.1 the parameters obtained so far from such experiments have been collected, together with the corresponding references.

Of particular advantage is the application of the LS technique to very thin films because, as mentioned in Sect. 8.2.1, it is currently the only method to pick up the standing modes in this case. For small d, the exchange term in (8.10) becomes dominant and we obtain a linear relation between ω and d^{-2} whose slope is given by $\gamma D(n\pi)^2$. For values of d below about ≈ 10 nm, the frequencies become too high for conventional Fabry-Pérot Brillouin spectroscopy (BS).

Table 8.1. Magnetic parameters as obtained by means of light scattering. For definition of parameters see Sect. 8.1. Quantities are given in SI and cgs units

Material	J [T] / 10^4 [G]	g	D 10^{-17} [Vs] / 10^{-9} [G cm^2]	K_b 10^3 [J/m^3] / 10^5 [erg/cm^3]	K_s 10^{-3} [J/m^2] / [erg/cm^2]	Ref.
Fe bulk	2.12	2.09				[8.62]
Fe	2.11	2.09				[8.46]
Fe	2.15	2.1	2.2	$K_1 = 4.7$		[8.63]
Fe		2.1	2.34			[8.45]
Fe	0.7		2.13	1.67		[8.60]
Fe	2.1			3.1		Fig. 8.12
Fe (100) on GaAs	2.0		2.61	$K_1 = 4.1$ $K_u = 0.8$[a]		[8.17]
Fe (110) on GaAs	1.48–1.57			$K_1 = 2.1,\ 2.5$ $K_u = 1.0 \ldots -3.8$[a]		[8.17]
Fe (110) on W	1.8				$K_s^\perp = -2.8$	[8.18]
					$K_s^\parallel = -0.024,\ -0.028$	[8.8]
...Ag/Fe...		≈ 2.0		$(4\pi DM_s)_{eff} \approx 0$[a]		[8.68]
...Cu/Fe...fcc Fe		1.42		$(4\pi DM_s)_{eff} = -6.5 \ldots -2.7$ kG[a]		[8.68]
				$4K_u^{(2)}/M_s = 0.23 \ldots 5.60$ kOe[a]		[8.19]
Bulk Fe/Au	2.1408	2.071			$K_s^\perp = 0.54$	[8.70]
Bulk Fe/Ag					$K_s^\perp = 0.79$	[8.70]
...Pd/Fe...	1.7				$K_s^\perp = 0.15$	[8.26]
...W/Fe...	2.01					[8.59]
Ni bulk	0.603	2.21	2.95			[8.62]
Ni	0.58	2.2				[8.63]
...Mo/Ni...	0.352–0.603					[8.91]
Ni$_{0.8}$Fe$_{0.2}$	0.9	2.2	2.09			[8.63]
Ni$_{0.8}$Fe$_{0.2}$	1.0	1.98	2.41			[8.77]
Co	1.68	2.16	2.7			[8.52]
...Nb/Co...	1.57–1.66	2.13				[8.58]
Co$_{0.8}$Cr$_{0.2}$	0.99–1.16	3.1–3.4	0.7	$H_k = 0.6 \ldots 1.0$ kOe[a]		[8.76]

[a] For a definition of these quantities see the original literature.

320   *P. Grünberg*

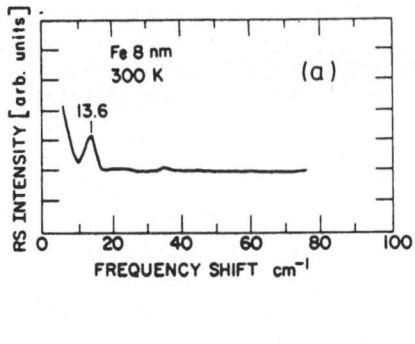

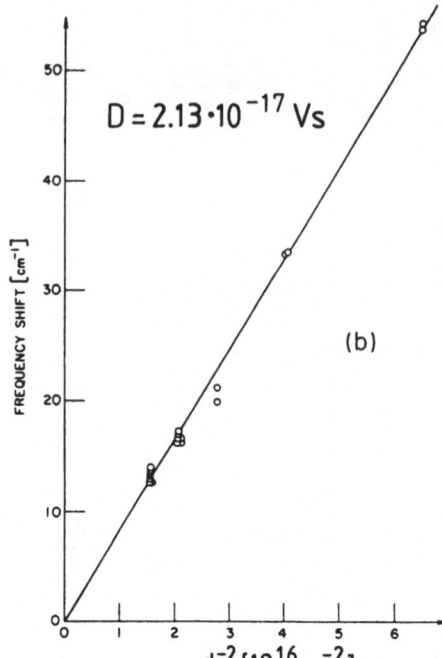

$$D = 2.13 \cdot 10^{-17}\ Vs$$

Fig. 8.8. (a) Raman spectrum from a 8 nm thick Fe film, showing second standing spinwave mode. (b) Mode frequencies obtained by Raman spectroscopy vs. d (1 cm^{-1} $\cong$ 30 GHz) [8.45]

Basically BS could also be extended into this frequency range but there is a practical reason for preferring Raman spectroscopy (RS). This is the simple fact that RS allows streak-like illumination of a sample to be employed. This is achieved by means of a cylindrical lense, whereas in BS the exciting laser light has to be focussed onto a very small spot. Hence heating effects at high incident powers are more critical in BS and this, in addition to convenience, is a good reason to prefer RS here. Even for RS, however, sampling times of a few hours are necessary in order to obtain spectra of the kind as shown in Fig. 8.8a. In Fig. 8.8b the evaluation based on (8.10) is displayed, where it has been assumed that the mode observed in the spectrum of Fig. 8.8a is the one with $n=2$. Interpreting it as the $n=1$ mode would lead to unreasonably high values of the exchange constant D. The $n=1$ mode may possibly be hidden in the Rayleigh tail. The value of D obtained from these experiments is compared in Table 8.1 with values from other sources.

b) Surface and Interface Anisotropies

Within the feamework of Sect. 8.1 the distinction between bulk and surface anisotropy is clear: bulk anisotropy enters in the equation of motion, surface anisotropy in the boundary conditions. Experimentally, such a clear distinction is not always possible. Consider, for example, an anisotropy introduced by surface strains. The strains and the associated anisotropy will certainly not be restricted to the surface but will reach to some extent into the bulk. It seems adequate to treat such a case by introducing a thickness dependent effective bulk anisotropy, an approach which has frequently been adopted, e.g. in [8.17–19]. *Rado* and *Hicken* [8.8] have recently reinvestigated earlier data by *Hillebrands* et al. [8.18] which were obtained this way and found that considering surface anisotropy properly as part of the boundary condition and not as an effective bulk anisotropy yielded practically the same result. Nonetheless care is needed, because this might not be generally true. A detailed theoretical treatment of surface anisotropy, adapted to the case of microwave absorption has been given by *Puszkarksi* [8.30].

A particularly interesting case which has been investigated by various groups and methods, including LS experiments, is the anisotropy of $\{110\}$-type Fe films on GaAs and on W. Figure 8.9 shows the frequencies of the DE mode as a function of the angle between the magnetization and the crystallographic [100]-direction for thin films of Fe on GaAs, grown epitaxially parallel to $\{110\}$. Results for two values of d are displayed. In these plots the highest frequency indicates the direction of the easy axis. Hence we see immediately that for the film with $d = 1900$ Å [part (a)] the easy axis is along [100] as expected from bulk Fe. For the film with $d = 90$ Å [part (b)], however, the anisotropy has changed; the easy axis is now along [110]. To describe this behaviour, in addition to the usual cubic anisotropy of Fe described by K_1, an additional in-plane anisotropy described by K_u had to be introduced. Here K_u is treated as a thickness-dependent bulk anisotropy. For $d = 90$ Å, K_u causes the easy axis to be along the in-plane [110] direction. This change of the easy axis from [100] to [110] for $d \leq 100$ Å was found previously by means of ferromagnetic resonance [8.65, 66]. It also appears for epitaxial Fe layers on $\{110\}$ W. These have also been investigated by means of LS and the spectra are shown in Fig. 8.10. Note the strong upwards shift of the DE mode for $d \leq 40$ Å. From the DE theory [8.15] we would expect a down-shift (compare Fig. 8.2b) as indeed seen in the spectra for $d \geq 40$ Å. A quantitative analysis reveals that the upwards shift is caused by superposition of an in- plane and an out-of-plane anisotropy. The parameters are quoted in Table 8.1 and agree with results from conversion electron Mössbauer spectroscopy (CEMS) [8.67].

Fe on GaAs has a very good lattice match (only 1.4% mismatch), whereas for Fe on W there is an appreciable mismatch of $\approx 10\%$. The peculiar anisotropy which causes the switching of the easy axis was therefore believed to be due not to lattice strains [8.67] but, at least in part, to growth steps parallel to the in-plane [110]-direction [8.18]. Apart from a clarification of certain sign conventions, the analysis of *Rado* et al. [8.8] revealed that the normal to the film plane is a hard (easy) axis for $\{110\}$ Fe on W (GaAs). However, the form anisotropy always holds the magnetization in plane where the switching effect occurs.

The anisotropy of $\{100\}$-type Fe surfaces and thin films has also been investigated recently by means of LS, assisted by various other methods [8.19, 68–71]. The results show that, for this orientation, very thin Fe films lose their bulk anisotropy. What remains is a strong anisotropy perpendicular to the surface which also shows some thickness dependence. This has been studied for

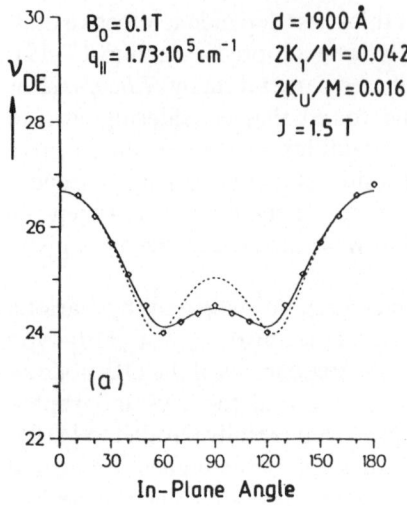

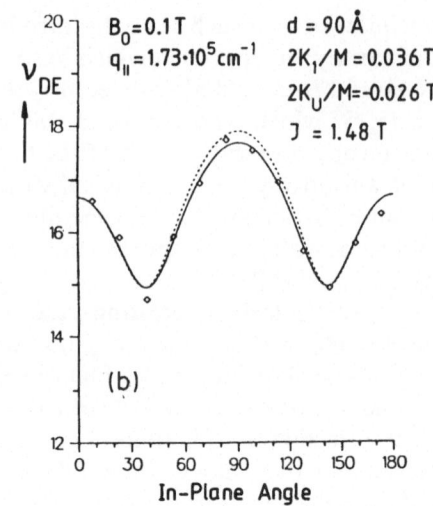

▲

Fig. 8.9a, b. DE mode frequencies from epitaxial (110) Fe films as a function of the angle between the magnetization and an in-plane [100] direction. Film thickness is $d=1900$ Å in (**a**) and $d=90$ Å in (**b**) [8.17]

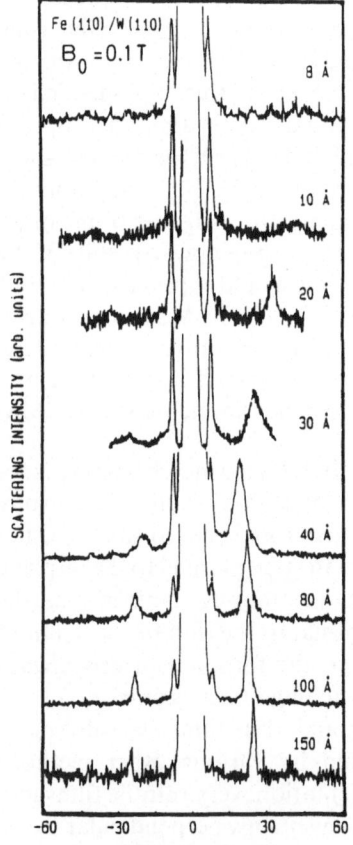

Fig. 8.10. DE mode in very thin films of epitaxial (110) Fe on (110) W. The narrow line close to the elastic scattering is a phonon [8.18]

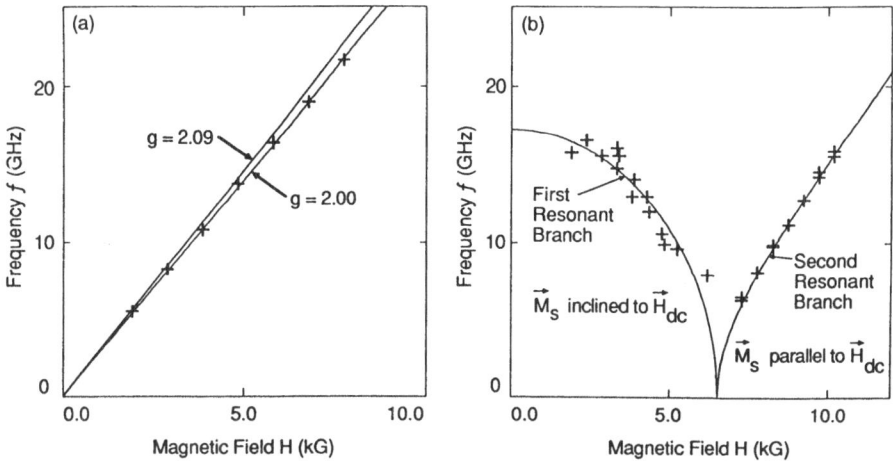

Fig. 8.11a, b. LS from the DE mode in very thin {100} Fe films: (**a**) 3 monolayers (ML) of bcc Fe sandwiched between Ag, (**b**) 3 ML fcc Fe sandwiched between Cu [8.68]

free surfaces, surfaces covered with Au and Ag, and for fcc-type Fe films sandwiched between Cu layers. Figure 8.11 shows plots of DE mode frequencies vs. external field from three monolayers of Fe, sandwiched between Ag layers and between Cu layers. The Fe between Cu has fcc structure. Below $B_0 = 0.65$ T, the magnetization tilts out of the sample plane, leading to an increase of the mode frequency. For decreasing B_0 the tilt increases. (At the same time the LS intensities decrease, as expected from the arguments given in Sect. 8.2.) Similar results are obtained from multilayers of Fe and Cu. The anisotropy of a {100}-type Fe surface has also been investigated by means of LS from a bulk crystal (whisker) covered with Au and Ag [8.70]. In order to increase the accuracy in the determination of the mode frequency, the laser was foccussed onto a spot containing two oppositely oriented domains, separated by a domain wall. Due to this the DE mode appeared on the S and aS side, thus improving the precision in the evaluation of the spectra. Comparison of these results with results obtained by means of FMR [8.71] and with thin-film data [8.69] show good agreement. For bcc Fe the anisotropy is strongest for the free surface, decreases upon coverage with Ag and further decreases upon coverage with Au. Slight oxidation is sufficient to reduce the anisotropy considerably. Thin films of fcc Fe sandwiched between Cu also show a strong perpendicular anisotropy. Data on surface anisotropy which have so far been obtained by means of LS are summarized in Table 8.1.

c) Mode Repulsion Experiments

Mode repulsion at branch crossings as indicated in Fig. 8.2c has produced some excitement ever since the discovery of the RW condition (8.5) from which it results. Since it depends on the surface anisotropy, it can be used for a determination of the latter. Group-theoretically, it has to do with the low

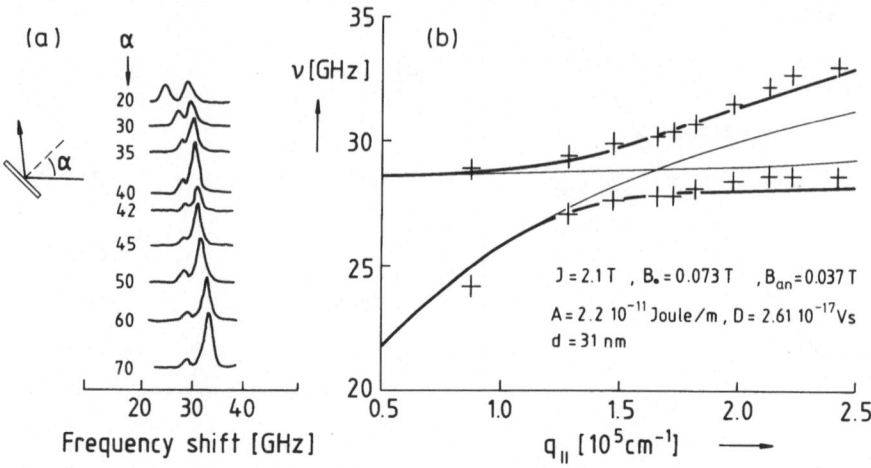

Fig. 8.12. (a) LS spectra from a thin Fe film for different values of $q_\parallel$ displaying mode repulsion at branch crossings (see also Fig. 8.2c). (b) Mode repulsion in comparison of theory and experiment [8.75]

symmetry of an in-plane magnetized film under inclusion of the dipolar interaction [8.35]. In such a low symmetry system all states belong to the same representation and they interact, which is the reason for mode repulsion. For a perpendicularly magnetized film there is a higher symmetry due to the presence of a mirror plane and there is repulsion (anticrossing) only between states with the same parity [8.35].

Experimentally, mode repulsion was first observed in MA experiments on YIG films for magneto-exchange branches with the same parity [8.73]. In LS experiments it was seen for the crossover of the DE mode and the first standing spin wave mode in Fe films [8.61, 74, 75] and films of Co_xCr_{1-x} [8.76]. An example is displayed in Fig. 8.12. We see here LS spectra taken for different values of the angle α between sample normal and incident light. This results in different values of $q_\parallel = [4\pi/\lambda_L] \sin\alpha$, where λ_L is the wavelength of the exciting laser light. In Fig. 8.12b the experiment is compared with the theory. For the crossing thin lines the RW condition (8.5) was neglected, whereas for the thick ones it was included, under the assumption $E_{surf} = 0$. The agreement of the experiment with the theory including the RW condition is another indication that $E_{surf} = 0$ is a good approximation here, as is expected for oxidized films (see end of Sect. 8.3.1b).

8.3.2 Double Layers

a) Modes in Double Layers with Dipolar Interlayer Coupling

The coupling of two ferromagnetic films via the dipole–dipole interaction of the precessing moments can be studied by means of LS from spin waves. The modes

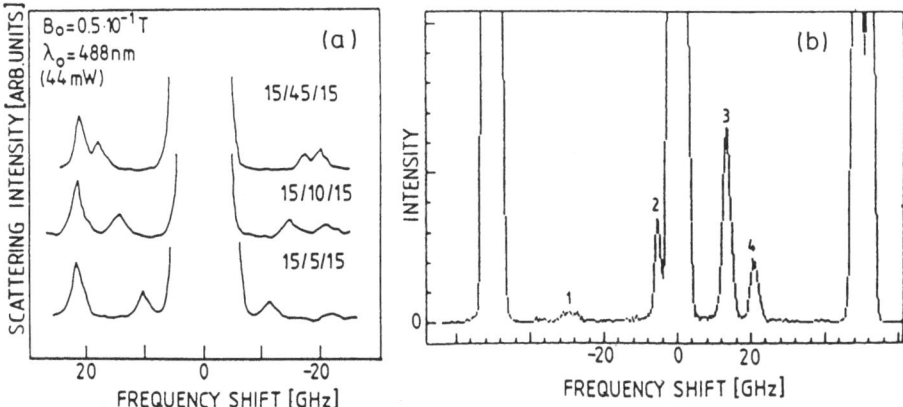

Fig. 8.13a, b. LS spectra displaying the dipolar coupling. (a) For parallel alignment as in Fig. 8.3a and various d_0 [8.61]. (b) For antiparallel alignment as in Fig. 8.3b [8.83]. In (a) thicknesses are given in the notation $d/d_0/d$ [nm]

seen by MA do not show such effects, at least not in the configurations considered here. Figure 8.13a displays spectra from double layers in parallel alignment with decreasing interlayer thickness d_0, from top to bottom. We see an increase of the splitting, which is due to the increase of the dipolar coupling when d_0 is reduced.

Dipolar coupling is a purely geometrical effect. One may therefore ask what new information can be derived from spectra such as Fig. 8.13a, other than can already be obtained from single films. Clearly the difference lies in the presence of an interface and, besides interlayer exchange (to be treated in the next section), new anisotropies can result. We will come back to this aspect in Sect. 8.3.3.

Before we proceed to the inclusion of interlayer exchange, however, a case should be mentioned where a spectrum that reflects the dipolar coupling provides new information which is otherwise hard to obtain. This is the case of a double layer with antiparallel alignment. It has recently been studied on double layers of permalloy [8.77] and good agreement between theory and experiment was established. Another example is displayed in Fig. 8.13b. This spectrum was obtained from a double layer of Fe in antiparellel alignment and shows the typical features expected from Fig. 8.3b. Note in particular the different mode frequencies on the S and aS sides. These are due to the fact that, classically, S and aS scattering on a collective excitation are correlated with opposite wavevectors and we have given a symmetry argument (see Sect. 8.1.2a) as to why these should be associated with different frequencies. As will be seen in Sect. 8.3.2c, such a spectrum can be of great value in identifying antiparallel alignment in magnetic double layers. Modes with a certain frequency and wave vector travelling only in one direction are called unidirectional or nonreciprocal and are of interest for microwave processing devices [8.78].

b) Modes in Double Layers with Positive Interlayer Exchange Coupling

We call the interlayer exchange coupling positive (negative) if it leads to ferromagnetic (antiferrogmagnetic) alignment of adjacent magnetic films and treat first the case of positive interlayer coupling. In the schemes shown in Fig. 8.4 we are then at positive A_{12}. Two experimental examples are displayed in Fig. 8.14. The thickness of the individual permalloy films was 100 Å, as assumed for the calculation in Fig. 8.4. The line marked with an arrow is the exchange mode. The lowest traces represent the exchange-decoupled cases where the position, given by (8.9), is marked by a dashed line. In the upper traces, which represent full coupling, it has shifted to the position given by (8.10). The other mode is the DE mode and does not shift, as expected. To give a feeling for the values of A_{12} that result for a certain interlayer thickness, the mode frequencies obtained from Fig. 8.14 for Pd interlayers are including in Fig. 8.4b and the corresponding thicknesses are indicated. The value for full coupling, A_{12}^c, is

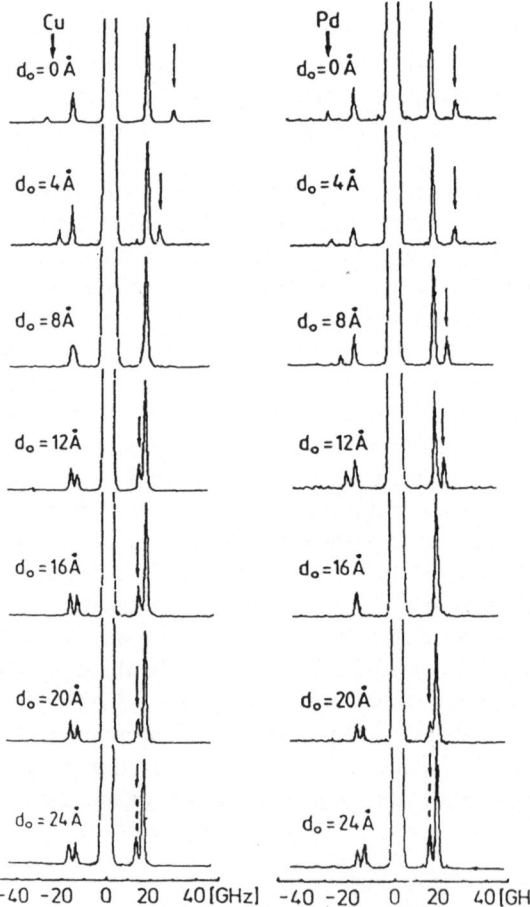

Fig. 8.14. LS spectra from double layers of $Ni_{0.8}Fe_{0.2}$ with Cu and Pd interlayers of different thickness d_0. The position of the exchange mode, marked with an arrow, depends on the size of the interlayer exchange [8.6]

calculated via (8.7) from the permalloy exchange constant $A = 10^6$ erg/cm and lattice constant $a = 3.54$ Å using $n_{latt} = 2$.

Closer inspection of Fig. 8.14 reveals that, for a given interlayer thickness, the exchange coupling of the permalloy films is stronger if the interlayer is of Pd than if it is of Cu. This can be compared with the exchange interaction of magnetic atoms in diluted alloys. Here the magnetic ordering temperature can be taken as a criterion for the strength of the exchange. It is known that the interaction of magnetic atoms diluted into Pd is also particularly strong; this is due to the large magnetic polarizability of the Pd. Comparison of layered structures and diluted alloys was performed in a number of other cases and showed that in all cases so far investigated, there is a good correspondence [8.79, 80]. While it is clear that a one-to-one relation of layered structures and diluted alloys would be an oversimplification, it is felt that the investigation of interlayer exchange can also contribute to the understanding of magnetic alloys. The layered structures have the great advantage that the interaction length, the interlayer thickness d_0, is, in principle at least, better defined than in the case e. g. of dilute alloys, where there is always an averaging over many such distances.

Interlayer exchange coupling has been investigated before (for a review see e. g. [8.81]). This was done by means of magnetic hysteresis and MA experiments. The conclusion was drawn [8.82] that most of the coupling is due to holes in the interlayer, giving rise to magnetic bridges. Meanwhile, conditions for the preparation and characterization of the films as well as the techniques to study the coupling have greatly improved and it seems worthwhile to look again into these questions.

c) Modes in Double Layers with Negative Interlayer Exchange Coupling

From Fig. 8.4b we expect a strong decrease of the frequency of the exchange mode when the interlayer exchange A_{12} becomes negative. For obvious reasons we call this type of coupling antiferromagnetic. An experimental example is displayed in the lowest trace of Fig. 8.15. The upper trace of this figure shows for comparison the fully coupled limit $(A_{12} = A_{12}^c)$ and the middle part the decoupled limit $(A_{12} = 0)$. The gross features expected from Fig. 8.4b are reproduced: for antiferromagnetic coupling the exchange mode drops below the value expected for $A_{12} = 0$. The samples quoted in Fig. 8.15 were epitaxial films with the film plane parallel to a $\{100\}$-type plane. They displayed the bulk anisotropy of bcc Fe, which has to be included for a more detailed analysis of these spectra [8.83, 84].

In double layers with antiferromagnetic coupling for small enough external fields the magnetizations of the two films are expected to align antiparallel. That this is indeed the case can also be shown by means of LS experiments, as was pointed out in Sect. 8.3.2a. A corresponding spectrum is shown in Fig. 8.13b. The scattering geometry under which this spectrum was obtained, revealed that the magnetizations in the double layer were antiparallel and perpendicular to the previously applied external field [8.83, 84]. This situation resembles the spin flop

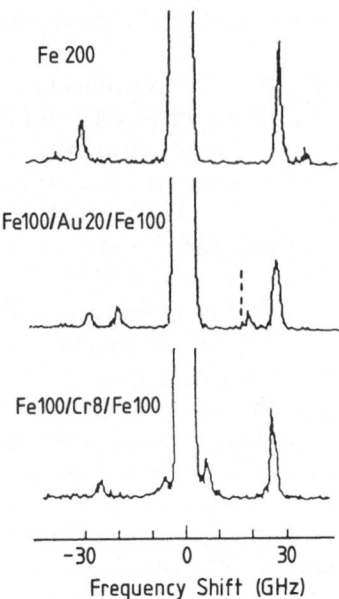

Fig. 8.15. LS from spin waves in double layers representing the fully coupled limit (*upper part*), decoupled limit (*middle part*) and antiferromagnetic coupling (*lower part*). Type of sample is also indicated, where the numbers give the thicknesses of the quoted materials in [Å]. Selected from [8.83]

phases of antiferromagnets. In principle, the sign of the magnetization directions in the two layers is also contained as information. Rotating both magnetizations into the opposite direction would interchange the S and aS side in the spectrum of Fig. 8.13b.

Figure 8.16 demonstrates how this can be used to interpret hysteresis curves. We see here the magnetization reversal of a sample with antiferromagnetic interlayer coupling as in the lowest trace of Fig. 8.15, and for comparison a sample representing the decoupled case as in the middle part of Fig. 8.15. In the decoupled case, the hysteresis is rectangular with a small coercive force, as also observed from single Fe films. In contrast, the curve from the sample with negative interlayer coupling is strongly sheared and for small B_0 the net moment is small. Without further information one would not know whether this is due to the formation of domains or, e. g., to antiparallel alignment. The LS spectra and the scattering geometry in which they were obtained [8.83, 84], characterize the antiparallel alignment for small values of B_0. This enabled the interpretation of the whole hysteresis curve, as shown by the arrows.

Negative coupling is also found between epitaxial Fe films in {110} orientation and between epitaxial {100} Fe films and polycrystalline permalloy films [8.85, 86]. The hysteresis curves can be rather complicated but LS was always a valuable tool to identify the alignment.

A peculiar behaviour is displayed by the exchange mode at fields B_0 where the magnetizations of the magnetic films cant. This is shown in Fig. 8.17. Here the exchange mode frequency, v_{em}, is plotted as a function of B_0, for B_0 along the in-plane [100] and [110] directions, which are the easy and hard axis of Fe in that plane. In both cases v_{em} goes through a minimum. Hysteresis curves from these

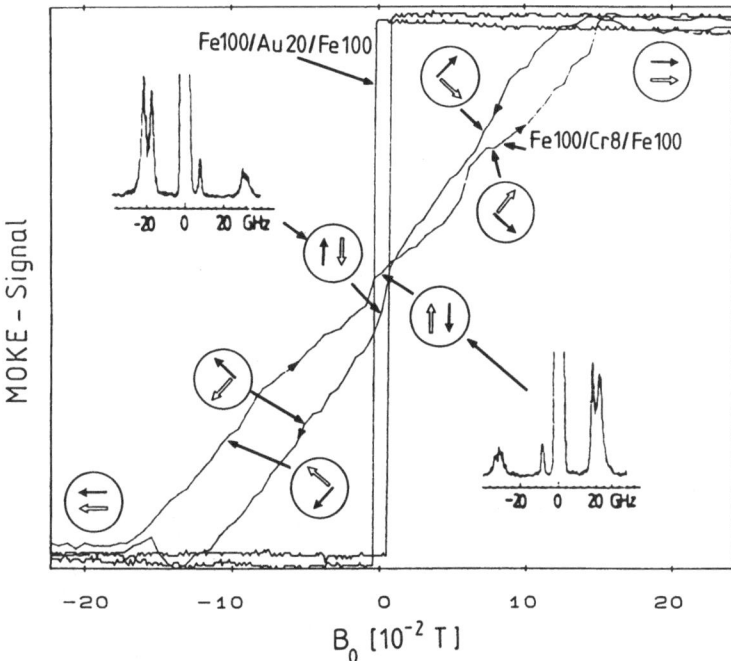

Fig. 8.16. Hysteresis curves of Fe double layers representing the decoupled case (same sample as in the middle part of Fig. 8.15) and the antiferromagnetic coupled case (*lower part* of Fig. 8.15). LS spectra to identify the alignment for small values of B_0 are also shown. Encircled pairs of arrows show alignment of M in the two films for different B_0 [8.86]

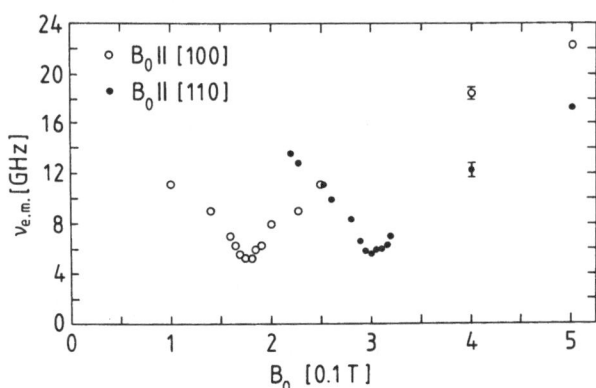

Fig. 8.17. Frequency of the exchange mode, ν_{em} as a function of B_0, for B_0 along the in-plane easy axis [100] and hard axis [110] [8.85]

samples revealed that the minima occur at those points where, for decreasing B_0, the moments of the two films start to cant [8.85]. For B_0 along the hard axis, this field is somewhat higher. *Hinchey* and *Grünberg* [8.39] found that the anisotropy causes a shift to higher frequency as observed for small values of B_0. Without the anisotropy and neglecting the dipole–dipole interaction, the frequency would fall to zero and remain there throughout the whole canted phase. This is the famous Goldstone mode behaviour [8.87]. It would be interesting to see how it is affected here by the dipole–dipole interaction.

We note a certain similarity between Fig. 8.17 and Fig. 8.11b. In both cases the frequency minima mark those values of B_0 where, upon decreasing B_0 the static magnetization starts to tilt away from the previous direction. In Fig. 8.11b the new direction is out of plane, in Fig. 8.17 it is characterized by the antiparallel alignment. In both cases the minima are due to a partial or complete balance of competing torques.

The microscopic origin of the antiferromagnetic coupling across Cr is not yet clear. In a localized picture of the antiferromagnetism of Cr one would expect, for films grown parallel to $\{100\}$ planes, that the coupling parameter A_{12} oscillates between positive and negative values depending on whether there is an even or odd number of atomic Cr planes between the two Fe films. This is due to the fact that according to the magnetic structure of bulk Cr the moments should be parallel within $\{100\}$, but antiparallel for adjacent planes. For $\{110\}$ planes there are as many moments pointing up as down and, ideally, any effect related to the antiferromagnetic structure should cancel out. Nevertheless antiferro-magnetic interlayer coupling was also found for Fe–Cr–Fe structures grown parallel to $\{110\}$ [8.85]. One might suspect that interface roughness or interdiffusion could alter the situation expected from the magnetic structure of bulk Cr to lead to the effects as they are observed. Ths is presently being investigated.

In this context it is interesting to note that negative interlayer coupling was also found in Gd/Y [8.88] and Dy/Y [8.89] layered structures, where it was shown to be due to the polarization of the conduction electrons in Y by the moments in the Gd or Dy [8.90].

8.3.3 Modes in Multilayers

Due to the possibility of fabricating magnetic multilayers, there has also been interest in the mode spectra of such systems. It has been noted [8.57, 29] that the collective modes described in Sect. 8.1.2 are the only true collective magnetic phenomena of such systems so far observed experimentally. Hence some of the motivation was simply to prove the existence of these modes.

Apart from this, magnetic parameters have also been deduced for multilayers of Mo/Ni [8.91], of Fe/Pd and Fe/W [8.26], and of Co/Nb [8.58]. For example, LS experiments have yielded new insight into the magnetic structure of Fe–Pd multilayers. Spectra are shown in Fig. 8.7. It was found from magnetometry that apparently, for very thin Fe films, the magnetic moment of the Fe films increases [8.92]. An analysis based on magnetometry in conjunction with LS [8.26] led to the conclusion that the magnetization of the Fe layers was in fact slightly reduced, as compared to the bulk Fe value but there was also an appreciable polarization of the Pd interlayers, leading to an increased total moment of the whole stack. The Fe – Pd interface was found to be characterized by a small out of plane anisotropy (see Table 8.1).

8.4 Modification of Magnetic Material Properties by Layering and the Role of Light Scattering in Their Characterization

Layering of magnetic materials can be used as a tool to modify the materials' properties. Various effects have now become known that occur as a result of the layering.

For example strong magneto-optic Kerr effects occur in layered structures of Fe and Co interspaced by metals like Cu or Au [8.93]. One can interpret this as being due to a particular mixing of the optical constants of the various materials. This is surprising because it means that one can enhance the Kerr effect of Fe or Co by layering it with a nonmagnetic metal. It could be shown that this effect is related to another interesting phenomenon, namely the enhancement of the Kerr effect at light frequencies where the real part of the dielectric constant is close to one. Generally, this happens close to the plasma edge and also explains the anomalously large Kerr effect of PtMnSb [8.94]. Needless to say, such an enhancement is of interest for magneto-optic recording.

Another effect, which has been known for a long time but is still not yet fully understood, is called exchange anisotropy [8.95]. It is due to exchange coupling across the joint interface between a ferro- and an antiferromagnetic layer and causes a shift of the hysteresis curve. Recently, structures were reported in which exchange anisotropy occurs in combination with a compensation point [8.96]. Compensation points describe temperatures at which the magnetizations of oppositely aligned sublattices – or in this case layers – are equal in magnitude. The overall magnetization then vanishes and the coercive force of the system becomes very large. Such compensation points are also relevant for magneto-optic recording.

We have seen in Sect. 8.3.1b that at interfaces anisotropies often occur that are perpendicular to the film plane. This effect was found to be particularly strong in layered structures of rare-earth and transition metals [8.97], where it is probably due to a pair anisotropy of the two kinds of atoms. This is also of interest for magnetic storage media.

In Sect. 8.3.2c new types of magnetization reversal in an external field were described, resulting from antiferromagnetic interlayer exchange. An application is not immediately obvious. Recently, on the other hand, it was reported that the antiparallel alignment in such structures can lead to strong magnetoresistance effects [8.98, 99]. This is of great interest for magnetoresistive field sensors.

In Sect. 8.2.3 and 8.3.2a we have described some new mode spectra as genuine properties of layered magnetic structures. Such modes are being considered for applications in signal processing devices [8.78]. A nonreciprocal behaviour, as displayed by the modes of a double layer with antiparallel alignment (Sect. 8.3.2a), is of particular interest here.

In addition, there are many effects in thin films which are useful for various applications, but which have to be considered more as bulk properties of the film material. They are due to a particular structure or composition. An anisotropy due to columnar growth or a compensation point due to a certain composition are examples. Lattice mismatch between film and substrate and strains are general phenomena and it has been known for a long time that they can result in interesting properties such as perpendicular anisotropies.

From these examples, we see that many possible applications can motivate research on layered magnetic structure. However, there are also fascinating phenomena like the Rudermann-Kittel like oscillations in Gd–Y and Dy–Y layered structures [8.88, 89] which, at present, are more of academic interest but nonetheless contribute greatly to the excitement in this field. In the structures described in Sect. 8.3.2b, such oscillations so far have not been found, although they should be there according to theoretical considerations [8.100, 101]. It remains to be seen whether the sample quality was not sufficient or whether there are more profound reasons.

In this article the usefulness of a particular probe for such investigations has been described. We have seen that LS from spin waves provides information on all relevant magnetic parameters mentioned above. The other method which is most closely related, and in many cases can be used as an alternative, is MA. While MA has some advantages, in particular with respect to resolution, we have seen that in some cases LS can be superior. A good example was the observation of standing spin waves in very thin films (Sect. 8.3.1a). A general advantage of LS is the fact that it operates with a frequency scan and not with a field scan, unlike the conventional MA experiment. In some cases a constant field B_0 is an important experimental condition, e. g. for the observation of the mode spectrum of a double layer with antiparallel magnetization where we are restricted to small values of B_0 (Sect. 8.3.2c). In Sect. 8.2.2b we have seen how LS and MA can be combined to produce a powerful tool for the investigation of spin wave relaxation.

Although the microscopic origin of many effects in layered magnetic structures is still rather unclear, we have seen that interesting modifications of materials properties can be obtained by the layering. The LS method can usefully supplement the information obtained by other methods for the characterization of these structures.

Acknowledgements. I would like to thank J. Barnas, B. Hillebrands and J.F. Cochran for proofreading this manuscript and for enlightening discussions.

References

8.1   J.R. Sandercock: *Light Scattering in Solids II*, ed. by M. Cardona, G. Güntherodt, Topics Appl. Phys. Vol. 51 (Springer, Berlin, Heidelberg 1982) Chap. 6
8.2   A.S. Borovik-Romanov, N.M. Kreines: Phys. Reports **81**, 351 (1982); *Spin Waves and Magnetic Excitations 1*, ed. by A.S. Borovik-Romanov, S.K. Sinha (Elsevier, Amsterdam 1988) p. 81
8.3   C. Patton: Phys. Reports **103**, 251 (1984)

8.4   P. Grünberg: Prog. Surf. Sci., ed. by S.G. Davison, Vol. 18, no. 1 (1985)
8.5   M.G. Cottam, D.J. Lockwood: *Light Scattering in Magnetic Solids* (Wiley, New York 1986)
8.6   P. Grünberg: Thin Film Techniques for Low Dimensional Structures, NATO ASI series, ed, by R.F.C. Farrow, S.S.P. Parkin, P.J. Dobson, J.H. Neave, A.S. Arrott (Plenum, New York 1987) p. 487
8.7   T. Wolfram, R.E. Dewames: Prog. Surf. Sci. ed. by S.G. Davison, Vol. 2, no. 4 (1972)
8.8   G.T. Rado, R.J. Hicken: J. Appl. Phys. **63**, 3885 (1988)
8.9   J.F. Cochran, J.R. Dutcher: J. Appl. Phys. **63**, 3814 (1988)
8.10 G.T. Rado, J.R. Weertman: J. Phys. Chem. Solids **11**, 315 (1959)
8.11 G.T. Rado: J. Appl. Phys. **61**, 4262 (1987)
8.12 F. Hoffmann: Phys. Stat. Sol. **41**, 807 (1970)
8.13 F. Hoffmann, A. Stankoff, H. Pascard: J. Appl. Phys. **41**, 1022 (1970)
8.14 M. Vohl, J. Barnas, P. Grünberg: submitted for publication
8.15 R.W. Damon, J.R. Eshbach: J. Phys. Chem. Solids **19**, 308 (1961)
8.16 L.K. Brundle, N.J. Freedman: Electron. Lett. **4**, 132 (1968)
8.17 G. Rupp, W. Wettling, R.S. Smith, W. Jantz: J. Magn. Magn. Mater **45**, 404 (1984)
8.18 B. Hillebrands, P. Baumgart, G. Güntherodt: Phys. Rev. B**36**, 2450 (1987)
8.19 J.R. Dutcher, B. Heinrich, J.F. Cochran, D.A. Steigerwald, W.F. Egelhoff: J. Appl. Phys. **63**, 3464 (1988)
8.20 R.E. Camley, T.S. Rahman, D.L. Mills: Phys. Rev. B**27**, 261 (1983)
8.21 P. Grünberg, K. Mika: Phys. Rev. B**27**, 2955 (1983)
8.22 K. Mika, P. Grünberg: Phys. Rev. B**31**, 4465 (1985)
8.23 P.R. Emtage, M.R. Daniel: Phys. Rev. B**29**, 212 (1984)
8.24 G. Rupp, W. Wettling, W. Jantz: Appl. Phys. A**42**, 45 (1987)
8.25 J. Barnas: J. Phys. C**21**, 1021 (1988) and ibid. p. 4097
8.26 B. Hillebrands, A. Boufelfel, C.M. Falco, P. Baumgart, G. Güntherodt, E. Zirngiebl, J.D. Thompson: J. Appl. Phys. **63**, 3880 (1988)
8.27 P. Grünberg: J. Appl. Phys. **51**, 4338 (1980)
8.28 P. Grünberg: J. Appl. Phys. **52**, 6824 (1981)
8.29 P. Grünberg: J. Appl. Phys. **57**, 3673 (1985)
8.30 H. Puszkarski: Prog. in Surf. Sci. **9**, 191 (1979)
8.31 M. Pomerantz, J.C. Slonzewski, E. Spiller: J. Appl. Phys. **61**, 3747 (1987)
8.32 J.F. Cochran, B. Heinrich, A.S. Arrott: Phys. Rev. B**34**, 7788 (1986)
8.33 K. Vayhinger, H. Kronmüller: J. Magn. Magn. Mater. **72**, 307 (1988)
8.34 K. Vayhinger, H. Kronmüller: J. Magn. Magn. Mater. **62**, 159 (1986)
8.35 K. Vayhinger: Diploma thesis (1985) University of Stuttgart, Germany
8.36 B. Hillebrands: Phys. Rev. B**37**, 9885 (1988)
8.37 J.F. Cochran, J.R. Dutcher: J. Appl. Phys. **64**, 6092 (1988)
8.38 B. Hillebrands: To be published
8.39 L. Hinchey, P. Grünberg: Submitted for publication
8.40 R.P. van Stapele, F.J.A.M. Greidanus, J.W. Smits: J. Appl. Phys. **57**, 1282 (1985)
8.41 E.L. Albuquerque, P. Fulco, E.F. Sarmento, D.R. Tilley: Solid. St. Commun. **58**, 41 (1986)
8.42 L.L. Hinchey, D.L. Mills: Phys. Rev. B**34**, 1689 (1986)
8.43 R.E. Camley: Phys. Rev. Lett. **45**, 283 (1980)
8.44 R.L. Stamps, R.E. Camley: J. Appl. Phys. **56**, 3497 (1984)
8.45 S. Blumenröder, E. Zirngiebl, P. Grünberg, G. Güntherodt: J. Appl. Phys. **57**, 3684 (1985)
8.46 R.E. Camley, M. Grimsditch: Phys. Rev. B**22**, 5420 (1980)
8.47 R.E. Camley, T.S. Raman, D.L. Mills: Phys. Rev. B**23**, 1226 (1981)
8.48 J.F. Cochran, J.R. Dutcher: J. Magn. Magn. Mater. **73**, 299 (1988)
8.49 P. Grünberg, F. Metawe: Phys. Rev. Lett. **39**, 1561 (1977)
8.50 R.E. Camley, P. Grünberg, C.M. Mayr: Phys. Rev. B**26**, 2609 (1982)
8.51 M. Grimsditch, A. Malozemoff, A. Brunsch: Phys. Rev. Lett. **34**, 711 (1979)

8.52 S.P. Vernon, S.M. Lindsay, M.B. Stearns: Phys. Rev. B29, 4439 (1984)
8.53 M.G. Cottam: J. Phys. C15, 1573 (1983)
8.54 W. Wettling, W. Jantz: J. Magn. Magn. Mater. 45, 364 (1984)
8.55 G. Srinivasan, C.E. Patton, P.R. Emtage: J. Appl. Phys. 61, 2318 (1987)
8.56 G. Srinivasan, C.E. Patton, J.G. Booth: J. Appl. Phys. 63, 3344 (1988)
8.57 M. Grimsditch, M.R. Khan, A. Kueny, I.K. Schuller: Phys. Rev. Lett. 51, 498 (1983)
8.58 G. Rupp, W. Wettling, W. Jantz, R. Krishnan: Appl. Phys. A37, 73 (1985)
8.59 B. Hillebrands, P. Baumgart, R. Mock, G. Güntherodt, A. Boufelfel, C.M. Falco: Phys. Rev. B34, 9000 (1986)
8.60 B. Heinrich, S.T. Purcell, J.R. Dutcher, K.B. Urquhart, J.F. Cochran, A.S. Arrott: Submitted for publication
8.61 P. Grünberg, M.G. Cottam, W. Vach, C.M. Mayr, R.E. Camley: J. Appl. Phys. 53, 2078 (1982)
8.62 J.R. Sandercock, W. Wettling: J. Appl. Phys. 50, 7784 (1979)
8.63 P. Grünberg, C.M. Mayr, W. Vach, M. Grimsditch: J. Magn. Magn. Mater. 28, 319 (1982)
8.64 P.X. Zhang, Y.Z. Pang, W. Zinn: Solid State Commun. 60, 449 (1986)
8.65 G.A. Prinz, G.T. Rado, J.J. Krebs: J. Appl. Phys. 53, 2087 (1982)
8.66 J.J. Krebs, F.J. Rachford, P. Lubitz, G.A. Prinz: J. Appl. Phys. 53, 8058 (1982)
8.67 U. Gradmann, J. Korecki, G. Waller: Appl. Phys. A39, 101 (1986)
8.68 B. Heinrich, K.B. Urquhart, J.R. Dutcher, S.T. Purcell, J.F. Cochran, A.S. Arrott, D.A. Steigerwald, W.F. Egelhoff: J. Appl. Phys. 63, 3863 (1988)
8.69 K.B. Urquhart, B. Heinrich, J.F. Cochran, A.S. Arrott, K. Myrtle: J. Appl. Phys. 64, 5334 (1988)
8.70 J.R. Dutcher, J.F. Cochran, B. Heinrich, A.S. Arrott: J. Appl. Phys. 64, 6095 (1988)
8.71 S.T. Purcell, B. Heinrich, A.S. Arrott: To be published
8.72 P.X. Zhang, Y.Z. Pang, W. Zinn: Solid. State. Commun. 60, 449 (1986)
8.73 R. Henry, S.D. Brown, P.E. Wigen, P.J. Besser: Phys. Rev. Lett. 28, 1272 (1972)
8.74 P. Kabos, W.D. Wilber, C.E. Patton, P. Grünberg: Phys. Rev. B29, 6396 (1984)
8.75 P. Grünberg: Unpublished
8.76 G. Srinivasan, C.E. Patton: IEEE Trans. MAG-22, 996 (1986)
8.77 P.X. Zhang, W. Zinn: Phys. Rev. B35, 5219 (1987)
8.78 see e.g. W.S. Ishak, K. Chang: Hewlett-Packard Journal, Feb. 1985
8.79 P. Swiatek, F. Saurenbach, Y. Pang, P. Grünberg, W. Zinn: Proc. 3rd Int. Conf. on Phys. of Magn. Mat., Szczyrk-Bila, Poland, (World Scientific Singapore 1987) p. 389
8.80 P. Swiatek, F. Saurenbach, Y. Pang, P. Grünberg, W. Zinn: J. Appl. Phys. 61, 3753 (1987)
8.81 A. Yelon: Physics of Thin Films 6, 205 (Academic, New York 1971)
8.82 O. Massenet, F. Biragnet, H. Juretschke, R. Montmory, A. Yelon: IEEE Trans. MAG-2, 553 (1966)
8.83 P. Grünberg, R. Schreiber, Y. Pang, M.B. Brodsky, H. Sowers: Phys. Rev. Lett. 57, 2442 (1986)
8.84 P. Grünberg, R. Schreiber, Y. Pang, U. Walz, M.B. Brodsky, H. Sowers: J. Appl. Phys. 61, 3750 (1987)
8.85 F. Saurenbach, U. Walz, L. Hinchey, P. Grünberg, W. Zinn: J. Appl. Phys. 63, 3473 (1988)
8.86 P. Grünberg, F. Saurenbach: Symposium on Layered Structures, Tokyo 1988 to appear in Bull. Mat. Res. Soc.
8.87 see e.g. R.M. White, T. Geballe: "Long Range Order in Solids", In Solid State Physics, Suppl. 15 (Academic, New York 1979) p. 37ff.
8.88 C.F. Majkrzak, J.W. Cable, J. Kwo, M. Hong, D.B. McWhan, Y. Yafet, J.V. Waszczak, C. Vettier: Phys. Rev. Lett. 56, 2700 (1986)
8.89 M.B. Salamon, S. Sinha, J.J. Rhyne, J.E. Cunningham, R.W. Erwin, J. Borchers, C.P. Flynn: Phys. Rev. Lett. 56, 259 (1986)
8.90 Y. Yafet, J. Kwo, M. Hong, C.F. Majkrzak, T. O'Brien: J. Appl. Phys. 63, 3453 (1988)

8.91 A. Kueny, M.R. Khan, I.K. Schuller, M. Grimsditch: Phys. Rev. B **29**, 2879 (1984)
8.92 F.J.A. den Broeder, H.J.G. Draaisma, H.C. Donkersloot, W.J.M. de Jonge: J. Appl. Phys. **61**, 4317 (1987)
8.93 T. Katayama, Y. Suzuki, H. Awano, Y. Nishihara, N. Koshizuka: Phys. Rev. Lett. **60**, 1426 (1988)
8.94 P.G. van Engen, K.H.J. Buschow, R. Jongebreur, M. Erman: App. Phys. Lett. **42**, 202 (1983)
8.95 W.H. Meiklejohn, C.P. Bean: Phys. Rev. **105**, 904 (1957)
8.96 D.J. Webb, A.F. Marshall, A.M. Toxen, T.H. Geballe, R.M. White: paper presented at INTERMAG 1987, Tokyo, Japan
8.97 N. Sato: J. Appl. Phys. **59**, 2514 (1986)
8.98 G. Binasch, P. Grünberg, F. Saurenbach, W. Zinn: Submitted for publication
8.99 N.N. Baibich, J.M. Broto, G. Creuzet, P. Etienne, A. Fert, A.R. Fert, S. Hadjoudj, F. Nguyen Van Dan: 12th Int. Coll. on Magn. Films and Surf., Le Creusot, France (1988)
8.100 A. Bardasis, D.S. Falk, R.A. Ferrell, M.S. Fullenbaum, R.E. Prange, D.L. Mills: Phys. Rev. Lett. **14**, 298 (1965)
8.101 K. Yosida, A. Okiji: Phys. Rev. Lett. **14**, 301 (1965)

Additional References with Titles

Chapter 2

Brazis, R., Safonova, L.: "Electromagnetic waves in layered semiconductor-dielectric periodic structures in dc magnetic fields". Proc. SPIE Intl. Congress Optical Sci. and Eng., Hamburg, Sept. 1988 (to be published)

Castillo-Mussot, M. del, Mochán, W.L., Barrera, R.G.: "Effect of plasma waves on the dispersion relation of conductor-insulator superlattices". In *Lectures on Surface Science* ed. by Castro, G.R., Cardona M. (Springer, Heidelberg, 1987) p. 28

Chu, H., Yia-Chung Chang: "Phonon-polariton modes in superlattices: The effect of spatial dispersion". Phys. Rev. B38, 12369 (1988-I)

Giraldo, J., Castillo-Mussot, M. del, Barrera, R.G.: "Electron-hole pair excitation in multilayered conducting heterostructures". Phys. Rev. B38, 5380 (1988)

Haupt, R., Wendler, L.: "Damping of polaritons in finite semiconductor superlattices". Solid State Commun. 61, 341 (1987)

Haupt, R., Wendler, L.: "Dispersion and Damping properties of plasmons polaritons in superlattice structures". Phys. stat. sol. (b) 142, 125 (1987)

Hawrylak, P.: "Plasmon and electron-hole-pair damping of excited vibrational and electronic states in quasi-two-dimensional electron systems". Phys. Rev. B35, 3818 (1987)

Ivchenko, E.L., Kosobukin, V.A.: "Exciton polaritons in semiconductors with a superlattice". Sov. Phys. Semicond. 22, 15 (1988)

Kushwaha, M.S., Halevi, P.: "Magnetoplasma modes in thin films in the Faraday configuration". Phys. Rev. B35, 3879 (1987)

Liu, Y., Sooryakumar, R.: "Discrete coupled plasmon-phonon modes in finite semiconductor multi-layers". Solid State Commun. 64, 1081 (1987)

Mochán, W. Luis, Castillo-Mussot, M. del, Barrera, R.G.: "Effect of plasma waves on the optical properties of metal-insulator superlattices". Phys. Rev. B35, 1088 (1987)

Nkoma, J.S.: "Surface phonon polaritons in semi-infinite semiconductor superlattices". Phys. stat. sol. (b) 139, 117 (1987)

Raj, N., Tilley, D.R.: "Polariton and effective-medium theory of magnetic superlattices". Phys. Rev. B36, 7003 (1987)

Santoro, G.E., Giuliani, G.F.: "Acoustic plasmons in a conducting double layer". Phys. Rev. B37, 937 (1988)

Sy, H.K., Song, L.M.: "Surface plasmons in type II superlattices". Phys. stat. sol. (b) 149, 595 (1988)

Teich, W.G., Mahler, G.: "Plasmons excitations in thin metal films". Phys. stat. sol. (b) 138, 607 (1986)

Wendler, L.: "Landau damped collective excitations of the quasi-two-dimensional polaron gas in double heterostructures". Solid State Commun. 65, 1197 (1988)

Wendler, L., Pechstedt, R.: "Coupled intra- and intersubband plasmon-phonon modes in double heterostructures". Phys. Rev. B35, 5887 (1987)

Yang, R.-Q., Tsai, C.-H.: "A note of the plasmon dispersion in semiconductor superlattices". Solid State Commun. 63, 1081 (1987)

Chapter 3

Abstreiter, G., Eberl, K., Friess, E., Wegscheider, W., Zachai, R.: "Silicon/Germanium strained layer superlattices". 5th Int. Conf. on MBE, Saporo, Japan, 1988, in press

Agulló-Rueda, F., Mendez, E.E., Hong, J.M.: "Doubly resonant Raman scattering induced by an electric field". Phys. Rev. B38, 12720 (1988)

Albuquerque, E.L., Fulco, P., Tilley, D.R.: "Lattice dynamics for superlattices". Phys. stat. sol. (b) 146, 449 (1988)

Basca, W., Känel, H. v., Mäder, K.A., Ospelt, M., Wachter, P.: "Inelastic light scattering from strained-layer Si/Ge superlattices". Superlattices and Microstructures 4, 717 (1988)

Bland, J.A.C.: "Photoexcited and forbidden LO mode Raman scattering in epitaxial InP(001) layers". Superlattices and Microstructures 4, 485 (1988)

Catellani, A., Sorba, L.: "Acoustic-wave transmission in semiconductor superlattices". Phys. Rev. B38, 7717 (1988)

Chu, H., Ren, S., Chang, Y.: "Long-wavelength optical phonons in polar superlattices". Phys. Rev. B37, 10746 (1987)

Dereux, A., Vigneron, J.-P., Lambin, P., Lucas, A.A.: "Polaritons in semiconductor multilayered materials". Phys. Rev. B38, 5438 (1988)

Dharma-wardana, M.W.C., Lockwood, D.J., Aers, G.C., Baribeau, J.-M.: "The observation of resonant phonons in ultra-thin superlattice structures". J. Phys. C, in press

Emura, S., Gonda, S.-I., Matsui, Y., Hayashi, H.: "Internal-stress effects on Raman spectra of $In_xGa_{1-x}As$ on InP". Phys. Rev. B38, 3280 (1988)

Fasol, G., Tanaka, M., Sakaki, H., Horikoshi, Y.: "Interface roughness and the dispersion of confined LO phonons in GaAs/AlAs quantum wells". Phys. Rev. B38, 605 (1988)

Friess, E., Brugger, H., Eberl, K., Krötz, G., Abstreiter, G.: "Confined optical modes in short period (110) Si/Ge superlattices". Solid State Commun., in press

Gammon, D., Shi, L., Merlin, R., Ambrazevicius, G., Ploog, K., Morkoc, H.: "Enhanced and quenched Raman scattering by interface phonons in semiconductors superlattices: what are the defects?". Superlattices and Microstructures 4, 405 (1988)

Gridin V.V., Beserman, R., Morkoc, H.: "Energy dependence of the electron-phonon coupling in a thin layer GaAs/AlAs superlattice". Phys. Rev. B37, 9061 (1988)

He, J., Djafari-Rouhani, B., Sapriel, J.: "Theory of light scattering by longitudinal acoustic phonons in superlattices". Phys. Rev. B37, 4086 (1988)

Holtz, M., Venkateswaran, U.D., Syassen, K., Ploog, K.: "Resonant Raman scattering in GaAs/AlAs thin-layer superlattices under high pressure". Submitted to Phys. Rev. B

Iikawa, F., Cerdeira, F., Vazquez-Lopez, C., Motisuke, P., Sacilotti, M.A., Roth, A.P., Masut, R.A.: "Optical studies in $In_xGa_{1-x}As/GaAs$ strained-layer superlattices". Phys. Rev. B38, 8473 (1988)

Iikawa, F., Cerdeira, F., Vazquez-Lopez, C., Motisuke, P., Sacilotti, M.A., Roth, A.P., Masut, R.A.: "Raman scattering from InGaAs/GaAs strained-layer superlattices". Solid State Commun. 68, 211 (1988)

Kash, J.A., Hvam, J.M., Tsang, J.C., Kuech, T.F.: "Localization and wave-vector conservation for optical phonons in $Al_xGa_{1-x}As$ and thin layer of GaAs". Phys. Rev. B38, 5776 (1988)

Kasper, E., Kibble, H., Jorke, H., Brugger, H., Friess, E., Abstreiter, G.: "Symmetrically strained Si/Ge superlattices on Si substrates". Phys. Rev. B38, 3599 (1988)

Kobayashi, A., Roy, A.: "Phonons in the alloy superlattice GaAs/AlGaAs". Phys. Rev. B35, 2237 (1987)

Lockwood, D.J., Baribeau, J.-M., Timbrell, P.Y.: "An annealing study of relaxation and interface quality in $Si - Si_{1-x}Ge_x$ strained-layer superlattices". J. Appl. Phys. 65 (1989)

Lou, B., Sudharasanan, R., Perkowitz, S.: "Anisotropy and infrared response of the GaAs-AlAs superlattice". Phys. Rev. B38, 2212 (1988)

Ren, S.-F., Chu, H., Chang, Y.-C.: "Phonon dispersion curves of GaAs/AlAs superlattices". Superlattices and Microstructures 4, 303 (1988)

Ren, S.-F., Chu, H., Chang, Y.-C.: "Anisotropy of optical phonons and interface modes in GaAs-AlAs superlattices". Phys. Rev. B 37, 8899 (1988)

Sapriel, J., He, J., Djafari-Rouhani, B., Azoulay, R., Mollot, F.: "Coupled Brillouin-Raman study of direct and folded acoustic modes in GaAs-AlAs superlattices". Phys. Rev. B 37 (1988)

Sapriel, J., Rao, E.V.K., Brillouet, F., Chavignon, J., Ossart, P., Gao, Y., Krauz, P.: "Raman scattering study of Al/Ga interdiffusion in ion-implanted and annealed GaAs-Ga$_{1-x}$As$_x$As superlattices". Superlattices and Microstructures 4, 115 (1988)

Shon, L., Inoue, K., Matsuda, O., Murase, K., Yokogawa, T., Ogura, M.: "Raman and photoluminescence investigations of ZnSe-ZnS strained-layer superlattices". Solid State Commun. 67, 779 (1988)

Tamura, S., Wolfe, J.P.: "Acoustic phonons in multiconstituent superlattices". Phys. Rev. B 38, 5610 (1988)

Wang, Z.P., Han, H.X., Li, G.H., Jiang, D.S., Ploog, K.: "Raman scattering from TO phonons in (GaAs)$_n$(AlAs)$_n$ superlattices". Phys. Rev. B 38, 8483 (1988)

Zachel, R., Friess, E., Abstreiter, G., Kasper, E., Kibbel, H.: "Band structure and optical properties of strain symmetrized short period Si/Ge superlattices on Si(100) substrates". Int. Conf. Phys. Semicond. Warsaw, 1988, in press

Chapter 5

Gumbs, G., Ali, M.K.: "Dynamical maps, Cantor spectra, and localization for Fibonacci and related quasiperiodic lattices". Phys. Rev. Lett. 60, 1081 (1988)

Holzer, M.: "Three classes of one-dimensional, two-tile penrose tilings and the Fibonacci Kronig-Penney model as a generic case". Phys. Rev. B 38, 1709 (1988)

Laruelle, F., Etienne, B.: "Fibonacci invariant and electronic properties of GaAs/Ga$_{1-x}$Al$_x$As quasiperiodic superlattices". Phys. Rev. B 37, 4816 (1988)

Lockwood, D.J., McDonald, A.H., Ares, G.C., Dharma-Wardana, M.W.C., Devine, R.L.S., Moore, W.T.: "Raman scattering in a GaAs/Ga$_{1-x}$Al$_x$As Fibonacci superlattice". Phys. Rev. B 36, 9286 (1987)

Ogadaki, T., Aoyama, H.: "Hyperinflation in periodic and quasiperiodic chains". Phys. Rev. Lett. 61, 775 (1988)

Ohsawa, K., Ninomiya, T.: "Band structure of one-dimensional quasicrystal and Lie algebra". J. Phys. Soc. Jpn. 57, 2448 (1988)

Seel, M., Trickey, S.B.: "Electronic structure of Fibonacci copolymers and superlattices". Solid State Commun. 66, 537 (1988)

Sokoloff, J.B.: "Perturbation theory for the electrical conductivity of quasiperiodic lattices". Phys. Rev. B 37, 7091 (1988)

Wang, C., Barrio, R.A.: "Theory of the Raman response in Fibonacci superlattices". Phys. Rev. Lett. 61, 191 (1988)

Xie, X.C., DasSarma, S.: ""Extended" electronic states in a Fibonacci superlattice". Phys. Rev. Lett. 60, 1585 (1988)

Villaseñor-González, P., Mejía-Lira, F., Morán-López, J.L.: "Renormalization group approach to the electronic spectrum of a Fibonacci chain". Solid State Commun. 66, 1127 (1988)

Chapter 7

Albuquerque, E.L., Fulco, P., Sarmento, E.F., Tilley, D.R.: "Spin waves in a magnetic superlattice". Solid State Commun. 58, 41 (1986)

Clemens, B.M., Eesley, G.L.: "Relationship between interfacial strains and the elastic response of multilayer metal films". Phys. Rev. Lett. 61, 2356 (1988)

Dodson, B.W.: "Atomistic analysis of the standard-modulus effect in metallic superlattices". Phys. Rev. B 37, 727 (1988)

Hillebrands, B.: "Calculation of spin waves in multilayered structures including interface anisotropies and exchange contributions". Phys. Rev. B 37, 9885 (1988)

Jankowski, A.F.: "Modelling the supermodulus effect in metallic multilayers". J. Phys. F 18, 413 (1988)

Jankowski, A.F., Tsakalakos, T.: "The effect of strain on the elastic constants of noble metals". J. Phys. F 15, 1279 (1985)

Wolf, D., Lutsko, J.F.: "Structurally induced supermodulus effect in superlattices". Phys. Rev. Lett. 60, 1170 (1988)

Subject Index

Topics in Applied Physics Founded by Helmut K. V. Lotsch